'Marvellous . . . A compelling read . . . One of the best science books of the year. By dint of a lot of flair and eclectic learning, Ball makes us look afresh at the most commonplace subject imaginable'
Graham Framelo, *New Scientist*

'The author's panoramic knowledge, conveyed through a clear and often delightful writing style, makes attractive reading for a technically literate, but not necessarily expert, audience . . . Ball's contribution is a delightful status report'
Frank H. Stillinger, *Nature*

'He is an engaging writer and his greatest success with this book is that even the most scientifically backward reader will have no difficulty digesting the explanations of awkward scientific concepts'
Dick Ahlstrom, *Irish Times*

'A superb and quenching portrait of this modest association of atoms that alone permits life among us'
Kirkus

'I strongly suggest that anyone with even a passing interest in the subject of water should read this book. However, if you have important deadlines to meet then beware, for you may find it hard to put down'
Austen Angell, *Physics World*

Philip Ball is a science writer and a consultant editor with the international scientific journal *Nature*. He communicates all areas of science to the public in national and international newspapers, on radio and TV. He is the author of *Designing the Molecular World: Chemistry at the Frontier*, which won the Association of American Publishers award for chemistry, *Made to Measure: New Materials for the 21*st *Century* and the bestselling *The Self-made Tapestry: Pattern Formation in Nature*.

H_2O

A Biography of Water

PHILIP BALL

PHŒNIX

A PHOENIX PAPERBACK

First published in Great Britain by Weidenfeld & Nicolson in 1999
This paperback edition published in 2000 by Phoenix,
an imprint of Orion Books Ltd,
Orion House, 5 Upper St Martin's Lane,
London WC2H 9EA

Second impression 2001

A CIP catalogue record for this book
is available from the British Library.

ISBN 0 75381 092 1

Printed in Great Britain by
The Guernsey Press Co. Ltd, Guernsey, C.I.

Contents

Illustrations

Preface

> Water was the matrix of the world and of all its creatures...Just as the noblest and most delicate colours arise from this black, foul earth, so various creatures sprang forth from the primordial substance that was only formless filth in the beginning. Behold the element of water in its undifferentiated state! And then see how all the metals, all the stones, all the glittering rubies, shining carbuncles, crystals, gold, and silver are derived from it; who could have recognized all these things in water?
>
> Paracelsus c.1531–5

Some substances become mythical. They transcend their physical and chemical materiality and manifest themselves in our minds as symbols, as qualities. In the collective unconscious of a culture, their material constitution becomes secondary to their symbolic value. Gold, to the alchemists, was more than a metal – it was perfection, the goal of a spiritual quest. It is not enough to describe fire as a luminescing gas, and there is something vital and irrevocable about blood that makes it no mere colloidal suspension. In general, a scientific account of mythical substances is bound to disappoint.

With water, that need not be so. Even when we remove its symbolic trappings, its association with purity, with the soul, with the maternal and with life and youth, when we reduce it to a laboratory chemical or a geological phenomenon, water continues to fascinate. At first glance a simple molecule, water still offers up profound challenges to science.

But why a biography? Because, like a person, water has immediate, evident and familiar characteristics that can be understood only, if at all, by a consideration of its deeper make-up, of the hidden factors that shape its behaviour. So I shall need to explore water's inner nature, the physics and chemistry of its unique personality. Moulded thus by physical forces, water leads a remarkable life in the wide world. I shall look at the influence water exerts on life, on the planetary environments of Earth and other worlds and stars, even on our own preconceptions about the possibilities of science.

This journey through water's place in creation requires that we explore its origins in two spheres: the first material, the second conceptual. For there is always more than one way to relate a life. There is, for instance, the story of the public persona, and the alternative narrative of the real facts (which, in any biography, will always be murky, ill-defined, known only to the subject and then only imperfectly). There is the question of genealogy and, if we want to be thorough about it, the question of racial ancestry, even of the origins of species. There is the story of the person's place in culture – can one recount Henry Ford's life without tracing our evolving cultural attitude to the motor car? In like fashion, we cannot avoid the conclusion that for humankind water is a force of social change – a precious resource to be treasured, nurtured and used wisely, for the alternative is deprivation, disease, environmental degradation, conflict and death.

With water, we have at least the good fortune that our impersonal subject permits us an unusually clear distinction between attitudes and facts. Our beliefs about what water is have undergone dramatic revision throughout history – and needless to say, as in all ages we believe our current vision to be the true one. This is a story that holds a mirror to ourselves, lets us see how our perception of the material world has turned somersaults over the centuries. But water has throughout this time been just what it is today: H_2O, a remarkable chemical compound with a private history. Some parts of that history are now reconstructed; some remain secret. But whichever way we tell it, water came from the Universe, and is out there still.

Yet how many of us, chemists aside, will bring to mind that banana-shaped molecule when we hear water invoked? Instead we see rivers, brooks, gurgling springs and wide oceans. We see a liquid, the archetype (we might imagine) of all that flows. Not for nothing is the science of flow called *hydro*dynamics. Water is not, of course, the only substance that forms a liquid – but very few do so for the conditions of temperature and pressure under which life is comfortably conducted. For this reason, it is hardly surprising that ancient natural philosophers concluded, when they saw metals melt, that these substances were imbibing the qualities of water itself. Water's uniqueness inheres largely in its liquid state, which is why I shall need to delve into the neglected world of liquid physics – only to find how thoroughly water disrupts the theoretical landscape.

Every famous figure attracts myths, and water has those in plenty. The ancient myths are familiar to us – the world called forth out of a primal ocean, the Biblical flood. But in Chapters 10 and 11 I explore the twentieth-century myths of hydrous natural science, the strange ideas and stories that have grown from water's genuine oddness. Here is a detour through pseudo-science into almost-science and even into honest-to-goodness science that simply got things wrong, or did the job badly. You have to stay on your guard if you want to get to grips with water in all its guises.

There is much more to be said about water: this biography is not exhaustive. There is room for exploring its poetics, its literary influences, as well as saying more about its social, historical and technological roles. Water needs many biographers, because in truth it is not a personality but more like a culture to itself, with laws, arts, and a unique history and geography. What I want to do is to relate the central secret of water's nature – to explain why it is so remarkable a substance, and why as a result it is the matrix of life.

I am always deeply impressed by how generously scientists and colleagues will offer advice, often in considerable detail, for no greater reward than my gratitude. I am indebted on this account to Peter Atkins, William Brock, Marie-Clare Bellissent-Funel, David Chandler, Chris Chyba, Pablo Debenedetti, Chris Doake, Jan Engberts, John Finney, Peter Fisher, Felix Franks, Laura Garwin, Michael Grätzel, Jim Kasting, John Rowlinson, Gene Stanley, John Stednick, Allerd Stikker, Chris Surridge, Warren White, and David Williams. The book has been improved immeasurably by the sound advice and patient attention of my editor Toby Mundy. And thank you Julia for encouragement, suggestions, support and everything else.

Philip Ball
London, February 1999

PART I

Cosmic Juice

1 The First Flood

Water's origins

> Surely this is a great part of our dignity...that we can know, and that through us matter can know itself; that beginning with protons and electrons, out of the womb of time and the vastness of space, we can begin to understand; that organized as in us, the hydrogen, the carbon, the nitrogen, the oxygen, those 16 to 21 elements, the water, the sunlight – all, having become us, can begin to understand what they are, and how they came to be.
>
> George Wald (Nobel Laureate in Medicine) (1964)

> Those stars are the fleshed forebears
> Of these dark hills, bowed like labourers,
> And of my blood.
>
> Ted Hughes, 'Fire Eater'

In the beginning there was water. While the earth was formless and empty, the Hebrew God was 'hovering over the waters'. There was no sky, no dry land, until God separated 'the water under the expanse from the water above it' and commanded that 'the water under the sky be gathered to one place'.[1] Then the world emerged – from an infinite, primeval ocean.

This is echoed in similar myths throughout the world. In central and northern Asia, North America, India and Russia a recurring motif is that of the Earth Diver: an animal or a god who plunges to the bottom of a primordial ocean to bring up a seed of earth. The Polynesian cosmogeny reproduces that of the Old Testament in extraordinary detail: the supreme being Io says 'Let the waters be separated, let the heavens be formed, let the earth be!' For the Omaha Native Americans, all creatures once floated disconsolately on a wholly submerged Earth until a great boulder rose from the deep. In Hindu mythology, the sound that embodied Brahma became first water and wind, from which was woven the web of the world. 'Darkness was there, all wrapped around by darkness, and all was Water indiscriminate' says the beautiful creation hymn of the *Rig*

Veda (3700 BC).[2] For the Maya of central America also, the deity Hurakan called forth the land from a universe of darkness and water.

Why does this idea of a watery beginning resonate throughout disparate cultures, without heed to the local particulars of geography or religious tradition? Ultimately its origin may be psychological: the land was knowable for ancient peoples, but the sea was a symbol of the unconscious – something mysterious, pristine, unfathomable. I know of no creation myths where the land came first and the seas followed in a subsequent deluge.

Yet land and sea are contemporaneous and complementary in some traditions. The Judaeo-Christian distinction between flesh and blood is a distinction between the earthy and watery aspects of the corpus of the world. In Norse mythology, the land is the flesh and bones of Ymir, the first giant, slain by Odin. His salty blood, gushing from the spear-wound in his heart, became the oceans. So too in Chinese myth are land and sea coeval aspects of a primal being, Pan-Ku the sculptor, whose medium was his own body.

But is there any truth of a more material nature in these myths – was the world once covered with water? And where has the water come from?

In the beginning

In myth, the origin of the Universe is seldom differentiated from the origin of the Earth. To look beyond the beginning of our world is to ponder the eternal: the Chaos of the Greeks, the abyss of fire and ice called Ginnungagap by the Norse, or the supreme deity Akshara-Brahma in Hindu tradition. Today, Earth's beginning is merely a local question, a moment of parochial interest in an already mature Universe. The real moment of creation goes back at least six billion years beyond that, and it is as fantastic as any myth.

Origins are seldom uncontentious. Current fashion sometimes has it that the idea of a cosmic Big Bang is best regarded as our latest cultural myth, as much a social construct as the slaying of Ymir. On the one hand it can only be arrogant to suggest otherwise; on the other, it's this particular kind of confidence that makes science possible. From a scientific perspective, the Big Bang is beyond question still the best model we have for the birth of the Universe, and rests on some formidable pillars. To address the question of why water is what it is, modern cosmology provides a consistent and explanatory framework in a way that Odin's murder of Ymir does not.

Imagine watching a movie of an explosion moments after it has happened. You see many fragments, rushing away from each other within a 'bubble' of expanding size. When, in 1929, the astronomer Edwin Hubble saw the galaxies of the Universe behaving in the same way, he was forced to the same conclusion as the one we would reach from the movie: this is the aftermath of an explosion. But Hubble was seeing it from the inside – we are riding on one of those fragments, the Milky Way galaxy. The Universe is getting bigger, as all the galaxies rush away from one another. The natural inference is that all the matter in the Universe was once focused into a much smaller volume, which went bang. Albert Einstein had deduced as much in 1917: when he applied his theory of general relatively to the Universe as a whole, he found the equations predicting that it had to be either expanding or contracting. That seemed to him then to be a crazy notion, and so he added a 'fudge factor' to remove the expansion. But Hubble's discovery persuaded him in 1931 that no fudging was needed after all.[3]

After just one million-billionth of a second, when according to some theories the Universe might have been just a metre across, the temperature would have been in the region of a billion billion degrees. In such extremes, there can be no atoms and molecules, no matter as we currently know it.

But as it expanded, the temperature of the Universe dropped rapidly. At the end of the first day of creation, it would have been about ten million degrees – about as hot as the centre of a star. The Universe today, with all its stars and supernovae and quasars, is but a dim, cool remnant of this cosmic fireball. In 1965 Arno Penzias and Robert Wilson at Bell Telephone Laboratories detected the faint afterglow that pervades the sky: a uniform background radiation of microwaves coming from all directions, indicating an average temperature of almost three degrees above absolute zero. This cosmic microwave background is all that is left of the Big Bang's fury.

The fabric of water

George Wald's view, quoted at the beginning of this chapter, is mine: understanding what we are composed of, and where that stuff came from, is part of our dignity. It demands, too, a greater humility to read the lives of stars, rather than divine providence, in our bones and blood. But bones and blood must come later; for now, I want to follow only the gestation of those protons and electrons, and from them the hydrogen, the oxygen – and the water.

For this is what we're after, these two elements: the H and the O, which unite so readily to create our eponymous subject. Water is H_2O, the only chemical formula that everyone learns: two atoms of hydrogen welded to one of oxygen. Their union is a molecule – a cluster of atoms. Chop up a block of ice, and keep chopping – and your finest blade, finer than the keenest surgical scalpel, will eventually reduce the fragments to these individual three-atom clusters. If you chop beyond that, you no longer have water. The H_2O molecule is the smallest piece of water you can obtain, the basic unit of water.

So here is a central aspect of water's character: it is a *compound*, an association of atoms, divisible into atoms of different natures. Yet water is so fundamental to the world that for millennia it was mistaken, naturally enough, for an *element*, something indivisible. Hydrogen and oxygen are elements, because they each contain only one kind of atom. But there is no 'water atom' – only a water molecule, made up of two different types of atom.

Before making bread one must make flour; and before water could come into the Universe, there had to be hydrogen and oxygen atoms. But before flour comes wheat – and atoms too have more fundamental constituents, Wald's protons and electrons.

As far as atoms are concerned, protons and electrons are like knives and forks at the dinner table: no matter how big the table, there are equal numbers of each. The difference between atoms of different elements – between an atom of oxygen and one of carbon, say – is simply that they contain different numbers of protons. In this regard, the underlying pattern of atoms is numerical, as Jacob Bronowski says in *The Ascent of Man*. An atom with one proton (and one electron) is hydrogen; an atom with eight of each is oxygen. At one level, chemistry is as simple as counting.

At another level, it is clearly not. For mere proton book-keeping offers no clue as to why hydrogen atoms join with oxygen atoms in the ratio of 2:1, nor why sodium (eleven protons) is a soft, reactive metal, chlorine (seventeen protons) a corrosive gas, silicon (fourteen protons) an inert, grey solid. To understand any of this we need to consider how the electrons are deployed: for there are deeper patterns in the arrangement of the electrons that determine the element's chemical properties.

Protons and electrons are not, as British physicist J.J. Thomson believed at the turn of the century, lumped together inside the atom as a heterogeneous blob. Rather, they bear to one another something like the relationship of the planets to the Sun, with the electrons orbiting a central, dense nucleus where the protons are. In this 'solar system' model of the atom, proposed by Thomson's protégé New Zealander Ernest Rutherford, if the nucleus of an atom were scaled up to the size of the Sun then the electrons would be more distant than Neptune's orbit by a factor of about ten. Yet we shouldn't take the model too seriously: electrons can't be pinpointed like planets, and do not follow well-defined, elliptical paths but instead occupy regions of space called orbitals. These regions, which have the shapes of spheres, lobes and rings centred on the nucleus, are best regarded as hazy 'electron clouds', rather like swarms of bees around a hive. From the manner in which an atom's electrons are distributed amongst the various available orbitals flows the whole of chemistry.

Moreover, atomic nuclei grasp their electrons not by the force of gravity but by electrical attraction: an electron is negatively charged, and a proton has a positive charge of equal magnitude. An atom, with equal numbers of both particles, is electrically neutral. Electrons, however, can be stripped away from atoms, rather as a passing star could pull a planet from a nearby solar system. The depleted atom then has an excess of protons over electrons, and so is positively charged. Atoms can also gain an excess of electrons over protons, and so become negatively charged. These charged atoms are called *ions*. This is why, even though protons and electrons are equally represented in a neutral atom, it is the number of protons that is the fundamental characteristic of an element. To pull a proton out of an atom, you have to dig it from the dense mass of the nucleus. That takes a huge amount of energy, and it converts the atom into a different element entirely.

Although hydrogen atoms have one proton and oxygen atoms have eight, oxygen is about sixteen times heavier than hydrogen. There is a third ingredient to the atom – a particle called the neutron, which has virtually the same mass as a proton but is electrically neutral. All atoms bar hydrogen have neutrons as well as protons in their nuclei, and generally speaking the nuclei contain equal numbers of each. The vagueness in this statement is, I fear, unavoidable, for two reasons. First, the number of neutrons tends increasingly to exceed the number of protons for heavier atoms: the proportions are pretty much 50:50 for light atoms like carbon, oxygen and nitrogen, whereas lead atoms have around forty per cent more neutrons than protons. Second, even atoms of the same element can possess different numbers of neutrons. Oxygen atoms can contain seven, eight, nine or ten neutrons to accompany their eight protons, while hydrogen atoms can contain no, one or two neutrons. These different forms of atoms of the same element are called *isotopes*. Most hydrogen atoms have no neutrons; but 0.000015 per cent of all of those in nature have one neutron. This heavier isotope is called heavy hydrogen, hydrogen-2, or deuterium.

This is, I appreciate, the stuff of dry chemistry textbooks, and I regret forcing it on you so soon. I hope it is of some consolation to learn that this is all you will need to know about atoms for the rest of the book. But they are the alphabet of chemistry, so we need to be at least on familiar terms with them. Besides, if we are to consider how the Universe cooked up water, we need to know which ingredients must go into the pot.

The soup goes cold

Water is but a simple dish: the recipe tells us to mix hydrogen and oxygen. The first ingredient is the easy one: it dropped right out of the Big Bang, once things got cool enough. That's to say, protons – the nuclei of hydrogen atoms – condensed out of the fireball about a millionth of a second after time and space were born.

But at this point the temperature would have been around a trillion degrees, which is too hot for protons to hold on to electrons. The Universe was then a soup of protons and electrons, seasoned with neutrons and other subatomic particles such as neutrinos, all swimming in a seething broth of X-rays. And for a good few minutes, that's how things stayed; the Universe was too hot to be interesting.

Although protons could not yet combine with electrons, they could at least team up with each other and with neutrons – for the force that binds protons and neutrons together in the nucleus, called the nuclear strong force, is many, many times stronger than the electrical force of attraction between protons and electrons. Just one hundred seconds into the Big Bang, with temperatures close to three billion degrees, protons and neutrons began to combine to form the nuclei of heavier elements – a process called nucleosynthesis. Fusion of these particles led to the formation of the nuclei of several light elements: helium-4 (an amalgam of two protons and two neutrons),[4] lithium (three protons and three or four neutrons) and boron-11 (five protons, six neutrons). About a quarter of the mass in the Universe is helium-4, formed by nucleosynthesis in the early days of the Big Bang.

The proportion of the Universe's total mass that comes from all other elements is tiny, however: about one to two per cent in all. In other words, around three quarters of the Universe's mass is hydrogen, and the rest is mostly helium. Once the temperature had dropped to around 4000°C, nuclei became able to grasp and retain electrons. Protons teamed up with electrons, and hydrogen atoms were born.

Atomcraft

If chemistry had relied solely on the Big Bang, the Periodic Table would be but a short, formless list of half a dozen elements – easier to grasp, perhaps, except that you wouldn't exist to appreciate it. By the time it had fashioned boron, the Big Bang had exhausted its atom-making vigour.

Fortunately for us, gravity came to the rescue. Within the diffuse clouds of matter synthesized in the Big Bang, gravity began the slow but inexorable task of galaxy-building. Where the gas was ever so slightly denser, the inward tug of gravity was that bit stronger. And so almost imperceptible variations in density gradually became accentuated, condensing into ever more compact blobs, like a sheet of rainwater on a windscreen breaking up into a network of droplets. These amorphous clumps became the precursors of vast galaxy clusters, within which smaller clumps condensed into separate galaxies – a hierarchical fragmentation right down to the scale of the nebulae that would ultimately become stars.

As the pull of gravity made matter collapse in on itself, the stuff heated up. Stars ignited and began blazing. One by one, the lights came on again throughout the Universe. The stars are more than mere fireballs – they are engines of creation, and out of their fiery hearts come the elements needed to make worlds.

Transmutation made real

> Astronomy is an indispensable art; it should be rightly held in high esteem, and studied earnestly and thoroughly.[5]

So said the itinerant physician and alchemist Paracelsus in the sixteenth century, unsuspecting all along that the stars possessed the art he himself sought: the ability to convert one element to another. Stars are the alchemists of the Universe.

In the interior of stars, hydrogen nuclei are fused together to generate heavier elements; this is the process of nuclear fusion, and it is how stars conduct nucleosynthesis. Young stars are made mostly of hydrogen, which fuses in three steps to generate helium-4 and a great deal of energy. Over its lifetime, a typical star burns about twelve per cent of its hydrogen to helium in this way.

One often hears that this transmutation of elements is a thoroughly modern idea, unrelated in more than a coincidental sense with the alchemists' belief that elements can be interconverted. But on the contrary, it is possible to follow a continuous thread of logic and supposition from Paracelsian metaphysics to Enrico Fermi's first atomic pile in Chicago in the 1940s.

In 1815 British chemist William Prout proposed that atoms of the heavier elements are formed by the clustering together of hydrogen atoms, making hydrogen the 'first matter' or *prote hyle* from which Aristotle had suggested all matter is composed.[6] Tempting though it is to suggest that in this way Prout anticipated the twentieth-century discoveries of nuclear fusion and the structure of the atom, the reason Prout's idea wasn't laughed out of court (although it was by no means uncontroversial) was in fact because the legacy of alchemy was still in the air. Indeed, no less a figure than the eminent British chemist and physicist Michael Faraday remained convinced of the doctrine of elemental transmutation throughout his life.

Prout's theory was elaborated by the French chemist Jean Baptiste Dumas in the 1840s. Dumas noted that the atomic weights of some elements, which by then were known with impressive accuracy, were certainly *not* whole multiples of the atomic weight of hydrogen, and therefore these elements could not be made of clusters of hydrogen atoms. Dumas proposed that the fundamental unit of matter might instead be some subdivision of the hydrogen atom, perhaps a quarter or a half. Unknown to Dumas, the discrepancies are actually a consequence of the fact that elements exist in nature as a mixture of isotopes, so that their average mass does not correspond to a whole number of protons. The link between these ideas and the chemistry of the extraterrestrial Universe was made by Norman Lockyer in the 1870s. During this and the preceding decade, astronomers detected the fingerprints of many earthly elements in the light emitted by the Sun and other stars. Lockyer, in parallel with the Frenchman Pierre Janssen, discovered a new element in 1868 purely from its distinctive imprint on the spectrum of sunlight – a series of dark bands where the element absorbs light of certain colours. Lockyer called the element helium (after *helios*, Greek for the Sun), and it was not found on Earth until twenty-seven years later.

Lockyer developed a theory of the 'evolution of stars and chemical elements' which drew explicitly on Dumas's elaboration of Prout's hypothesis. He proposed that heavy elements are made from lighter ones inside stars as they cool from a blue-white brightness to a red dimness – a progression inferred from the observed colours of different stars. The British chemist William Crookes developed a similar hypothesis in the 1880s, based on the observation that gases subjected to high voltages could be decomposed into a plasma, a mixture of ions and electrons. Crookes considered plasmas to be a 'fourth state of matter' consisting of subatomic particles akin to those postulated by Prout and Dumas. He constructed an exotic scheme for the evolution and transmutation of elements from this plasma, which he assumed to be the stuff of stars.

Enter oxygen

In 1919 physicist Francis Aston, working at the Cavendish Laboratory of Cambridge University, developed a device that enabled him to measure the relative masses of atomic nuclei with great precision: the 'mass spectrograph', which we would now call a mass spectrometer. He found that even the nuclear masses of individual isotopes are generally not exactly whole multiples of hydrogen's; they are somewhat lighter, although typically by a margin of only a fraction of one per cent. The tiny difference in mass reflects the fact that a huge amount of energy is released when protons and neutrons combine to form heavier nuclei: the energy accounts for the 'missing mass', and is calculated according to Einstein's famous formulation $E=mc^2$. For the first time, Aston realized the vast energy lurking within the nuclei of atoms. When Ernest Rutherford, the director of the Cavendish, demonstrated in 1919 that a nuclear transmutation process could be induced by artificial means, scientists realized that it might be possible to extract this energy technologically – for better or worse. French physicist Jean Perrin proposed in the same year that the Sun and other stars might derive their energy from the fusion of hydrogen to heavier elements. In other words, nuclear fusion might not be just a *consequence* of the furious solar environment, as Lockyer had supposed, but the *cause* of it. Arthur Eddington added his approval in 1920: 'What is possible in the Cavendish Laboratory may not be too difficult in the Sun.'

In the mid-1930s the Russian George Gamow put Perrin's idea on firmer footing, suggesting that hydrogen is transformed to heavier elements by capturing a succession of protons or neutrons. German physicist Hans Bethe showed in 1939 that a tiny dose of carbon is needed to stimulate this process. A newly formed star condensing from a gaseous nebula typically contains about one per cent carbon, primarily in the form of the isotope carbon-12. This can provide the seed for the six-step sequence of nuclear reactions that converts hydrogen-1 to helium-4. The carbon-12 is recycled: consumed at the beginning of the sequence, but regurgitated at the end. By definition, it acts as a catalyst. This means that a tiny amount of carbon can facilitate the fusion of a lot of hydrogen.

At first glance, this cycle doesn't seem to get us very much further, since its net result is to transform hydrogen to helium – and we've seen that this can happen anyway, without the help of carbon. But in the intermediate steps of the cycle, other elements are formed: three different isotopes of nitrogen, and one of oxygen (the rare isotope oxygen-15). In Bethe's so-called C-N-O cycle, oxygen makes its entrance onto the cosmic stage.

The C-N-O cycle provides a significant fraction of a star's energy output. In fact, for stars several times more massive than the Sun it becomes a more important power source than the direct hydrogen-to-helium reactions. Because a star is constantly reiterating this cycle, it maintains a steady amount of carbon, nitrogen and oxygen in its atmosphere. Clearly, however, this can't be the whole story either. The C-N-O cycle generates only oxygen-15, while the isotopes we see in nature are mainly oxygen-16, -17 and -18. And what about all the even heavier elements?

Bethe supplied part of the answer. He showed that at particularly high temperatures a new set of nuclear reactions becomes possible, in which oxygen-16, oxygen-17 and fluorine-17 also take part. But this side-branch of the C-N-O cycle requires a pinch of oxygen-16. Newly formed stars in the present-day Universe acquire this oxygen isotope from the interstellar material from which they condense; but where did it come from in the first place?

The answer was provided in 1957 by Margaret and Geoffrey Burbidge, William Fowler and Fred Hoyle, in a paper that still defines today most of what we know about the nucleosynthesis of heavy elements in stars. The reason a star doesn't just keep collapsing once gravity has pulled it together from gas and dust is that the intense radiation produced by nuclear fusion gives the gas buoyancy, rather as a burner supplies buoyancy to the air in a hot-air balloon. But in the autumn years of a star's life, when it has burnt up most of its hydrogen and its fusion engine grows cooler, this buoyancy is lost and the star begins to contract. Gravitational collapse generates heat in the star's dense core, which is now mostly helium-4. At the same time, the star's outer atmosphere of gas expands and cools to a red glow, and it becomes a red giant star. In the hot, dense core, the star starts to burn helium. The nuclei fuse to make new elements whose masses grow in leaps of four: boron-8, carbon-12, oxygen-16, neon-20, magnesium-24, silicon-28 and beyond. Oxygen-18, meanwhile, is formed from fusion of helium-4 with nitrogen-14.

Eventually the helium in the star's core is used up too, and so the star has to resort to burning whatever it has left, which is mostly carbon and oxygen. This requires still higher temperatures and pressures, which are conveniently supplied when the diminishing fuel reserves permit further contraction, raising the core temperature to around a billion degrees. At this point, carbon-12 and oxygen-16 undergo fusion to generate a series of elements of about twice their mass: sodium-23, silicon-28, phosphorus-31 and sulphur-32. Thereafter, silicon-28 fuses with the helium nuclei produced in other reactions to make elements up to and heavier than iron. In their final evolutionary stage, such stars have a concentric-shell structure with a core of the heaviest elements and successive shells rich in silicon, in carbon/oxygen/nitrogen, helium and finally hydrogen.

And there is more. Stars larger than around four times the mass of the Sun may end their lives in spectacular fashion: as supernovae, which explode with a brightness that momentarily outshines the entire galaxy in which they reside. When such a star has finally exhausted its supply of nuclear fuel, there is nothing more to prevent the catastrophic collapse of the dense core under its own gravity. The in-rush of matter in the core generates a shock wave, and the star becomes unstable. In an awesome rebound, the outer envelope of the star is cast off and out into space while the star's core implodes to unspeakable densities whose inner region is a liquid of neutrons. Here atomic nuclei are unable to retain their separate identities but instead become crushed to a featureless miasma, and most of the protons combine with the electrons to produce a preponderance of neutrons. So the supernova becomes a dark, compact neutron star surrounded by an expanding shell of matter rich in a variety of elements. Such is the energy of the supernova's outburst that new nucleosynthesis reactions are triggered, enriching the debris with very heavy elements such as thorium and uranium.

These elements – the whole Periodic Table of them – are scattered through space. As a result of supernovae, the void between the stars is sprinkled with the raw material from which worlds are made. Walt Whitman anticipated this process in 1855 in an inspired poetic leap of imagination: 'a leaf of grass is no less than the journey-work of the stars'.[7] And Ted Hughes, in 'Fire Eater', reads the origins of earth and water in the firmament.

Wet space

Oxygen is the third most abundant element in the Universe – albeit a very poor third to hydrogen and helium, whose primordial generation in the Big Bang ensures that they constitute almost all of the fabric of creation. But helium is unreactive, a cosmic loner. And so should we after all be surprised that water, the combination of the Universe's most popular reactive elements, is so pervasive? This molecule, the matrix of life, is the product of the Universe's two most generous acts of creation: the Big Bang, which started it all and gave us a cosmos made mostly of hydrogen; and stellar evolution, which reformulates this element, whose very name means 'water former', into oxygen and all the other elements that make up the world. Within the imponderable expanses of interstellar space, these two elements unite – and there in the making is the river Nile, the Arabian Sea, the clouds and snowflakes, the juice of cells, the ice plains of Neptune, and who knows what other rivers, oceans and raindrops on worlds we may never see.

Every supernova sends a potent brew of atoms and molecules spewing out into the cosmos. But the cosmos is a big place, and even the creative might of an exploding star is a drop in the ocean. The space between the stars of our galaxy is emptier than the best human-made vacuum; and yet there is enough finely dispersed matter out there to make around ten billion more stars, one twentieth of the number in the luminous drapery of the Milky Way. This tenuous stuff is mostly hydrogen, but it has been delicately seasoned over the aeons with other elements and molecules, a dizzy menu of them. You'll find plenty of hydrogen molecules (H_2, two atoms hand in hand) out there, but also carbon monoxide, hydrogen cyanide, methanol and ethanol, ammonia, formaldehyde, and yes, water. There's solid matter too: tiny grains of silicate minerals, specks of soot and diamond – often with a coating of ice. All you need, in fact, to make a planet.

In some parts of the galaxy the gas and dust between the stars is clumpy, forming vast 'molecular clouds' which can block out the starlight beyond to give us fantastic sights like the Horsehead Nebula in Orion. In these clouds, stars may form as the matter condenses under its own gravity. That water is abundant in these regions was discovered in 1969 by astronomer Charles Townes and coworkers. The watery signature was barely legible: just a single bright peak in a microwave spectrum of cold interstellar gas. Molecules in interstellar space are usually detected from the lines that they strip out of the spectra of light from more distant objects – each type of molecule absorbs light at characteristic colours. But the water that Townes saw was not absorbing the microwave radiation – it was emitting it. The water was glowing! Improbable as it might seem in the deep freeze of space, molecules in interstellar clouds can be pumped full of energy. The molecules get 'hot' by undergoing collisions in dense regions of the clouds, and they cool again by emitting radiation. The molecules can synchronize their emission: the radiation emitted by one excited molecule can 'tickle' a second into emitting too, and before long a whole slew of hot molecules are casting off their excess energy. Much the same processes are responsible for light emission in some lasers. Because the 'light' from these collisionally pumped molecular clouds is in the microwave region of the spectrum, they are not cosmic lasers but *m*asers (from Microwave-Amplified Stimulated Emission of Radiation). What Townes and colleagues saw was the first known astrophysical water maser. These extensive astrophysical objects are now known to be regions where the gas is collapsing to form new stars. Water sends out a signal of star formation to the Universe at large.

Star formation: it's what every world needs. To make a planet, you first have to make a sun.

The great flood

The ancient Greeks guessed well about our planet's origin, for they believed that Mother Earth – Gaia – arose from a primordial Chaos. Chaos is the etymological origin for the word 'gas', and it was from gas and dust that the Earth was formed, along with the Sun and our sister planets. In an inspired guess, Immanuel Kant proposed as much in 1755.[8]

As a clump of gas collapses within a molecular cloud, it rotates and flattens out into a disk. While most of the matter gathers into the central core and is incorporated into the nascent star, some is left further out in the disk, where it provides the material for the formation of a planetary solar system. Several disk-like embryonic stars have been seen elsewhere in the galaxy. Some of the disks are punctuated by ring-shaped voids, thought to be the tracks engraved through the dust by newly formed, circulating planets. This happened in our own stellar disk – the solar nebula – about 4.6 billion years ago, when the Earth was one of those orbiting blobs.

But planets do not arise fully formed from globules of condensed solar nebula. We know this because there is far less of certain gases – neon, argon, krypton – in today's atmosphere than is thought to have been distributed through the solar nebula. Because these gases are chemically unreactive, we would expect them to remain as abundant as ever they were if our planet and its atmosphere was just a clump of pristine solar nebula with its elements rearranged.

No, planet formation is less stately and more traumatic than this. The accretions of gas and dust in the solar nebula formed smaller rocky bodies called planetesimals that range in size from boulders to moon-sized asteroids. These swarming planetesimals engaged in fearsome collisons that smashed each other to rubble – but the rubble from each collision then cohered into a single, larger object through the tug of its own gravity. Rather like companies, larger planetesimals grew at the expense of smaller ones until the disk was swept free of debris and only the planets remained, the multinational conglomerates of the solar system. The inner planets – Mercury, Venus, Earth and Mars – are relatively small, dense, rocky orbs. But out beyond the asteroid belt, where some of the smaller debris escaped capture, the planets were able to retain vast envelopes of gases and liquids: here we find the gas giants Jupiter and Saturn, and the frozen worlds of Uranus, Neptune and Pluto.

Earth was not a good holiday destination in those early days. The heat generated during its formation from colliding planetesimals created a global inferno. And around 4.5 billion years ago the Earth seems to have collided with a planetesimal about the size of Mars. Were this to happen today, you might as well cancel the papers. Global nuclear war would be a picnic in comparison – an impact this size would almost shatter the planet, and would certainly extinguish all life. As it was, it sheared off enough material to form the Moon, boiled away any atmosphere that the Earth then possessed, and left the planet a ball of molten rock (magma) for millions of years, its surface awash with a fiery ocean from pole to pole.

Yet collisions were not wholly destructive. On the contrary, they ultimately gave the planet an atmosphere, water – and the possibility of harbouring life. In the part of the solar nebula where the Earth condensed, volatile substances like water and carbon dioxide were rare commodities – only further out, where the temperature was low enough for them to condense and freeze, could they become a major component of planetesimals. These colder bodies could sequester a coating of ice from the gas and dust, just as snowflakes high in our atmosphere sweep up water vapour from the air. Blundering in and out of the nascent inner solar system, such objects most probably added water to the rocky mixture that was becoming the Earth.

To test whether this idea holds water, so to speak, planetary scientists today study the composition of meteorites. These cosmic boulders – well, they are more like pebbles on the whole, and some are no bigger than grains of sand – are mostly the leftovers of planet formation, the bits that never quite got incorporated into planets. It's likely, then, that the mixture of elements and compounds of which they are comprised reflects the composition of early planetesimals. They are still raining down on us from the skies, albeit in far smaller numbers than when the world was young. Many meteorites do indeed carry a bountiful crust of ice – not just water ice, but also frozen carbon dioxide, ammonia and other volatile compounds.

Meteorites called carbonaceous chondrites, which are rich in carbon compounds, can contain up to twenty per cent water, either as ice or locked up in the crystal structures of minerals. The most abundant type of meteorites, ordinary chondrites, carry much less water – around 0.1 per cent of their mass. Yet even this would be more than enough to fill the oceans, if the Earth was formed primarily from planetesimals with this composition.

But meteorites are not the only objects still wandering amongst the planets. There are itinerants the size of mountains out there, and they could deliver huge quantities of water to the Earth and its neighbours in a flash. I'm talking about comets, the unruly rabble of the outer solar system. Comets mostly originate in a roughly spherical cloud of objects stretching way beyond the orbit of the most distant planet Pluto, perhaps more than halfway to the nearest neighbouring star system. This halo, called the Oort cloud, contains around a million million comets, whose immense, looping orbits bring them occasionally sweeping through the inner solar system – as we saw in spectacular fashion with comet Hale-Bopp in 1997. They consist mostly of volatile gases condensed into ices, of which by far the most abundant is water. Mixed in with the ice is a scattering of mineral dust, making comets immense dirty snowballs. Generally they are a few hundred metres to several kilometres across, and so contain an awesome amount of water. Halley's comet, for instance, is a potato-shaped lump about eight by sixteen kilometres in size, with a mass of about one hundred trillion kilograms – most of which is ice. A typical comet is still larger, containing around one thousand trillion kilograms of water. A million comets like this would be enough to supply all of Earth's oceans.[9]

I'm glad to say that comets do not collide with Earth with anything like the frequency of small meteorites: the last major collision may have been sixty-five million years ago, possibly hastening the dinosaurs' demise. But comets swarmed through the solar system in far greater numbers when the Earth was forming, and would have crossed paths with the planet far more regularly, bringing oceans on their backs.[10] It seems that the gravitational tug of the outer planets Uranus and Neptune, as well as nearby stars, helped to rearrange the orbits of comet-like planetesimals in the Oort cloud so that they would pass more often through the inner solar system. Meanwhile these and the other giant planets, particularly Jupiter, eventually swept up most of the debris from the solar system and so quietened down the game of cosmic billiards by about a billion years after the planets had formed. Had this not happened, huge impacts might have delayed the appearance of life on Earth for billions of years. So we may have our neighbours to thank not only for our oceans but also for the life that spawned in them.

But I'm jumping the gun, for the oceans did not appear until many millions of years after the planet was formed. Four and a half billion years ago the Earth was still a molten magma ball, seething from the collision that ejected the Moon. As the planet cooled, its constituents separated like curdled milk. Within about fifty million years, the iron of which much of the Earth was comprised had sunk to the core, and the lighter elements – silicon, aluminium, calcium, magnesium, sodium, potassium and oxygen, along with some remaining iron – formed a rocky crust at the surface, just as slag floats on top of molten iron in a smelter.

Amongst all this rocky stuff were the volatile compounds delivered by collisions as the planet accreted – hydrogen, nitrogen, hydrogen sulphide, carbon oxides, water. While the Earth was molten, these volatile compounds dissolved in the magma, but as the molten rock cooled and solidified, the vapours were released in a process called degassing. The atmosphere that resulted from degassing was very different from today's, consisting mostly of carbon dioxide, nitrogen and water vapour.

Hydrogen is too light to be retained by the Earth's gravitational field, and was gradually lost from the early atmosphere into space. For this reason, the Earth is steadily losing its water too, albeit very slowly. The Sun's ultraviolet rays split water in the upper atmosphere into its constituent hydrogen and oxygen atoms, a process called photolysis. The hydrogen then escapes into space. This water-splitting costs the planet the equivalent of a small lake's worth of water each year. That sounds like a lot – and it certainly would be if it all came from a single lake! But averaged over the amount of water on the planet, the loss is probably quite small: photolysis may have reduced the Earth's water reserves by just 0.2 per cent since the planet was formed.

The day the rains came

Those formative years were steamy times on Earth, for all the water was in the sky. And then one day, somewhere between 4.4 and 4.0

billion years ago, the temperature had fallen far enough for water to condense. Clouds massed in the sky, and the oceans rained down. Sadly, I have to confess that this would not truly have happened so suddenly, one fine day in the Hadaean era – but I like the image. Yet however you look at it, there's no avoiding the conclusion that a deluge must eventually have ensued that leaves the Biblical version looking like an April shower. This was the Original Flood, and had anyone been there to witness it I don't think an Ark would have done them much good.

Far from eradicating life, this deluge set the stage for life's entry. It turned the face of the world blue, and created a planet that exists, in atmospheric scientist James Lovelock's words, as 'a strange and beautiful anomaly in our solar system'.[11]

2 Blood of the Earth

Seas and rivers of the world

Whence flow the Seas? Whence have free Springs their head? Whence do the far extended Rivers rise?

Titus Lucretius Carus (56 BC), *De Rerum Natura*

...as man has within him a pool of blood wherein the lungs as he breathes expand and contract, so the body of the earth has its ocean, which also rises and falls every six hours with the breathing of the world...

Leonardo da Vinci, *Notebooks*

We live on a blue planet, and seem more or less determined to disguise the fact. Our maps – North America and Asia stretching out to one another like Michelangelo's divine fingers in an attempt to bridge eastern and western land masses – give no clue that, seen from some angles, the globe is nearly all sea. Standard cartographic projections appear designed to maximize land area at the expense of the waters, to hide away the awesome glaze of the Pacific Ocean. Over two thirds of the planet's surface is covered by liquid water, and over one twentieth by ice. We call our home Earth – but Water would be more apt.

This is only human nature. Living in London, New York, Tokyo or even Lower Chevening, one forgets that other places are not the same. Wherever we are, we all too easily assume that our environment is representative. If extraterrestrial beings were to drop by on Earth to collect a random sample of its wildlife, we tend to imagine them hovering over the plains of Texas, plucking up cowboys and housewives. But it is more likely that they would collect a tank of algae from the Pacific.

Long adapted for land life, we have never truly come to terms with the dominance of the seas. These huge watery plains are hardly less scary today than they were to our ancestors, who populated their nether reaches with fabulous beasts and feared the chasms that lay just out of sight over the horizon. Even the deserts hold fewer terrors.

We have colonized the dense, steamy valleys of Amazonia, the frozen Arctic wastes and the starkness of the Siberian steppe – and yet we remain a little nervous even about what lies in the depths of the Scottish lochs, let alone the unplumbed abysses of the great oceans. In some ways we know more about the Moon, Venus and Mars, reassuringly free of liquid water, than about the ocean floor.

So we have a curious relationship to water on Earth. It nurtured and sustained civilization – yet the fresh waterways that fed the cultures of ancient China and Egypt, Mesopotamia and the Indian continent make up barely a tenth of a thousandth of all the liquid water on the planet. Just about all of the rest is salty, and lethal to the thirsty adventurer. Water giveth and water taketh away, in floods that resound through the legends of many cultures, in hurricanes and other wild faces of nature. Water has carried explorers far afield, yet it swallows up our puny vessels without a trace. The water gods, exemplified by Poseidon in Greek mythology, are ambiguous creatures with aspects both benign and terrible. Throughout the Book of Job, the contingencies of nature's waters are a continual metaphor for the trials of humankind.

The naïve psychological perspective associates water with life, and I will later say much that reinforces this intuition. But at a mythological level, the natural waters of the Earth offer humankind a journey into death. The Styx is the conduit to Hades, the Ganges even today a repository of the deceased. The Nile and the Tigris were not only holy in Near Eastern belief but the dwelling place of the dead, ruled by demigods with the power of resurrection. From the association of streams and rivers with death and rebirth comes the Christian practice of baptism.

But it is also via the broad oceans that water is associated with our passage beyond the borders of life. For the earliest seafarers this link was all too real. Death at sea has a special, mythical status: the drowned pass on to an altogether more fathomless fate than those whose corporeal being is returned to the shallow earth. In ancient cultures, children who died in birth were often carried to the river or the sea for fear that their disease would harm the fertile ground. The Ship of the Dead is a potent and recurring symbol: the *Flying Dutchman*, the *Marie Celeste*. Our enduring fascination with the doomed *Titanic* taps into this rich seam, while disasters far worse have faded to obscurity. Today there are new spectres abroad in the planet's liquid, as the final chapter will show.

Water is the agent of geological, environmental and global change. It confers fecundity in parched regions, while its passing turns grassland to desert. It spells the difference between blue skies and grey. The ebb and flow of oceanic heat bring peculiarities and extremes in climate – the benevolent warmth of the Gulf Stream, the jumpy pulse of El Niño – events which threaten drought or downpour depending on where you are, even the transition to ice-age conditions. Ice itself may be not just the harbinger but part of the very cause of these glacial spells, during which the Earth is refrigerated for thousands of years.

For all its fluidity, water is also one of the main shaping agents of nature. It makes rugged corrugations in highlands, carving out the intaglio of river valleys. It eats away at coastlines to generate underhangs, caves and eventually collapse, and to shift entire beaches down the coast. On its course from mountain to sea it may leave exquisite rock sculptures in its path. Cycles of freezing and thawing split apart the firmest of rocks, reducing slopes to rubble or heaving up stones from beneath the ground in fantastic, geological 'fairy rings'. And in tongues of ice water scours the Earth into broad valleys and shifts huge boulders over great distances.

In this and the following chapter I look at these many faces of natural water: the agent of planetary-scale life and death, the sculptor, the occult lubricant of volcanism and the agent of mineral formation, the brushwork of the skies. Water is what makes our planet unique.

The water wheel

But my brothers are as undependable as intermittent streams,
as the streams that overflow
when darkened by thawing ice
and swollen with melting snow, but that cease to flow in the dry season,
and in the heat
vanish from their channels.

Job 6: 15–17

Every day, every passing second, water is on the move. The rivers flow, the oceans perform their slow and elegant gyrations, the clouds congeal and weep. Each 3100 years, a volume of water equivalent to all the oceans passes through the atmosphere, carried there by evaporation and removed by precipitation. Yet only a thousandth of one per cent of the planet's total water resides in the atmosphere at any moment, enough to deposit just one inch of rain if it all fell uniformly throughout the world. This constant overturn of water between the reservoirs on land, in sea and sky is called the hydrological cycle (Fig. 2.1), and it is as crucial for life on Earth as is the presence of liquid water in the first place.

Leonardo da Vinci recognized that the Earth recycles its fluids. He appreciated that evaporation creates the clouds: 'the heat of the sun...calls up their moisture from the expanses of the sea'. But he believed that rainfall alone is not quite sufficient to account for the mighty torrents that pour from the mountains to the lowlands, and instead supposed that rivers are fed largely by water drawn up from the sea through 'the body of the mountain' by the 'natural heat' of the earth. His cycle from sea to mountain-top had therefore no pressing need to include evaporation:

> ...so therefore, one may conclude that the water passes from the rivers to the sea, and from the sea to the rivers, ever making the self-same round, and that all the sea and the rivers have passed through the mouth of the Nile an infinite number of times.[1]

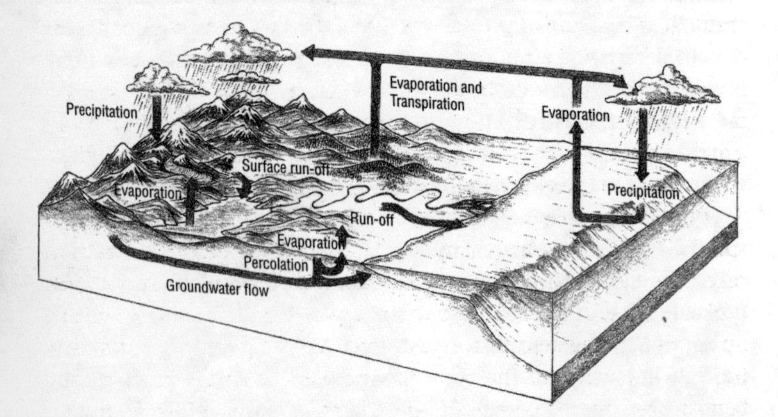

Fig. 2.1 The hydrological cycle carries water on an unending journey through streams, rivers and oceans, the atmosphere, the ice sheets, living systems and the deep Earth.

Not infinite, in fact – but certainly a huge number. And the Earth's natural heat, its internal volcanism, plays no role in the affair; rather, it is the Sun alone that powers this churning of the global water cycle.[2] The French lawyer and amateur geologist Pierre Perrault put Leonardo right in 1674, showing that evaporation and precipitation rather than an 'internal distillation in the earth' is 'sufficient to make springs and rivers run throughout the entire year'.[3] He estimated that the amount of rain that falls in the upper Seine valley is five times greater than the amount that the river bears away – one of the first ever uses of quantitative methods in the Earth sciences.

Most of the water that falls as rain has found its way into the sky from the sea surface: the Sun's heat removes from the oceans the equivalent of one metre depth each year – 875 cubic kilometres in total every day. A further 160 cubic kilometres evaporates each day from the land surface. Of course, this rate of evaporation varies widely with the seasons and with geographical location: because the tropics are warmer, the rate of evaporation there is at least four times greater than at the poles.

Evaporation from the ground and from plants (a process called transpiration – see page 223) removes water to the atmosphere, while precipitation, generally as rain and snow, supplies it to the land. The difference between precipitation and evaporation defines the amount of fresh water available for lakes, streams and other reserves on land. This 'run-off', which is mostly returned to the oceans through rivers, adds up to about 100 cubic kilometres globally per day. In deserts, evaporation is about equal to precipitation and there is essentially no run-off. In the Amazon basin, about half of the rainfall ends up as run-off, and most of this finds its way into the great Amazon river, which delivers an awesome one-fifth of the total fresh water input to the global oceans.

The various cogs of the hydrological cycle turn at a wide range of speeds. Rainfall in a river's upland source region can take weeks to reach the sea, while water vapour evaporated from the sea surface typically takes about ten days to fall again as rain. For water locked up as ice (in the so-called cryosphere), the cogs may grind slowly indeed. The water at the base of the polar ice sheets has typically been frozen for hundreds of thousands of years. Most mountain glaciers melt and recede by a few kilometres every decade under present-day conditions, while the sea ice in the polar seas expands and retreats seasonally.

The very existence of a hydrological cycle is a consequence of water's unique ability to exist in more than one physical state – solid, liquid or gas – under the conditions that prevail at the surface of the planet. Volcanic areas excepted, the Earth's surface never gets hot enough to boil water; but it evaporates readily nonetheless, since the amount of water vapour in the air is generally well below the 'saturated vapour pressure', the maximum humidity of air before water droplets start to condense. That's why the oceans are, to a greater or lesser degree, always 'steaming'. When moist air cools, the water vapour may condense back to the liquid state, producing the pearly billows of clouds or the dank blankets of mountain mist. This cycle of evaporation and condensation has come to seem so perfectly natural that we never think to remark on why no other substances display such transformations. Almost all of the non-aqueous fabric of our planet remains in the same physical state.[4] The oxygen and nitrogen of the air do not condense; the rocks, sands and soils do not melt (except in the furnace of the deep Earth) or evaporate. If these substances are transformed at all, it is often through the agency of water, which will dissolve many gases and minerals alike.

The freezing of water, meanwhile, can send it on a millennia-long detour from the cycle of evaporation and precipitation. Yet the ability of water to enter the solid state is also a crucial aspect of the overall cycle. When water is frozen during the ice ages, the world's seas recede, the climate becomes drier, deserts expand and ecosystems may be utterly transformed.

We will later see that this propinquity of water's changes of state and the planet's surface conditions is an anomaly, a peculiar outcome of water's unique molecular character. It is only because water is different that rain washes the streets, that brooks gurgle and rivers roar down from the mountains, that the surf crashes on the rocks.

The hydrological cycle emphasizes the dynamic nature of the Earth's environment: it is constantly repeating and renewing itself. Substances other than water are cycled by geological and biological processes too. Carbon from atmospheric carbon dioxide gets woven into the fabric of plants, may be thence consumed by animals, settles as dead organic debris to the ocean floor, is carried into the deep Earth at the convergence of tectonic plates, and is recycled into the atmosphere by volcanic emission of gases. Nitrogen from the air is converted by bacteria into nitrogen-containing nutrients in the soil, and is thereby taken up into living cells before being converted back to the nitrogen molecules of air by other microbes feeding off dead organic matter. These cyclic sequences of chemical and biological transformation of the elements are called biogeochemical cycles.

Water is the lubricant for biogeochemical cycling. Because it is such a superb solvent, and because it is itself in constant flux, it helps to convey other substances hither and thither, between different ecosystems and different climates. Carbon dioxide in the atmosphere dissolves in the surface waters of the sea to provide a carbon source for marine photosynthesis, and in turn this biological growth in the ocean's upper layer drives the rest of the ocean's carbon cycle. Essential nutrients pervade the seas in soluble form: nitrate, phosphate, sulphate, and metals such as iron. The swift churning of the hydrological cycle helps to drive the cycling of these other substances: rain and rivers flush inorganic nutrients out of the minerals of the rocky Earth and carry them to the sea. There is little exaggeration in saying that it is water, in the end, that makes the world go round.

Only such a dynamic environment, constantly changing yet constantly repeating itself, can support life. At the same time, life itself can come to exert an important and often dominant influence on these natural cycles, something that is particularly evident in the carbon and nitrogen cycles. Biogeochemical cycles create feedbacks that can enhance or damp out disturbances to the environment, such as changes in the intensity of the Sun's radiation or episodes of unusually vigorous volcanic activity. This gives the planet the potential to regulate itself, to maintain stable cycles and a relatively constant environment in the face of changing circumstances. The extent to which this really does happen, and in particular the degree to which living organisms play a part in it, is the central issue in the debate over James Lovelock's Gaia hypothesis, the idea of a self-regulating Earth.

Deep blue

> Have you journeyed to the springs of the sea
> or walked in the recesses of the deep?
>
> Job 38:16

What lies over the ocean's rim? Since humans first took to the sea, this question has been irresistible. The Phoenicians and Vikings crossed the Atlantic long before Columbus and Magellan in the heyday of European seafaring, and Chinese mariners reached the east coast of Africa in the fifteenth century, well before Portuguese colonists. By 1700 maps of the Atlantic Ocean were almost as accurate as today's. But although dragons may have been banished from beyond the world's end, it was largely the promise of distant lands (and resources), not the allure of the blue waters, that stimulated these explorations. There was little systematic effort to look at the seas for their own sake until the celebrated voyage of the British research vessel HMS *Challenger* from 1872 to 1876, which circled the globe and took depth soundings and ocean water samples in an attempt to look at the oceans as a part of the planet's geography, rather than as a highway to foreign exploitation.

What we have learnt since then is sobering. Around half of the Earth's solid surface is between three and six kilometres below sea level: the places where we live are like the tips of icebergs. The deepest parts of the ocean – the trenches – can plummet to over eleven kilometres, more than two kilometres deeper than Mount Everest is high. The floors of the great oceans are scarred down their middle by rugged, submerged ridges several kilometres high. These mid-ocean ridges mark the borders of tectonic plates: here magma wells up from the mantle, cooling at the sea bed to solidify into fresh ocean floor, while the plates move apart on either side.

The global conveyor

All the oceans of the world are interconnected. Yet the lumbering continents and the high points of the ocean floor modulate the degree of connection, enabling us to define distinct water masses, 'basins' separated by shallow shelves. Once mariners considered that there were seven seas to sail: the Atlantic, Pacific, Indian and Arctic Oceans, the Mediterranean and Caribbean Seas and the Gulf of Mexico. Today we recognize only three different ocean basins – the Atlantic, Pacific and Indian – although a fourth ocean, the Southern or Antarctic, is loosely designated as the southernmost portions of these three, encircling Antarctica itself. The world's seas, including the Mediterranean, the North Sea, the Red Sea, the Arabian Sea and the East and South China Seas, are large bodies of water that lie on the margins of the oceans and are typically separated by narrow gaps or straits, such as the Strait of Gibraltar, or by high ridges on the sea bed.

The waters in these oceans do not simply sit there bobbing up and down; they are constantly passing in concerted masses from one place to another, both vertically and horizontally, like shoppers riding the escalators in a crowded multi-storey department store. It is a stately procession, a sluggish reflection of the jets, streams, fronts and vortices of the ever-active atmosphere. But unlike the movement of air masses, the circulation of water between the oceans is constrained by the essentially arbitrary distribution of the continents. These have found their way to their current positions through continental drift, the movement of the tectonic plates driven by the even more sluggish convective motion in the Earth's mantle. There is thus a poetic recurrence of movement here throughout the classical elements of air, water and earth, at a successively slower pace. We might with only a little poetic licence find the sequence completed with the overturning of 'fire' – molten iron – in the Earth's core.

Because the tectonic plates go right on drifting and colliding in their ponderous manner, there is nothing fundamental or fixed about the pattern of the oceans. In the Jurassic period, about 170 million years ago, there were only two major oceans. The Panthalassa Ocean occupied virtually the whole planet between longitudes 30°W (now the mid-Atlantic) and what is now the international date line (mid-Pacific), and the Tethys Ocean stretched between latitudes 30°N (level with North Africa today) and 30°S (which now passes through southern Australia) in the eastern part of the globe. The Panthalassa Ocean became the Pacific, while the Tethys was gradually squeezed out of existence as the Indian continent collided with Asia to throw up the Himalayas.

Today there is a strong asymmetry in the distribution of land and sea between the two hemispheres. Almost two thirds of the global ocean is in the Southern Hemisphere, and a remarkable ninety-five per cent of all land points have antipodes – equivalent points in the other hemisphere – in the sea. Only in the Southern Ocean can one sail around a complete line of latitude without encountering land.

The ocean's surface currents, which shift the top one hundred metres or so of water, are driven mainly by winds. They propel the sea surface just as blown air will create flow in a cup of coffee. So these ocean surface currents, which are crucial to the distribution of fish and other marine organisms and to the transport of heat around the globe, are at the mercy of the atmosphere. A confusing convention has developed whereby ocean currents are defined in terms of the direction they are going, while air currents – winds – are designated by the direction from which they come. So a westerly wind drives an eastward ocean surface current, both moving in the same direction. Westerly winds prevail between latitudes 30° and 60° in both hemispheres; closer to the equator, down to latitudes of 15°, the easterly trade winds drive westward currents. Around the equator itself the winds are weak – the doldrums – and predominantly easterly.

This zonal (east–west) and almost hemispherically symmetrical driving of ocean surface currents is modified by two factors: the continents get in the way, and the Earth is spinning. Except in the Southern Ocean, the currents are hemmed in by land masses to the east and west, and so are deflected northwards and southwards close to the coasts to trace out huge closed loops called gyres (Fig. 2.2). This gyration is accentuated and modified by the effect of the Earth's rotation, through an influence called the Coriolis force. Named after the nineteenth-century French engineer Gaspard Coriolis, this force acts on an object that moves within a rotating system.[5] You can feel this force if you try to walk in a straight line out towards the edge of a rotating platform: the Coriolis force impels you to veer away from the line. On Earth, this force causes a current to diverge away from its initial course towards the right in the Northern Hemisphere and the left in the Southern Hemisphere.

To see what effect this has on ocean currents, consider the North Pacific gyre (see Fig. 2.2). The easterly trade winds drive a westward current in the southern part of the gyre (around 15°N), while the westerlies at around 30°N generate an eastward current at this latitude. But as the flow is deflected to the right by the Coriolis force, it becomes squashed up against the Asian coast to the west but stretched out and broadened as it proceeds towards the North American coast. What this means is that the western part of the gyre becomes a very intense northwards flow, which is called the Kuroshio, while the southward limb of the gyre off California is much more dispersed. In the Kuroshio current, the speed of the flow can reach up to a metre per second, which is three to ten times faster than that typically observed elsewhere in the oceans.

This same phenomenon occurs in the North Atlantic, where the intense northeasterly flow from the Gulf of Mexico and the Florida Straits along the eastern North American seaboard corresponds to the Gulf Stream. This narrow flow eventually becomes a more diffuse northeasterly current, the North Atlantic Current, which travels towards the British Isles and Norway bringing warm water and a milder climate to western Europe than is experienced at comparable latitudes on the North American continent. A similar focusing of the subtropical gyres occurs in the Southern Hemisphere, leading to the intense Brazil Current in the South Pacific and the Agulhas off south-eastern Africa in the South Indian Ocean (Fig. 2.2).

All of this is generalization, and thus simplification, that considers only the annually averaged flows. Only the easterly trade winds are steady throughout the year. Moreover, inconstant flows at smaller scales embellish the gyre systems with vortex-like eddies. The inconstancy of the oceans is illustrated most strikingly in the Indian Ocean, where the surface currents rotate in different directions at different times of the year in response to changes in the Asian monsoon winds. From around June to September, the southwest monsoon over India drives a clockwise circulation centred somewhere near the equator; from November to March, the northeast monsoon over southeast Asia reverses this flow.

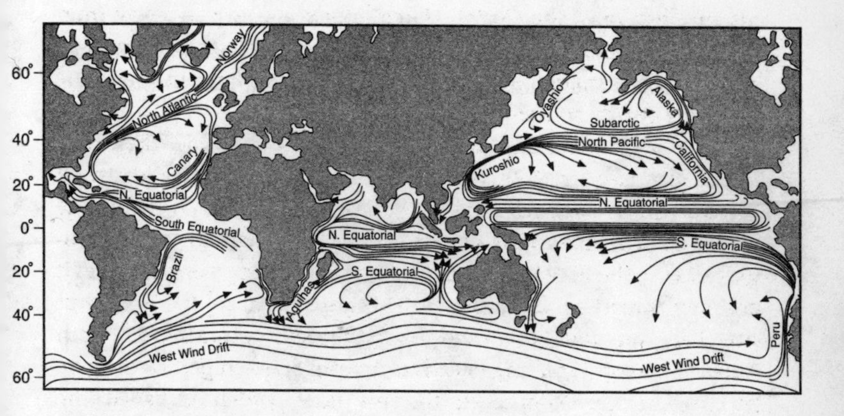

Fig. 2.2 Big wheels keep on turning. Global ocean surface currents are driven by winds. The obstruction by the continents and the force generated by the Earth's rotation mould the currents into circulating 'gyres'. The circulation in the Indian Ocean reverses direction between winter and summer owing to seasonal changes in the Asian monsoon winds.

Salt power

Winds can't drive ocean circulation at depths of much below a hundred metres. Deep circulation has another origin: it is driven largely by differences in water temperature. This churning, at depths of between one and five kilometres, carries warm water into colder seas, and so redistributes heat around the planet. Deep-water circulation forms a conveyor-belt flow which links all three of the world's oceans via the Southern Ocean (Fig. 2.3). To see how this flow is sustained, let's pick up a ride at the sea surface in the equatorial Atlantic Ocean. The upper part of the conveyor belt here is travelling northwards, and consists of water that has been warmed in the tropics. As it travels polewards, this water mass cools and becomes more dense. At the same time, evaporation from the sea surface leaves the surface water more salty, since the escaping water vapour does not take the salt with it. This enhanced saltiness (salinity) also increases the density of the sea water. So as the current progresses polewards it becomes heavier than the water below, and it sinks in the vicinity of the Labrador Sea, south of Greenland.

This cool, salty, dense water mass becomes North Atlantic Deep Water, which plunges to depths of around a thousand metres and is then carried in a return flow along the lower part of the conveyor belt back towards the equator. It passes from north to south across the entire Atlantic Ocean before reaching the Southern Ocean, where it rises towards the margins of the Antarctic continent. During the

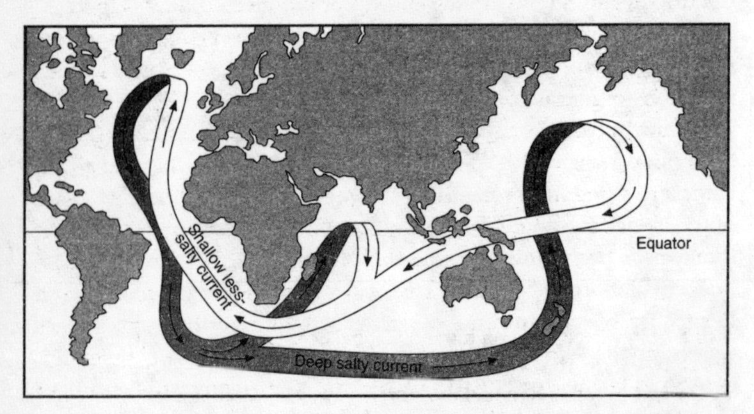

Fig. 2.3 The global conveyor. Circulation in the deep oceans bears warm, less salty water along its upper belt and cold, salty water on the lower belt.

Antarctic winter, a portion of this water mass freezes as sea ice in the Weddell Sea, leaving behind even colder and saltier water (since freezing, like evaporation, removes relatively pure water: the ice excludes the salt). This dense water sinks right to the ocean floor as Antarctic Bottom Water, the densest water in the oceans. But most of the deep current coming into the Southern Ocean from the Atlantic is borne eastwards as a salty flow called Antarctic Circumpolar Water, before turning northward into the Pacific Ocean off Australia. Here it warms up as it flows into the tropical Pacific Ocean, and the warmer, less dense water rises in the central North Pacific Ocean on the ascending branch of the conveyor belt, there to be carried back westward towards the Atlantic Ocean.

These differences in temperature and salinity mean that the deep oceans are not simply one homogeneous mass of water – they can be divided up into distinct reservoirs, which follow different pathways and mix rather little. The global conveyor belt of deep circulation is called the thermohaline circulation, since it is driven by changes in heat (Greek: *thermos*) and salinity (*halos*). The North Atlantic limb of the conveyor belt carries huge amounts of heat polewards in the warmer surface flow, which is released at high latitudes as the water cools and sinks. The heat delivered to the high-latitude North Atlantic Ocean in this way is about a quarter to a third of that delivered by direct sunshine. So the thermohaline circulation has a strong influence on climate. During the ice ages, it is believed that the circulation was far weaker, because the temperature difference between the tropics and high latitudes was less pronounced than it is today.

As the last ice age was coming to an end and the global climate was warming, around 12,000 to 10,000 years ago, there was a sudden reversion to glacial conditions around 10,500 years before present. This shift in climate, called the Younger Dryas event, seems to have been mind-bogglingly rapid: some estimates indicate that the global climate reverted from something like present-day conditions to those of the ice age in around fifty years. Some oceanographers believe that it may have been caused by a partial shutting down of the thermohaline circulation, as the extensive northern ice sheets melted and flushed fresh water into the ocean. This dilution of the salty surface waters in the North Atlantic Ocean would have made them less dense and less susceptible to sinking, and so the conveyor belt might have ground almost to a halt. The message is sobering, and is reinforced by similar rapid climate shifts that have shown up in climate records from the still more distant past: the deep circulation of the oceans may be a sensitive switch that could plunge the world into a deep freeze if disturbed.

Rhythms of the Moon

Superimposed on the large-scale average flows in the oceans are currents due to the daily ebb and flow of the tides, which can raise sea level by as much as fourteen metres on some coasts. Because the rise and fall of the tides has always been so important to many aspects of community life in coastal regions – from river transport to fishing to land access and availability – their cycles have been monitored with intense interest since antiquity, and their association with the phases of the Moon has been long known. The Japanese poet Shosammi Sueyoshi implies this connection with characteristic delicacy in the *Shinkokinshu* anthology (1205):

> The cries of the night
> Sanderlings draw closer
> To Narumi Beach;
> As the moon sinks in the sky
> The tide rises to the full.[6]

But it was not until 1687 that Isaac Newton provided a mechanical explanation for the 'Moon's floods'. Newton realized that the gravitational pull of the Moon would distort the Earth's shape and modify the profile of the oceans.

The gravitational pull of the Moon draws out a bulge in the ocean on the moonward side which is highest at the closest point on Earth to the Moon (the zenith).[7] But there is more to the tides than this, because of the Moon's rotation around the Earth. To be more precise, the Earth and the Moon are rotating around their common centre of mass – the point where a fulcrum would balance them on a set of cosmic scales. Because the Earth is much more massive than the Moon, the centre of gravity of the two is inside the Earth, 4729 kilometres from the planet's centre on the moonward side.

The rotation about this point creates a centrifugal force on the surface of the Earth. At the place on Earth that is furthest from the Moon (the nadir) the centrifugal force, dominating over the tug of the Moon's gravity, produces a second tidal bulge. The Moon rotates around the Earth in just over a terrestrial day – twenty-four hours and fifty minutes – dragging with it the tidal high points of the zenith and nadir and so creating a twice-daily (semi-diurnal) high tide. Because of the fifty-minute difference between a terrestrial day and the lunar rotation period, high tides come twenty-five minutes later on each successive half day.

Although the dominant influence on the tides, the Moon is not their sole instigator. The Sun exerts tidal forces too, which operate in just the same way but are a little under half as strong. The main (semi-diurnal) solar tide repeats every twelve hours. And both individually and acting together, the Sun and Moon establish still more subtle rhythms in the rise and fall of the seas, like the overtones of a bowed violin string. When the Earth, Moon and Sun line up at full and new moon, the solar and lunar tides reinforce each other and the waters rise to their highest extent. These are the spring tides. The low neap tides, on the other hand, happen when the line between the Earth and Moon is perpendicular to that between the Earth and Sun – when the Moon waxes through its first quarter or wanes through its third.

If the Earth was all water, it would be possible to predict the coming and going of the oceans' tidal bulges with mathematical precision. But the continents and the topography of the sea floor hinder their progress, and the Coriolis force deflects tidal currents in opposite directions in the two hemispheres. As a result, the arrival times of high tides at any point in an ocean basin cannot be predicted on the basis of astronomical calculations alone. It is a complex and largely empirical matter to map out the phases and amplitudes of the tides in any ocean basin – and even then these are subject to seasonal and longer-term variability. While Newton no doubt envisaged the tides as clockwork-like cycles, the truth is, as usual, more complicated, and fishermen are still better served by books of tidal tables than by pocket calculators.

The great arteries

> When the river rages,
> he is not alarmed
>
> Job 40:23

The great rivers of the world have a profound resonance even for those who have never set eyes on them. The names alone are enough to conjure up dark tales of exploration and adventure, romance and intrigue: the Congo, the Amazon, the Nile, the Volga and Seine and Danube. Perhaps it is too facile a connection, but I cannot help but wonder whether the stirrings that these names evoke are an echo of the ancient significance of the world's waterways, which offered travel in an age before airlines, and abundance in an age before supermarkets and global agro-industry. Historian Simon Schama sees still deeper roots:

> ...to see a river was to be swept up in a great current of myths and memories that was strong enough to carry us back to the first watery element of our existence in the womb. And along that stream were borne some of the most intense of our social and animal passions: the mysterious transmutations of blood and water; the vitality and mortality of heroes, empires, nations, and gods.[8]

The water that bathed and nurtured the roots of human civilization was fresh, not salt. All of the four oldest great civilizations sprung up by rivers and their fertile flood plains: Mesopotamia bracketed by the Tigris and the Euphrates (in modern Iraq); the Harrapan culture on the Indus (in what is now Pakistan); China on the mighty Yangtse and Yellow rivers from the brow of the Tibetan plateau; Egypt on the Nile. The fundamental nature of this dependence on water is reflected linguistically in Persian, in which the first word of the dictionary is *ab*, meaning 'water'. Herein lies the root of the word 'abode', from the Persian *abad*; and derived therefrom is *abadan*, 'civilized'. Quite literally, water constitutes the beginning of civilization.

Today rivers remain a source of plenty: of water for domestic supplies, for cooling, cleaning and other industrial purposes, for agricultural irrigation, for energy generation via hydroelectric power. For many people of the world they supply the staple protein intake in the form of fish, and their significance as routes of trade and transportation remains hard to overstate. They supply profound inspiration to artists and poets, and to scientists also. The metaphor of a river recurs throughout myth and literature. The primal significance of rivers is made explicit by the French poet Paul Claudel in a hypnotic recitation of names:

> Knowing my own quantity,
> It is I, I tug, I call upon all of my roots, the Ganges, the Mississippi,
> The thick spread of the Orinoco, the long thread of the Rhine,
> the Nile with its double bladder...[9]

Tales of epic exploration on the world's great rivers abound. The Nile, steeped in Egyptian myths of death, resurrection and fertility, its waters attributed a healing potency even as late as the seventeenth century, is the archetype of all rivers. The source of the Nile became an almost legendary Holy Grail. Caesar offered to abandon his wars in return for a glimpse of the springs from which it flowed, while for nineteenth-century explorers, obsessed with the Western idea of rivers as lines of power, the urge was to 'penetrate directly to the source'.[10] But the problem was that, as Claudel hints, this mighty river has no unique point of origin. The Blue Nile flows from the Ethiopian highlands, a fact known to the ancient Greeks. At Khartoum in Sudan it converges with the White Nile, whose source lies deep in the central African continent. John Speke identified this source as Lake Victoria in 1858, and in 1860 he established the Kagera river in Burundi as the Nile's southernmost point. David Livingstone found the source of the Congo on an expedition launched in 1866, but his quest for the still disputed source of the Nile was curtailed by his death from malaria in 1873. The Amazon, which rises in the Peruvian Andes just 160 kilometres from the Pacific Ocean, owes the name by which it is now known to the eurocentrism of the Spanish explorer Francisco de Orellana, who was purportedly assailed by a tribe of female warriors during his trek down the river in the 1540s. Thus did the river acquire the incongruous name of the tribe's counterparts in Greek myth.

Other rivers have long been major arteries of commerce. The Rhine, which makes a journey of 1320 kilometres from the Austrian Alps to the North Sea at Rotterdam, is navigable as far as Basel in Switzerland, and is linked by canal to the Ruhr industrial region of Germany and recently via the Rhine-Main-Danube waterway to the Danube – Central Europe's longest river – and thence to the Black Sea. But it is perhaps in the twin threads of the Tigris and the Euphrates, which empty into the Persian Gulf, that we can see the most profound role of rivers in human history. For in the land 'between two rivers', which is how the Greek name of Mesopotamia translates, nomads settled on the fertile flood plain in 8000 BC to become possibly the first farmers and herders in human history. There followed a succession of great civilizations: Babylonia in the lower valley from around 5000 BC, settled by tribes from near the coast of the Gulf; Sumeria in southern Mesopotamia from around 3100 BC; and Assyria in the northeast from 2000 BC. The need for coordinated irrigation to support the Babylonian settlements spurred the development of one of the earliest governmental structures, and also of industry – for coordinated engineering was required – and foreign trade in raw materials. The Sumerians made canals along the Tigris at least as far back as 2400 BC. In this fecund land we can discern how water brought culture and learning, social order and technological advancement, to the ancient world.

A force of nature

> As water wears away stones
> and torrents wash away the soil,
> so you destroy man's hope.
>
> Job 14:19

The geological role of rivers and streams[11] has many faces. They typically carry off around thirty per cent of the rain or snow-melt that falls on the areas that they drain. These areas, called drainage basins (also watersheds in the USA or catchments in the UK), are defined by the topography of the land, being typically bordered by ridges beyond which the next tiny stream feeds ultimately into another river. The shape of a drainage basin is determined by the river network itself as it incises channels into the landscape. The channel heads slowly cut back into the bedrock by washing away material through erosion. The result is typically a highly branched network, like the tree-root profile of the Amazon.

But the shapes of rivers can differ markedly. The Nile is the world's longest river – 6650 kilometres from the source of the White Nile in Burundi to its outflow in the Mediterranean Sea – but is rather thin and straight, with a catchment never exceeding 2000 kilometres in width. The Amazon is just 200 kilometres shorter, but it sprawls more widely to encompass a catchment of over twice the area of the Nile's: seven million square kilometres, an astonishing five per cent of the world's total land surface area. In part these differences can be ascribed to the differing climates of the two regions: much of the Nile runs through dry, parched lands for most of the year, whereas the Amazon's moist rain forest experiences extensive precipitation, giving it a wider source area and a greater annual flow than the Nile.

Of the different shapes that rivers and streams can adopt, two of the most common are called meandering and braided. Rivers flowing down low slopes over predominantly silty or clay terrain – conditions fulfilled on many flood plains – tend to wander in broad, symmetrical curves with a roughly constant wavelength. The word 'meander' derives from the River Maiandros (as the ancient Greeks called it) or Menderes (as it is now known) in Turkey, a famously twisty example. Meandering rivers change their course over time like a writhing snake, sometimes fast enough to complicate surrounding agriculture: the Mississippi can shift its tracks by up to twenty metres a year. Shifting meanders may leave behind them oxbow lakes where bends have approached close enough to fuse. These fusion events take place because the outside edges of each loop are always pushing farther out, since erosion is greatest at these points. The current is slowest, meanwhile, at the inside edges of meanders, which can consequently become clogged with deposits of silt. Braided rivers, on the other hand, follow a complex, interwoven network of paths separated by islands and spits. They seem to be the result of high sediment transport in the river water, a factor underlined by the similarity in appearance with small streams of water flowing over flat sandy beaches to the sea.

The migration of rivers and streams over a valley floor creates a flat flood plain, across which the waters rush if the river bursts its banks. Flood plains receive fresh doses of sediment during each flood; it is to such rich deposits, seasonally renewed and moistened, that the Nile valley owes the fertility that nurtured early Egypt. But flood-plain settlements live in risk too. Preferential deposition of sediment at the channel edges during floods can create natural levees around a river that help to confine it but may also allow it to rise above the level of the flood plain. If the river bursts these levees in flood, the result can be catastrophic.

Streams and rivers are a major shaping force of geology. They redistribute sediments to the tune of around sixteen billion tons each year – a figure that has risen dramatically since prehistorical times owing to human activities such as agriculture and dam building. These sediments may gradually extend the borders of the continents as they are dumped at vast river deltas. Rivers carve highlands into rugged landscapes, wearing away solid rock by a variety of processes. Sand and small stones carried by the flow grind away at the river bed, slowly carving out delicate flow-forms by abrasion. Larger rocks and boulders carried by more violent flows may crack and splinter the channel's boundaries. And ever the universal solvent, water erodes by chemical action too, dissolving minerals and releasing their elements into biogeochemical cycles.

Going underground

> Can reeds thrive without water?
>
> Job 8:11

Even the most arid of terrains is not always as dry as you'd think. Deserts have their oases, and regions miles from the nearest stream or river can be supplied with water from a deep well. How does this water appear from out of the parched earth?

A very small but, for human purposes, highly significant fraction of the world's water resides in hidden places. Around two thirds of the rain water that falls is returned directly to the atmosphere by evaporation and transpiration; and most of the rest is run-off, feeding streams and rivers. But a small amount permeates into the ground, draining through the spaces between soil grains until it reaches an impermeable layer of bedrock or dense clay. Here the water flows down the slope of the impermeable layer, winding its way through the soil's pores or through cracks and fissures in overlying rock. This is *groundwater*, and the permeable material through which it flows is called an *aquifer* – a 'water-bearer'. The channels of the aquifer are saturated with water, and the upper limit of this saturated region corresponds to the water table.

Strictly speaking, only the water in the saturated region is groundwater; above this, water that penetrates but does not saturate the soil is called vadose water, and this is the stuff that sustains most land plants and biological activity in the soil. If the water table rises close to the surface, the soil becomes waterlogged, and the only plants that can grow there are those tolerant to having their roots permanently doused in water, such as reeds and sedges. These saturated ecosystems are the Earth's wetlands: bogs, fens, marshes and swamps, all of them rich habitats and crucial to the biogeochemistry of the land masses.

The water table is not as clearly defined as the surface of a stream, since the water level peters out gradually. Moreover, its depth depends on the nature of the pore space in which the water sits, for in very narrow vertical pores the water level rises higher than gravity alone would permit. The water is sucked up the pores by 'capillary action', a consequence of the molecular forces of interaction between the pore walls and the liquid. It's this same effect that pulls up a curved meniscus when water meets the side of a glass beaker. The height of the 'capillary rise' is greater for narrower pores. In clay soils, where the grains are very small and the pore spaces between them are narrow, capillary action can raise the water level by over three metres.

A well can be created by drilling below the water table to reach an aquifer. If the water finds its own way out, for example through fissures in an aquifer's impermeable base that allow the water to emerge from a hillside or valley slope below the aquifer, the result is a spring. Some aquifers are confined between two layers of impermeable rock, and so there is no water table: the water cannot find its own level, but is forced to stay below the capping layer. Then it can become pressurized, particularly at depressions where the water flow is channelled into a basin. A hole drilled through such an aquifer releases this water under pressure, so that it surges spontaneously to the surface in a so-called artesian well.

Groundwater dissolves a rich concoction of minerals from the rocks through which it flows. It contains dissolved carbon dioxide from the atmosphere and from decomposition of plant matter, which turns it weakly acidic. If this water comes into contact with chalk and limestone – both comprised primarily of calcium carbonate, in combination with some magnesium carbonate – the acid reacts with the insoluble carbonates to generate relatively soluble bicarbonates, and the groundwater becomes a weakly alkaline bicarbonate solution. This is 'hard water', which causes scaling of water pipes and furring of kettles: the 'scale' is made up of calcium and magnesium carbonates, which precipitates again when the water is boiled and left to cool. 'Soft water', on the other hand, comes from aquifers that pass through rocks such as slate and granite, which contain little or no calcium and magnesium.

Dissolved minerals give spring water its health-sustaining properties. The 'healing waters' of spa towns such as Bath in southwest England have been celebrated at least since Roman times. The minerals impart a certain saltiness to the spring water, and this brew is all the more potent when the water is warmed during its passage through the earth, increasing its dissolving power. At warm springs such as those at Bath, the water penetrates deep enough through rock fissures to be heated by the ground's natural thermal gradient – the Earth gets warmer with depth at a rate of around 20–40°C per kilometre.

The temperature of thermal springs may be no greater than a comfortable 20°C or so, although at Bath it reaches almost 50°C. The hottest of hot springs, however, such as those that attract bathers in Iceland and Japan, are warmed by different means. These countries sit over or close to the borders of tectonic plates, where hot magma rises up through the Earth's mantle to manifest itself as volcanism at the surface. In these regions, the near-surface magma chambers make the geothermal gradient much steeper than it is elsewhere, and groundwater can become heated to boiling point, forming steaming pools of water or boiling mud where it reaches the surface. If a hot groundwater reservoir is largely enclosed except for a narrow shaft to the surface, the pressure caused by heating can trigger explosive ejections of water in the form of geysers (from the Icelandic *geysir*, meaning gusher). The pressures developed in geyser plumbing systems can be enormous: between 1900 and 1904 a new geyser appeared and then subsided at Waimangu in New Zealand that spurted to at least 1600 feet in a welter of mud and stones.

If more water is extracted from an aquifer for human use than is naturally fed into it by rain, the water table gets lowered. This can induce subsidence as the previously saturated ground dries and becomes more compacted. London subsided by an average of six to eight centimetres between 1865 and 1931 as a result of the lowering of its water table. There is now a rumour, however, that at least some parts of London's water table are rising again because the big breweries, which consumed much of the water in the city's aquifers, have moved out of town (or out of business). Londoners are warned of an impending threat of flood rising from the deep, because the capital no longer makes its own beer.

Crystal gazing

By comparing the workings of the Earth to those of the human body, Leonardo da Vinci was merely perpetuating a long tradition. The same theme was pursued in the sixteenth century by the German Georg Agricola, who had the benefit of a greater geological knowledge than Leonardo. In his classic study of minerals and metals *De Re Metallica* (1556), Agricola supposed that mineral veins are formed when groundwater percolating through the rocks, carrying dissolved salts, is 'baked by subterranean heat to a certain denseness' and so forms metal ore deposits. Agricola had in mind the transformation of one classical element to another: water to earth. But recouched in modern terms, his idea becomes a demonstrable truth: many ores are formed when groundwater dissolves minerals in the soil and rock and then precipitates them elsewhere in new combinations of elements.

Groundwater that percolates down to a source of heat, such as is offered by nearby volcanic activity, becomes a 'hydrothermal fluid'. Because water has a stronger dissolving power when warmer, these fluids may be supersaturated when they cool down again, and the dissolved mineral ions that they hold will recrystallize. Many hot springs are encrusted in rich mineral deposits where the water has cooled at the surface. If the water cools significantly before it reaches the surface, minerals can be precipitated instead in the fissures through which the fluid travels, creating mineral veins.

René Descartes disputed Agricola's hypothesis that groundwater from rain is the source of the deep fluids from which minerals form; he suggested instead that they originate from molten rock. In fact, he was right too: magma is indeed another source of hydrothermal fluids. As a body of magma cools deep in the Earth, igneous rocks crystallize out and the fluid that is left behind becomes increasingly enriched in volatile compounds such as carbon dioxide and water, as well as in ions that do not fit comfortably into the solidifying mineral's crystal lattice. These hydrothermal fluids form mineral veins as they are squeezed through fissures and lose their heat.

Hydrothermal fluids can also be generated when rocks are bent and squeezed by plate-tectonic motion. The folding and compression that results can expel and heat water from hydrous rocks, in which water molecules are inserted amongst the crystal's arrays of atoms. The minerals deposited this way are chemically related to the deformed source rocks: 'dewatering' of sandstone, for instance, can lead to deposits of quartz, since both are forms of silicon dioxide.

Not all mineral deposition requires hydrothermal fluids. Mineral salt deposits called evaporites are formed when a standing body of water at the Earth's surface evaporates, precipitating its dissolved mineral load as the solution gets increasingly concentrated. Evaporite minerals are typically found in salt marshes and around lakes, particularly the seasonal lakes called playas that accumulate and evaporate with the changing seasons in semi-arid prairies. Playas can range from just a few metres across to several kilometres, as in the case of Utah's Great Salt Lake. Streams that feed into the lake leach minerals from surrounding rocks and soils, enriching the lake with dissolved salts. The evaporite minerals that can be deposited this way include halite (sodium chloride, or common salt), gypsum (calcium sulphate, from which plaster of Paris is made) and borax (sodium borate).

A particularly important source of commercial metal ores arises from the formation of salt-rich solutions or brines, when evaporite minerals on a lake bed are dissolved by the water that percolates through the bottom sediments. As the sediments get compacted under their own weight and become deeply buried by the continual deposition of fresh material above, the water is expelled from the sediments' pores and is warmed by the geothermal gradient, and so becomes a good solvent for the evaporite salts. The resulting brines are typically rich in chloride and sulphate ions, and also in metal ions such as copper, lead and zinc. They may travel some distance through porous rock before the metals precipitate out as sulphide ores such as galena (lead sulphide) and sphalerite (zinc sulphide). Such metal ore veins are called Mississippi Valley Type deposits, after their occurrence in this river valley.

Evaporation of isolated bodies of sea water can create analogous deposits called marine evaporites. In 1970 scientists on board the *Glomar Challenger*, a research vessel equipped to drill cores in the sea bed for studies of marine geology, were astonished to find layers of evaporite deposits about a kilometre thick beneath sedimentary rocks in the Mediterranean Sea. Such massive evaporite formation could have been caused only by one thing: the evaporation of the entire Mediterranean Sea in the distant past. This seems to have happened around six million years ago, when a period of particularly dry climate lowered the sea level and cut off the connection to the Atlantic through the shallow Straits of Gibraltar, leaving the Mediterranean a land-locked sea. Without replenishment from the global oceans, the Mediterranean then slowly dried out and became a desert laden with huge tracts of gypsum and other evaporite salts –

an episode called the Messinian salinity crisis. The floor of this desert reached two thousand metres below global mean sea level, and the rivers feeding into it from Europe and Africa faced a huge plunge at the dried-up coast, carving great gorges into the rock. About five million years ago the dam at Gibraltar seems to have been breached, and the resulting waterfall, as the Atlantic fed back into the dry basin, would have made Victoria Falls look like a leaky tap. The torrent was one hundred times larger than that at Victoria, feeding about a hundred cubic kilometres of Atlantic water into the basin every day to fill up the Mediterranean again in about a century.

Lost oceans

Water pervades the Earth in subtle ways, and its influence goes way beneath the abyssal depths of the oceans. It gets hot down there, ever more so the deeper you go. At the base of the ocean crust, which is seven kilometres deep on average, the temperature is already around 150°C. By 660 kilometres down, in a region called the transition zone where the upper mantle is apparently distinguished from the lower mantle by a change in the mineral structure of the rock, the temperature is a furious 1500°C. It's not the kind of place you'd expect to find water.

And indeed, most geologists believed until the late 1980s that the deep Earth is a dry place. But recently that belief has been challenged in dramatic fashion. Some researchers now think that there may be enough water hidden in the Earth's bowels to refill the oceans thirty times. The water would be locked up in the crystal lattices of hydrous mantle minerals.

How might water get to such depths? Surface water is carried down into the Earth at subduction zones, where one tectonic plate is pushed down beneath another. The hard upper portion of the subducted plate plunges as a solid slab into the softer, hotter mantle below. A subducted slab typically contains a variety of hydrous minerals, such as clays and mica formed by the compression of sedimentary material deposited on top of the plate at the ocean floor. As the slab sinks into the mantle, the increasing temperatures and pressures squeeze the water out of the crystals, a process called dehydration. This liberated water then plays a critical role in the behaviour of the wedge of mantle lying above the subducted slab, because as it penetrates the mantle rock it drastically lowers the rock's melting temperature. As the mantle wedge becomes saturated with water released from the slab, rock that was previously stable in solid form at these high temperatures and pressures can melt into magma.

Because the rock is less dense when molten than when solid, the magma is buoyant within the mantle. It rises from the top face of the slab, where the water is released, and emerges in spectacular fashion through volcanoes at the top of the mantle wedge. If the two plates are converging beneath an ocean, the result is an arc of volcanic islands behind the subduction zone: an oceanic island arc, such as Japan (Fig. 2.4). If the subducted slab descends below the margin of a continent, a continental volcanic arc is formed instead, as in the Peruvian and Chilean Andes. The striking and hazardous volcanic activity in the Pacific 'Ring of Fire' is really a primal elemental conspiracy of fire, earth and water.

Subduction zones create a deep-earth limb of the hydrological cycle, in which water is drawn downwards from the ocean within hydrous rocks, released by dehydration of slabs, and recycled through magma either back into lithospheric rock or out into the atmosphere through steaming volcanoes. But some geologists are

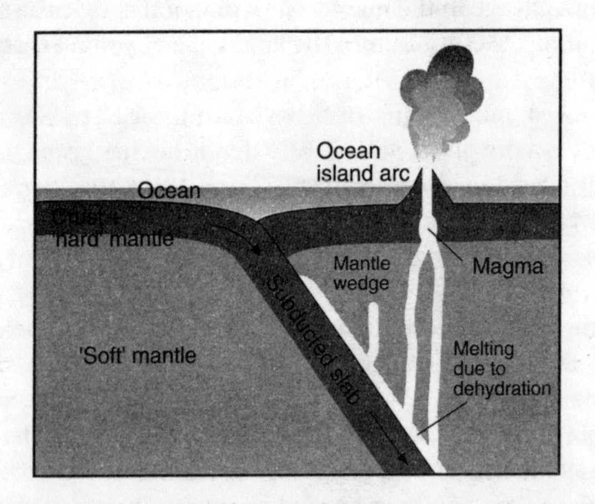

Fig. 2.4 Water in the deep Earth creates volcanoes where tectonic plates converge. The water is squeezed out of minerals in the descending slab and stimulates melting of the rock in the wedge above the slab. When two plates meet within the ocean (as shown here), this creates a series of volcanic islands like Japan. At the convergence of an oceanic and a continental plate, the result is an arc of volcanic mountains at the fringe of the continent, like the Andes in South America.

wondering whether water can get stuck down there: whether it could be transferred from dehydrating slabs into hydrous minerals that remain stable and do not melt at such depths.

It is no easy matter to investigate the composition of the deep Earth; just about all we know about it is indirect. When deep rock is exhumed in the form of molten magma, its composition is invariably changed from that at the depths at which it originates, owing to the gradations of temperature and pressure that it experiences during its ascent and to its interaction with the surrounding rock. In recent decades, however, it has become possible to study the Earth's mantle experimentally in the laboratory, using instruments that subject materials to tremendous pressures. These devices, called diamond anvil cells, squeeze rocks between diamond teeth to pressures that can equal those at the base of the mantle, where it meets the molten iron of the core.

Such studies have traditionally used 'dry' rocks, in which no water molecules are locked into the crystal structure, since it was assumed that the high temperatures and pressures of the mantle would drive water out. But in the late 1980s it was found that hydrous olivine, a mineral ubiquitous in the mantle, subjected to conditions like those in the mantle's transition zone will undergo a transformation to a related mineral called wadsleyite without giving up its water. In other words, if wadsleyite exists in the transition zone, it could potentially serve as a reservoir for trapped water. Some estimates suggest that as much as three quarters of the mantle between 400 and 525 kilometres depth could exist in the form of wadsleyite, which provides an immense possible reserve of water – over a dozen oceans' worth.

The consequences of a 'wet' mantle could be dramatic. 'Wet' magma might be responsible for catastrophic releases of volcanic rock in the past, now crystallized in the form of huge rocky plains called continental flood basalts. Flood basalts are created by plumes of magma that originate very deep in the mantle. As hot magma in a plume rises towards the surface, the pressure around it falls, since there is less weight of rock above, and so its melting point drops – just as the boiling point of water is lower on high mountains, where the air pressure is less. If such magma contains water – or picks it up en route – the melting point would be even lower, and so wet magma might undergo massive melting to erupt as vast outpourings of basalt. Water-induced melting of magma might explain some puzzling characteristics of flood basalts in Brazil and Namibia. This remains a tentative possibility, but it underlines the potential importance of water even in the realm of fire and earth.

Water the destroyer

> If he holds back the waters, there is drought,
> If he lets them loose, they devastate the land.
>
> Job 12:15

Nearly every culture has a legend of a catastrophic flood that destroyed all the world except for a chosen few. Noah is familiar throughout the Western world, and Atlantis too; but similar tales are to be found in cultures from China to Peru, New Guinea to native North America. Noah's voyage to the peak of Mount Ararat has its origins in the Assyrian and originally Sumerian legend of Gilgamesh, dating back at least as far as 2100 BC. But whereas Noah's god sought to cleanse the Earth of a corrupt and sinful people, it seems that Enlil, the chief god of Gilgamesh's pantheon, called down the heavy rains simply because the people who lived on the broad Earth were too noisy and disturbed his sleep. Only Utanapishtim, king of the city of Shurippak on the Euphrates, and his family and entourage were spared destruction, since the king was forewarned by Ea, god of wisdom, to build an ark.

It isn't hard to see how such legends might have arisen in the pre-Biblical Middle East, where the Sumerian cities along the Euphrates and the Tigris rivers were in constant danger from the massive floods of which these rivers are evidently capable. But the ubiquity of such legends testifies to the uneasy relationship of ancient civilizations with their water supplies. It is an uneasiness that persists today, when huge centres of population remain at threat from flooding.

There are, broadly speaking, two types of flood: those caused by rivers bursting their banks, and those that occur in coastal areas due to high sea swells, water waves and hurricanes. Because of changes in human activities and demography, the risks and damage caused by these floods are increasing in global extent. Today around forty per cent of deaths from acts of nature, as well as forty per cent of the societal costs of natural disasters, are the result of floods. Between 1965 and 1985, well over half of the federally declared disasters in the USA were floods caused by storms and rain. In that country alone, floods cause damages of typically $2 to $4 billion, and the loss of roughly 200 lives, each year. And although the figures vary widely, the trend is upwards. In the USA at least, this is not so much because the frequency of floods is increasing (although there is some possibility of that) but because property development is increasingly encroaching into threatened areas.

However, North American casualty figures are tiny compared with those in less developed countries. The USA has a well-established infrastructure for dealing with flooding: forecasts and warnings are issued and disseminated efficiently by fifty state offices of the National Weather Service's River and Flood Forecasting Service. In contrast, the facilities available to countries like India, Bangladesh and China, where huge numbers of people live in constant danger of floods, are far more limited. Thus, for example, when high rains from Hurricane Agnes caused massive flooding in the eastern USA from Georgia to New York that resulted in $3.5 billion worth of damage, only around 120 lives were lost; and although the Mississippi flood of 1993 was the worst for twenty years, causing damage of over $15 billion and rendering 50,000 people homeless in Iowa alone, only 487 were killed. This contrasts with the floods in northern Pakistan and India in 1992 which killed more than 2000 people and left hundreds of thousands homeless. And the combination of the Ganges, Brahmaputra and Megna rivers in Bangladesh leaves most of this low, fertile country – most of which is less than three metres above sea level – susceptible to flooding to depths greater than ninety centimetres every year.

Yet even this pales in comparison with the river floods of China, which are on a scale unsurpassed in recorded history. Ancient records tell of the Yangtse, Wei and Yellow rivers bursting their banks around 2297 BC to flood almost the entire north China plain, turning it into a virtual inland sea worthy of any legend. When the Yellow River flooded in 1332, around seven million people died from drowning and countless more from the resulting famine. As late as 1887, a similar event killed six million.

Rivers generally flood when high rainfall feeds them with more water than their channels can accommodate. In desert areas, rainfall tends to be brief but intense, accentuating the danger of very abrupt 'flash floods'. This danger abounds too in mountainous areas, where steep slopes focus the water into fast flows. The risks of flooding are increased when a high proportion of the fallen water finds its way to streams and rivers rather than soaking into the ground, and this is accentuated by the removal of vegetation – something that commonly takes place when land is claimed for human habitation. Covering natural soils with concrete and tarmac, or installing storm drains, are other human interventions that increase flood dangers.

Ice and snow also pose the threat of river flooding: the former by blocking constrictions in river channels (as seasonal ice breaks up, for instance), and the latter from melt water. Often these two effects operate concurrently. Such was the case, for instance, in 1936 when a flood in New England due to snowmelt and ice jams killed 107 people.

But river floods have not always been detrimental. In ancient Egypt the annual flooding of the Nile provided arable land that enabled the rich culture of the Nile delta to thrive. It's notable that Egyptian civilization is one of the very few that has no legend of a deluge, presumably because the Nile is so vast that it was always able to act as a buffer against intense meteorological events. At the same time, its bountiful floods could always be relied upon to distribute fertile silty soils across the delta, because of the insurance provided by its 'twin bladders' in Ethiopia and central Africa: if one source was unusually depleted one year, the other would generally compensate. The flooding commenced in southern Egypt around mid-August, and reached the northern end by mid- to late September. By October in the south and November in the north, it was all over, and the waterlogged flood plains were ready for crop planting. From around 3100 BC, the arable season was extended, and the area expanded, by the use of artificial irrigation systems: channels were dug, deepened and diverted to spread the waters further and to hold them for longer. So reliably was the extent of flooding linked to crop yields that the carefully monitored flood levels were used to determine the year's taxation rate.

Coastal floods have several causes, and effects fully as devastating as those of river floods – all the more so given that population densities in many countries, such as the USA, are commonly anomalously high around the coasts. Unusually high tides can be dangerous, but seldom catastrophically so except when they coincide with severe weather conditions. In 1953, for instance, high tides and storm-induced sea swells in the North Sea caused coastal floods in eastern England and in the Netherlands, killing over two thousand people in all. If high river outflows due to intense rains coincide with high tides, the river waters can be hindered from discharging into the sea and the result can be flooding of estuarine regions.

But the main flood dangers for coastal areas arrive in the furious form of tropical cyclones. These most awesome of atmospheric phenomena arise over the oceans in the tropics, between about latitudes 23° and 5°, from June to November. In the USA and the Caribbean they are known as hurricanes (from the West Indian *huracan*: big wind), and in the western Pacific as typhoons (from the Chinese *taifun*: great wind). They take the form of a huge vortex of inward-spiralling winds, up to a thousand miles across, which spins anticlockwise in the Northern Hemisphere and clockwise in the Southern owing to the Coriolis effect. In the very centre of the hurricane is a column of relatively clear, low-pressure air about ten to fifty kilometres across – the fateful eye of the storm, staring out into space. Surrounding this is a wall of thunderclouds, driving rain and furious winds blowing at up to two hundred miles per hour as they spiral inwards and upwards.

Tropical cyclones are initiated by the evaporation of water from a sea surface warmer than 27°C. The high rate of evaporation from such a warm sea creates extremely moist air, from which heavy rain condenses. When water vapour condenses, it releases heat, called 'latent heat' (see page 146). So rain formation warms the air further, and it rises by convection. This in turn sets up a region of very low air pressure, which sucks more moist air inwards. The structure is self-enhancing: the more it rains, the more latent heat is released and the more the consequent rising air and low pressure enhance the inward flow of moist air primed for condensation. Slowly the winds, spun up by the Coriolis effect, grow into a fearsome edifice that releases in a day about as much energy as an industrialized nation typically uses in a year.

The terrible winds of a hurricane wreak havoc when the vortex runs up against a coastline. But much of the hurricane's destructive power stems from its effect on the sea surface. The low pressure in the centre pulls the water into a bulge up to ten feet high; and more seriously, as the hurricane moves forward the winds push the water ahead of it into a great wave called a storm surge. This surge typically reaches a height of about twenty-five feet, and surge waves as high as forty feet have been recorded. The coastal flooding induced by the storm surge is responsible for ninety per cent of the deaths caused by hurricanes.

The Caribbean and eastern American seaboard are repeatedly lashed by hurricanes in the tropical Atlantic Ocean. In 1900 a hurricane unleashed a coastal wave on the Texan town of Galveston, causing flooding that killed around ten thousand people in the most deadly natural disaster ever to strike the USA. Belize City in Central America was similarly devastated by Hurricane Hattie in 1961, prompting its relocation fifty miles inland. And two of the biggest US hurricanes in recorded history struck in the past ten years – Hugo in 1989 and Andrew in 1992 – between them causing almost $30 billion of damage.

But it is probably Bangladesh in the Indian Ocean that suffers most from tropical cyclones, as they are called there. In 1737 a forty-foot storm surge swept along the coast of the Bay of Bengal, drowning around a million people. 1876 saw another devastating storm-surge flooding event, and in 1970 a wave fifteen feet tall inundated Bangladesh (then East Pakistan), flooding a million acres of rice fields and causing at least two hundred thousand deaths. Over half as many people again lost their lives this way in 1991. The problem is not just that Bangladesh is densely populated and lacking in the resources to erect defences or effect evacuations; the land is so low that even relatively small storm surges, a couple of metres high, can have terrible consequences. Truly, tropical cyclones represent one of the most awful conspiracies of natural waters in the skies and the oceans.

The great wave

> Your forces come against me wave upon wave.
>
> Job 10:17

As water-induced natural disasters go, few can be as terrifying as the giant coastal waves known as tsunamis, which afflict the Pacific coasts of the Far East in particular. That such colossal waves appear to be emblematic in Japanese culture, most famously in Hokusai's painting *The Great Wave*, is scarcely surprising, as they have afflicted this sea-bound country more profoundly than any other. *Tsunami* is a Japanese word, meaning 'long harbour wave', and as such it would once have been applied to any giant wave striking the Japanese coast. But today the term is restricted to those that have their origin in seismic outbursts: earthquakes or volcanic eruptions.

Arising suddenly out of calm seas, tsunamis have something of the supernatural about them, and so it is no wonder that they have been implicated in several near-legendary events. The influence of a tsunami has been detected in the account by the Greek historian Herodotus of how the city of Potidea was rescued from a Persian siege in 479 BC. During the siege, says Herodotus, the sea retreated, and the Persians took advantage of this to attack the exposed seaward side of the city, only to be drowned when the sea came flooding back. Recession of the sea is commonly mentioned in historical reports of the approach of a tsunami.

William Hobbs, who related this tale in 1907, was also moved to speculate that a tsunami might have been responsible for the parting of the Red Sea that allowed Moses and the Israelites to evade their Egyptian pursuers. Later commentators suggested that it was perhaps not the Red Sea that the Israelites crossed but the Sea of Reeds on the Mediterranean coast, and that the tsunami was produced by the eruption of Santorini volcano in the Aegean Sea in the fifteenth century BC – an outburst that may have inundated the northern coast of nearby Crete, triggering the collapse of the Minoan civilization. Tsunamis are not uncommon in the eastern Mediterranean – at least 141 have been identified since the second millennium BC – and I do not dare to investigate how many theories about the fate of Atlantis pursue similar lines of argument.

Japan has been hit so hard by tsunamis for the simple reason that it sits in the most seismically active area of the world. The death count from these occurrences is staggering: a single tsunami in 1896 claimed the lives of 27,000 Japanese people, and another 15,000 were killed in 1792 by a giant wave on the Shimabara peninsula.

The association of tsunamis with earthquakes has been recognised at least since the fifth century BC, when it was mentioned by the Athenian Thucydides. There are several ways in which earthquakes (or volcanic eruptions, which also cause ground tremors) set up a giant sea wave. Landslides occurring either on land or in sediments beneath the ocean surface can displace large volumes of water; such a landslide following an earthquake in Lituja Bay in Alaska in 1957 created a wave fully 200 feet high, which swept away all of the trees up to 530 metres on the Alaskan shore. Explosive eruptions can create a shock wave; an explosive tsunami of this sort following the eruption of Krakatau volcano in 1883 drowned over 36,000 people along the Javanese and Sumatran coasts. And the sea floor itself can push up a wave as it is displaced by slippage at an earthquake fault line; this is the cause of most tsunamis, such as that following an earthquake in Prince William Sound in Alaska in 1964.

Tsunamis can travel over tremendous distances in the ocean before their energy is dissipated. Some have been known to cross the Pacific Ocean from one coast to another: a magnitude-9 earthquake on the northwest seaboard of North America in 1700, for example, produced a wave that traversed the Pacific and caused damage and destruction in Japan, while an earthquake in Chile in 1960 sent tsunamis to Hawaii, Japan and Alaska. Small wonder that mid-Pacific islands like Hawaii have to be on the lookout in all directions. Fortunately, tsunamis move through the sea much more slowly than the seismic waves that cause them, which are transmitted rapidly through the firmer medium of the Earth's crust and mantle. So tsunami detection networks can rely on seismic monitoring to provide a warning signal. The difficulty is that it is not straightforward to tell whether or not a particular seismic event has generated a tsunami – and the ocean is an awfully big place in which to search. So warning systems have to balance safety against the problems of large-scale evacuation through excessive caution.

Water can be harnessed, but not tamed. We can make it serve us in little ways; its capacity to dominate and overwhelm us is far greater. Japanese poet Ki no Tsurayuki captured this ambivalence in the *Tosa Nikki* (AD 936):

> Seeking to fathom the mind of the raging god, we cast
> A mirror into the stormy sea. In that his image is revealed.
>
> Surely this cannot be the god whom we commonly associate with such gentle things as 'Limpid Waters', 'The Balm of Forgetfulness', and 'Pines along the Shore'? We have all seen with our own eyes – and with the help of a mirror – what sort of a god he is.[12]

The importance of appeasing this terrible god infuses the history of seafaring. Water, whose mirror-like surface is itself an invitation to reverie, tranquillity and calm –

> A pattern of wave ripples, woven – it seems –
> On a loom of green willows reflected in the stream.[13]

– this same water may, if the mood takes it, rise up and swallow us whole.

3 Storehouses of the Hail

Natural waters in the sky and ice

I am the daughter of Earth and Water,
And the nursling of the Sky;
I pass through the pores of the ocean and shores;
I change, but I cannot die…
I silently laugh at my own cenotaph,
And out of the caverns of rain,
Like a child from the womb, like a ghost from the tomb,
I arise and unbuild it again.

Percy Bysshe Shelley, 'The Cloud'

Suddenly I saw the cold and rook-delighting heaven
That seemed as though ice burned and was but the more ice.

W.B. Yeats, 'The Cold Heaven'

Not all of the planet's water flows. Water is unique on Earth in showing all three of its faces – solid, liquid and vapour – to the stars. Our blue gem is streaked and blotched with the pearly brilliance of cloud and ice. In this chapter I consider our planet's whiteness – the role played in nature by water vapour and ice. We've seen already that these states of water are but arcs on a great wheel, the hydrological cycle, which connects water in all its worldly niches. Chopping up the wheel in this manner can never be other than artificial, a convenience made possible by the visual discrimination that exists between sea and river, sky and snow-fields. Yet we shall see that there is at least one good reason to consider together the 'white waters' of the atmosphere and the poles, even though one of them is tenuous and the other of crystalline hardness. To sunlight streaming from space, a white surface is a reflective surface, never mind that we fly Jumbo jets through the one and need ice-breakers to penetrate the other. This means, as we shall see, that clouds and ice have central roles in the balancing act that determines the Earth's climate.

The soft fleece of the clouds is in fact a fabric of liquid and frozen water, its milky brightness derived from the same physical principle as that which leaves milk itself white: the droplets are of a similar size to the wavelengths of visible light, so they scatter sunbeams just as the obstacles on a pinball table scatter the pinballs. But clouds are merely an occasional reminder of the invisible truth that the air is full of water vapour – a transparent presence which, when the conditions so dictate, conjures up the damp shrouds of mountain mists, the cloying blankets of fog and the billowing, bruised towers of a thunderstack. Were we able to see at infrared wavelengths, the skies would be much more richly hued from the water vapour, which filters a part of the infrared out of sunlight. Using instruments called spectrometers that *do* see at these wavelengths, space travellers could use this special and nameless hue as an atmospheric indicator of surface water on other planets.

Clouds have lives as short as those of insects. Ice, meanwhile, can live as long as the mountains, or nearly so. Our planet's ice has many tales to tell, most of them very ancient. There is ice on this world that was a hundred thousand years old when the oldest known Palaeolithic paintings were being daubed on cave walls. And the scars that ice leaves in the landscape were the trigger for a revolution in geological science as profound as Copernicus's heliocentric theory was to astronomy. Locked away in frozen water is a key to our future too, since the great planetary ice sheets hold immense reserves of water that, if released, would transform the face of the Earth. A burning issue for today's Earth scientists is the question of how securely the floodgates are locked.

Something in the air

> When I made the clouds its garment
> and wrapped it in thick darkness...
>
> Job 38:9

To water, falling as it does from the heavens, ancient Chinese scholars ascribed a tendency towards 'downwards' motion – in contrast to the 'upwards' movement of fire. 'Water is that which soaks and descends', said Tung Chung-shu in the second century AD. But what comes down must first go up. Water is a component of the Earth's atmosphere.

The fraction of all the Earth's surface water that resides up there is minuscule: about a thousandth of one per cent, or just 0.035 per cent of all the fresh water. Yet atmospheric water is one of the principal mediators of terrestrial weather and climate, and is far and away the most important source of fresh water for human use. While the instantaneous water content of the atmosphere seems negligible, the constant traffic through this conduit of the hydrological cycle, with its attendant transformation from salty to fresh, is a vital component of the Earth's self-regulating environment.

Cloudbursting

> He wraps up the waters in his clouds,
> yet the clouds do not burst under their weight.
>
> Job 26:8

It's tempting to blame on Wordsworth and Shelley the misguided perception that clouds are distinct entities traversing the skies as jellyfish ply the seas: wandering lonely, or piloted 'over earth and ocean, with gentle motion'. Although individual clouds may be carried small distances by air currents, on the whole they are ephemeral sculptures of water and ice that condense out of clear skies, stick around for a while, and then disperse again like Marley's ghost. We'd do better to regard clouds as processes, like waterfalls, than as discrete objects.

A cloud forms when water vapour in a parcel of air condenses into tiny particles of liquid water or ice. Condensation occurs when moist air rises and cools, which can render it supersaturated with water vapour. The altitude at which supersaturation gives way to condensation is called the condensation level. There is a limit to how much water vapour air can hold, just as there is only so much salt a glass of water can dissolve. An air parcel that can accommodate no more water vapour without some of it condensing into liquid is said to be saturated, like a salt solution that can dissolve no more of the stuff.

And just as warm water will dissolve more salt than cold water, likewise warm air holds more moisture than cold. That's why moist yet undersaturated air can become supersaturated when it cools. But condensation of water droplets in saturated air won't happen unless the air contains tiny, suspended solid particles, designated 'condensation nuclei', which provide surfaces for the water to condense onto. Over the continents, condensation nuclei are mostly airborne particles of mineral dust, although in industrialized areas there are high levels of human-made particles too, such as specks of sooty carbon from fossil-fuel burning.

These particles originate from the land surface, and don't get carried far out over the oceans before they are washed out of the air in rain, or fall into the waves under gravity. But the seas have their own sources of condensation nuclei. Some are flecks of crystalline sea salt, which are formed when droplets from breaking waves evaporate in the air. These particles pollinate the sea hazes that linger over the ocean surface. Other oceanic condensation nuclei are manufactured by marine life. Certain types of phytoplankton produce a gas called dimethyl sulphide (DMS), apparently as a by-product of their metabolism. The distinctive odour of DMS is responsible for the invigorating smell of the coastal sea. In the atmosphere the gas is converted to sulphate, which can coalesce with the sodium and magnesium ions borne aloft in sea salt to form crystalline particles that absorb water vapour and dissolve into the seeds of cloud droplets.

For a cloud to shed rain, the droplets have to grow large enough to fall quickly through the air. But their growth is rather slow – depending on how big the condensation nucleus is, it may take a droplet over an hour to reach a size of twenty micrometres. After it reaches this size, further growth occurs primarily by coalescence with other droplets rather than by the condensation of more liquid on the droplet's surface. A droplet acquires its limiting descent speed – the free-fall or 'terminal' velocity – when the pull of gravity is balanced by the frictional resistance due to its passage through the air. A fine drizzle consists of droplets of typically 200 micrometres or so across, which reach free fall at a leisurely half a metre or so per second. Cloud droplets about a millimetre across become fully fledged raindrops, heading earthwards with a terminal velocity of about nine metres per second.

Sky sculptures

> Have you entered the storehouses of the snow
> or seen the storehouses of the hail?
>
> Job 38:22

Our archetypal idea of a cloud – fluffy white billows, like the head of a cauliflower – corresponds to the type known as cumulus, from the Latin word for 'lumpy'. These are formed at low altitudes when warm, moist air starts to rise in an updraft. When air close to the ground is warmed by the heat given off from the surface, it becomes less dense and more buoyant than the cooler air above, and so it begins to ascend – a process called convection. If the air reaches the condensation level, the water vapour condenses into the nacreous bulbs of a cumulus formation. The updrafts from surface heating of land or sea in summer produce 'fair-weather' cumulus clouds which scatter light showers from a sky that just an hour before was an unbroken blue canopy.

Most cumulus clouds are warmer than 0°C throughout, and so their particles are all liquid droplets – the defining feature of so-called warm clouds. Beyond their edges the air is appreciably drier – that is, undersaturated in water vapour – and so droplets that stray beyond the billows evaporate rapidly. For this reason, the edges of cumulus clouds are bright and well-defined.

Stratus ('layered') clouds are more static: stable and sprawling, they create unbroken cloud cover in an overcast sky. They are formed when an updraft fails to penetrate a more stable layer of air above, forcing the cloud to spread laterally. Low-altitude stratus clouds are warm (completely liquid) clouds. But both stratus and cumulus may also be found at higher altitudes, engendered by updrafts originating not at ground level but higher in the atmosphere. These are called altocumulus and altostratus, and are 'mixed clouds' in which the temperatures become cold enough for the formation of some ice particles amongst the water droplets. The temperatures in mixed clouds range from 0°C to –39°C.

The highest clouds are wispy, ethereal wraiths – cirrus, meaning 'feathery'. They have temperatures below –39°C, and are composed wholly of ice. The fuzzy edges of these clouds are a consequence of the relatively slow evaporation of ice crystals that have drifted beyond the main mass.

When sunlight passes through atmospheric ice crystals, it is refracted: the rays are bent as the light enters the crystal through one face and exits through another. A light beam that follows such a trajectory is always deflected from its initial path by at least 22°, and as a consequence, sunlight travelling through icy clouds cannot reach the eye within a cone of 22°. So the sky within this cone around our line of sight to the Sun appears darker, and the circular edge of the cone is traced out in a bright circle of light – a halo around the Sun. This is just one of the many spectacular optical effects produced when sunlight passes through ice crystals in the atmosphere. The ice crystals can become aligned with one another if they possess flat plates which act like parachutes. This transforms a halo into arcs and spots.

Any particular cloud stack does not necessarily correspond to a single cloud type exclusively, nor persistently. In the tropics, for example, strong convective updrafts can lift moist air to the top of the troposphere (the lower sixteen kilometres or so of the atmosphere), forming towering cumulonimbus ('raining cumulus') clouds capped with a cirrus anvil of ice particles. Such clouds are often the harbingers of fierce tropical squalls. Changes in air circulation can modify a cloud type: the onset of convection inside a layered altostratus cloud can puff it up into altocumulus, and conversely, a high-rising cumulus can spread laterally to become a stratocumulus formation if it meets a ceiling of stable air.

Precipitation in a typical mid-latitude stratus-type cloud stack can take several forms, not all of which reach the ground. Snow falls from the uppermost (cirrus) regions, only to melt into rain as it passes through warmer air below. Drizzle trails behind the honest-to-goodness rain in a stratus 'tail', where droplets have not had time to grow so large. Ice crystals growing in a mixed (ice/water) region of a cloud at sub-zero temperatures can purge the air of water droplets, which freeze instantly when they collide with ice particles. In this way, ice particles can accrete more ice at their edges, a phenomenon called riming. Riming can lead to the accumulation of thick ice on solid surfaces such as aircraft wings and television aerials. The result may be an 'ice storm', like the one that decimated power lines around Montreal in the winter of 1997–8 and led to widespread and prolonged power cuts.

Probably the most dramatic form of precipitation occurs in tall cumulonimbus clouds, where ice particles may be cycled many times through the mixed layer, accumulating more ice on each pass like snowballs rolling downhill. The particles are carried upwards into the frozen layer by the convective updraft at the core of the cloud stack, and then fall earthwards under gravity, only to be seized again by the updraft. After several cycles the ice particles may have grown to awesome sizes, and they are eventually ejected from the column to fall as hail. Most hailstones are less than a centimetre across, but monsters as big as thirteen centimetres have been recorded. A ten-centimetre hailstone weighs about half a kilogram, and will fall to a terminal speed of around 145 miles per hour. That's a projectile big and bad enough to dent the roof of a car, and it goes without saying that its effect on a poor bystander can be lethal.

Global sunscreen

How warm is your house in winter? That depends on two factors: how much heat you pump into it, and how much you lose. The same consideration determines the warmth of our planet, which both gains and loses heat by radiation through space.

Earth's radiation balance is the transaction underlying all considerations of climate change. The Earth is warm not just because it is close to the Sun, as those privileged few who have walked in the shadowed valleys of the Moon will testify. The relative warmth of our planet is due to the fact that the atmosphere retains some of the Sun's heat, rather than letting the solar radiation bounce back out into space as it does from the Moon. Without an atmosphere, the Earth would be around 35°C colder on average: below water's freezing point almost everywhere. The Sun's rays penetrate the atmosphere to heat up the Earth's surface, which then re-radiates the heat. But whereas sunlight is most intense in the visible part of the spectrum, the energy it carries is 'degraded' by being absorbed and then re-emitted: it is emitted as longer-wavelength radiation, in the infrared part of the spectrum. We can't see infrared radiation, but we can feel its effect: an electric bar heater warms from infrared rays even before the element glows red.

The atmosphere retains heat because, whereas clear air is transparent to visible light, it absorbs infrared radiation. The molecules of several of the gases that are mixed with the air in very small quantities – in particular carbon dioxide, methane, water vapour and nitrous oxide – have chemical bonds that vibrate at the frequencies of infrared radiation. Like undamped piano strings resonating to a sung note, these molecules soak up some of the radiation re-emitted from the warm surface of the Earth. This means that some of the energy of the Sun's rays gets trapped in the atmosphere, which is warmed as a result. This effect is called the greenhouse effect – misleadingly, since greenhouses don't work this way – and the gases in the atmosphere that cause it are known as greenhouse gases. All of those I've just listed have natural sources, and so there is a *natural* greenhouse effect to which we owe our clement environment.

Concerns about global warming are due to the increase in concentrations of greenhouse gases in the atmosphere as a result of human activities such as fossil-fuel burning, which releases carbon dioxide and other gases, and agricultural practices, of which methane in particular is a product.

Although it is the most important contributor to the natural greenhouse effect, water vapour is not normally classed as a greenhouse gas. The term is usually reserved for those gases whose atmospheric concentrations are directly influenced by human activities, and we don't have very much direct influence on the amounts of water vapour in the atmosphere. However, any excess greenhouse warming caused by increased levels of carbon dioxide, methane and so forth most certainly will have an effect on water vapour in the atmosphere. Global warming would lead to increased evaporation from the oceans, and so might be expected, on average, to increase the humidity of the air. At the simplest level, this might be regarded as a 'positive feedback' effect: warming puts more water vapour into the atmosphere, which absorbs still more infrared radiation and so warms the planet still further.

Fortunately, there are moderating influences too – negative feedbacks. For instance, some plants grow more rapidly in air enriched with carbon dioxide. So as the amount of atmospheric carbon dioxide increases, plants may work increasingly hard to pull it back out of the atmosphere and 'fix' it in their tissues.

Some of the most important potential negative feedbacks involve the biosphere – life on Earth. In his original Gaia hypothesis, James Lovelock proposed that the planet might regulate itself by biospheric feedbacks to stay at a temperature conducive to the continuance of life. In other words, life itself maintains the environment to its advantage, just as the body has feedback systems for maintaining homeostasis (a constancy of temperature and of such things as blood alkalinity) in the face of external change. This idea has been much debated, yet there is still no consensus on whether it is basically right. Some believe it requires a rather optimistic balance of feedbacks. Conceivably, self-regulation might be possible over a narrow range of climate change but not beyond. Certainly, there is no obvious reason why things *have* to be this way, for all that the idea has a romantic attraction. But either way, Lovelock's hypothesis has helped to put the biosphere at centre stage in our understanding of the planetary environment.

The hydrological cycle, however, introduces some of the biggest uncertainties in this understanding. The feedbacks that involve atmospheric water vapour seem to be highly complex, and at present it remains unclear even whether they are positive or negative overall. Take the case of evaporation. Just because there is more evaporation in a warmer climate doesn't have to mean that there is more atmospheric water vapour. It could just mean that there is more rain, so that the extra water that evaporates is just as soon removed again. The truth is assuredly more complicated than this, because rainfall depends on many local meteorological factors, and how these might respond to a warmer climate is far from clear. The likelihood is that global warming would create greater rainfall in some parts of the world but greater aridity in others.

Perhaps the most difficult aspect of the problem concerns the feedback effects that arise from clouds. Clouds increase the planetary albedo – they make it more reflective to sunlight. Decks of bright white clouds hanging over land and sea send some of the Sun's rays bouncing back to space before they can penetrate the atmosphere, reducing the difference between heat in and heat out, and cooling the Earth. On the other hand, a cloudy sky retains more heat from the warmed surface than does a clear sky, as you'd be aware after a chilly night in the desert. So within the lower atmosphere, clouds are a warming influence. Which is greater – the warming effect or the cooling? It is now generally agreed that the influence on albedo is dominant – that clouds act mainly as a sunscreen, exerting a net cooling effect.

But to deduce how cloud cover might affect climate change you need to be able to predict how cloud formation will alter if the world warms up. That is a difficult business, because it demands prediction not only of how air moisture will change but how atmospheric circulation patterns will be altered – a complicated business even for the purposes of predicting next week's weather, let alone the next century's climate. If clouds exert a net cooling influence, then an increase in cloud formation in a warmer world will constitute a negative feedback on global warming. But *will* the skies be more cloudy?

These complications bedevil one of the key predictions of the Gaia hypothesis, which embodies perhaps its most fertile contribution to environmental science. Plankton, you will recall, are active players in the formation of marine clouds: they generate dimethyl sulphide (DMS), which is converted in the atmosphere to sulphate particles that act as condensation nuclei for clouds. Lovelock, along with atmospheric scientists Robert Charlson, Meinrat Andreae and Stephen Warren, suggested in the early 1980s that this might be a vital component of the self-regulating influence of the biosphere. In a warmer world, they said, plankton growth might be stimulated, leading to more DMS production, more cloud condensation nuclei, more marine clouds – and so ultimately to a cooling effect. The question is: does warming indeed create more DMS, and is this manifested as greater cloud cover? These questions have preoccupied many atmospheric scientists throughout the late 1980s and 1990s, but the answers remain open. All we can really say right now is that the production of marine DMS and its atmospheric effects are more complex than anyone anticipated.

Understanding cloud formation processes is much more than an exercise in the study of nature's beauty; it goes to the very heart of predictions about the future of our planet.

The deep freeze

> The breath of God produces ice
> and the broad waters become frozen.
>
> Job 37:10

A drive to the coast during the height of the last ice age, about 18,000 years ago, would have taken most of you a fair while longer than it would today. Global mean sea level then was around 120 metres lower than it is now, and you could walk from Asia to Indonesia, from Australia to New Guinea, without getting your feet wet. A considerable proportion of the water that now fills the ocean basins was then bound up as ice in huge sheets that stretched over most of eastern Canada (the Laurentide ice sheet) and from Scandinavia to northern Germany and the British midlands (the Fennoscandinavian ice sheet) (Fig. 3.1). The cryosphere – the component of the Earth's water locked up in solid form in the ice caps, sea ice and icebergs, and glaciers – may today account for only two per cent of the total water volume at the planet's surface, but its influence throughout the planet's history is not to be forgotten. And even in these warmer times, that two per cent represents over three quarters of the Earth's fresh water.

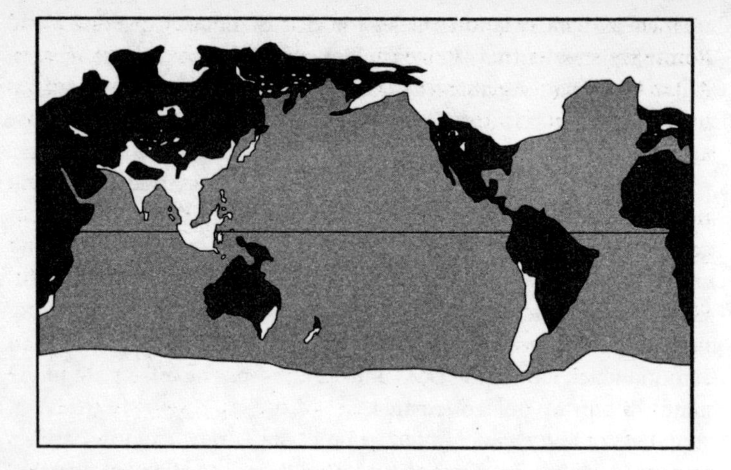

Fig. 3.1 At the height of the last ice age, around 18,000 years ago, ice sheets (white) smothered most of Europe and North America, as well as southeast Asia and southern South America. The outline of the land masses is a little different from that today, because of the lower sea level.

Ice covers about 5.7 per cent of the Earth's surface, although the seasonal melting of sea ice around Greenland and Antarctica ensures that this figure fluctuates periodically. We should also include the permafrost in the soil of Arctic tundra as a part of the cryosphere, as well as the winter snow that covers high mountains and northern forests and plains. Although temporary, this snow may have an important influence on the local climate. When soil and vegetation are covered by snow their albedo increases about fourfold, which means that they absorb less solar radiation. As snow melts in the spring, the decrease in albedo warms the ground quickly, providing a positive feedback on snow melting that allows a spring thaw to advance rapidly, removing snow cover within a matter of a few days.

Most of the world's ice is stored in and around Antarctica. Here the continental ice sheet reaches up to 4800 metres in thickness, with an average thickness of 2100 metres. The main continental ice sheet is divided in two by the Transantarctic Mountains; and there are smaller ice caps and glaciers on the Antarctic Peninsula to the far west. The East Antarctic Ice Sheet is the largest, and rests on bedrock above sea level. Most of the West Antarctic Ice Sheet, meanwhile, sits on rock below sea level. Beyond the land mass, great ice shelves tens of metres thick extend over the sea, most notably the Ross and Ronne ice shelves that bridge the Ross and Weddell seas in western Antarctica. These ice shelves persist all year round; but seasonal sea ice extends much further, floating as pack ice that breaks up in the austral summer.

The Antarctic and Greenland continental ice sheets are maintained not by the freezing of surface water but by the regular application of coats of new ice, in the form of snow. As snow gets buried by fresher snow falling on top, it gets compacted together under the weight. First it is compressed to a form of ice called firn, which is halfway between loosely packed snow and dense ice – rather like the stuff in a firm snowball. Firn is riddled with pockets of air trapped between the grains of ice. At greater depths most of these pockets are squeezed shut; but some remain as bubbles of air trapped in the dense ice. This trapped air represents a buried sample of the atmosphere from earlier times – a crucial resource for climate scientists.

The ice at the base of the East Antarctic Ice Sheet, mostly around three kilometres deep, is at least 250,000 years old: it fell as snow all that time ago, and has been locked up ever since. By drilling cylindrical cores of ice from the ancient ice sheets in Antarctica and Greenland, scientists can obtain a record of how the atmosphere has changed over this time period. They can determine how the amounts of greenhouse gases such as carbon dioxide and methane have fluctuated over the centuries, and so obtain clues about the way that natural processes respond to, or drive, climate change. Layers of trapped particles of dust and salt, meanwhile, can be the fingerprint of volcanic debris scattered by massive eruptions. And the ratios of different isotopes of oxygen and hydrogen in the ice itself record changes in global temperature.[1] Polar ice is a time capsule that holds vital secrets about our future.

If the sea-borne parts of the ice caps are frozen seas, we can regard mountain glaciers as icy rivers. For all that they are frozen solid, however, they are not immobile. The creaks and groans that emanate from their depths are a signal of their slow, creeping advance. Glaciers have their roots in high places and extend down into valleys, and so the ice flows downhill. This motion, which happens at a snail's pace and so is not readily discerned at the foot of the flow, seems to have first been noticed in the sixteenth century, and in 1751 Johann Georg Altmann concluded that it was caused by gravity. For some time it was believed that the ice slides as a solid mass over the rocky bed. But solid ice will freeze to bedrock, just as it will stick to your skin when it is very cold. Slippage across the underlying terrain, called basal slip, takes place readily only when the interface of ice and rock is lubricated by melt water. The freezing point of water is lowered by high pressure, so the tremendous compression at the base of a glacier can induce melting. Basal slip lubricated in this way is quite common for the glaciers in alpine mountains, where temperatures don't stay as cold as they do at the poles. The melt water beneath the ice is discharged into glacier-fed rivers, which supply many hydroelectric plants in Europe and Scandinavia. Rivers have been created artificially for this purpose in some glaciers by drilling tunnels into the ice.

In the polar ice sheets of Greenland and Antarctica, basal slip is restricted to particular channels within the ice mass, called ice streams, where the ice flows much faster than elsewhere. The speed difference between ice streams and the rest of the ice sheet can be tremendous. In the West Antarctic Ice Sheet, ice streams twenty-five to eighty kilometres wide can flow at speeds of up to six metres per day, over two hundred times faster than the flow in the rest of the ice. The cause of these ice streams is still not clear, although it is likely that they too are lubricated by melt water. The streams seem to be underlaid by loose, muddy sediments which may also relieve friction against the bedrock: but whether these sediments are primarily a cause of fast flow or a consequence of it (being formed by the grinding of the ice) remains unknown.

Where basal slip is not possible, ice sheets ooze downhill by another, more sluggish means: by a creeping motion of the ice itself, which acts under pressure like a highly viscous fluid or like toothpaste squeezed from a tube. This flow, called creep, is the main cause of glacier motion in very cold regions. A combination of creep and sliding carries the tongue of a glacier down an alpine mountain valley at speeds typically ranging from a centimetre to as much as a metre per day. Polar glaciers are generally more stately, advancing at a typical rate of several metres a year.

For all that mountain glaciers are constantly advancing, some of them appear to be withdrawing back up the slope because of accelerated melting at the front edge due to the increasingly warm climate of the past century or so. It appears that on average this is causing the world's mountain glaciers to retreat (Fig. 3.2).

Rewriting the Book

When glaciers retreat in times when the climate is relatively warm, the ponderous masses of ice leave tell-tale signs of their passage. A valley that has been scoured out by a glacier looks very different from one carved by flowing water. River valleys are typically V-shaped, whereas glacial valleys, once filled with a tongue of ice from wall to wall, are broader and shallower with a U-shaped profile (Fig. 3.3*a*). The rivers that now flow down glacial valleys are fed by tributary streams issuing from 'hanging valleys', which intersect the cliff-like slopes some distance above the valley floor so that the streams drop down in waterfalls (Fig. 3.3*b*). This discontinuity between the tributary height and the river height is not seen in normal river networks, and is a consequence of the fact that the tributaries that fed previously into the glacier did so at its top, way above the bedrock.

An advancing glacier is a bulldozer, pushing huge boulders before it. When the glacier retreats, these rocks are left behind as 'moraines'. The feature that distinguishes many glacial moraines from piles of boulders that have simply rolled down a valley's sides is that they are composed of a different rock type from that on which they sit, having been carried from a remote region where the bedrock geology was different. Such 'erratic' boulders of granite amidst the limestone mountains of the Swiss Jura were what led Swiss geologist Louis Agassiz to propose in 1837 that the glaciers of these mountains once extended much further than they did in his day.

It was not a new idea: the Scottish geologist James Hutton had suggested something similar in 1795. Despite Hutton's eminence, his idea had been given short shrift because it conflicted with one of the most fundamental beliefs amongst geologists of his time. They generally maintained that many unusual geological features, including moraines, were created by the catastrophic Biblical flood. No one doubted that this flood had occurred – a precise date was even assigned from studies of the Scriptures. James Hutton opposed this 'catastrophist' or diluvial theory with a 'uniformitarian' point of view, whereby the Earth's features were wrought by slow geological processes happening over vast periods of time.

This idea was not only unconventional; it verged on the blasphemous. For it required that the Earth had been formed much longer ago than the six millennia or so commonly argued on scriptural grounds, in order to give the slow forces enough time to do their

Fig. 3.2 Mountain glaciers have been melting over the past several hundred years. In the early nineteenth century the Argentière glacier in the French Alps, visible in the background, reached almost to the church in the foreground. (Andrew Goudie, University of Oxford)

work. One of Britain's leading geologists, William Buckland, proclaimed in 1819 that the evidence for the diluvial theory was incontrovertible on geological grounds alone, quite aside from the firm authority of the Bible.

Although a personal friend of Agassiz, Buckland received the Swissman's idea with much scepticism – until he accompanied him to the sites that had led Agassiz to his conclusions. By 1840, Buckland was won over, and he soon persuaded Charles Lyell, another of Britain's leading lights in geology.

Agassiz supposed that the high point of glacier advance at some time in the past had been a unique event – in other words, that there had been only one occasion in the past when the Earth was much colder. But once the geological signatures of glaciers became accepted as such in the 1840s, it emerged that there had been more than one such advance – more than one ice age. In 1842 the French mathematician Joseph Alphonse Adhémar performed calculations which implied that changes in the nature of the Earth's orbit around the Sun might be responsible for the recurrence of ice ages.

The characteristics of this orbit are today known to change in three major respects. First, the angle of tilt of the Earth's rotation axis – which currently stands at 23.5° to the plane of the orbit – varies periodically with a period of 40,000 years. This angle is called the obliquity. Second, the orientation of the rotation axis with respect to the Sun (the inclination) wobbles with a period of about 20,000 years, a phenomenon called precession. Third, the elliptical shape of the orbit – the eccentricity – changes cyclically every 100,000 years, becoming sequentially fatter and thinner.

But not all these changes were taken into account by Adhémar: he argued that the precession alone was responsible for an ice-age cycle. In 1864 the Scotsman James Croll developed the idea further by taking eccentricity changes into account. Croll's calculations seemed to provide a more tenable basis for astronomical control of climate, but they didn't turn out to match the geological record of when the ice ages actually occurred. It was not until 1920 that the sums were done correctly, by the Yugoslavian Serb Milutin Milankovitch. His subsequent refinement of the 'astronomical theory' of ice ages in 1930 still holds up today. According to this theory, cyclic changes in the Earth's three orbital parameters, now called Milankovitch cycles, alter the distribution of solar radiation that the Earth receives, and their net effect triggers long-term climate change. The onus was then

a

b

Fig. 3.3 Valleys carved by glaciers have a characteristic U shape (*a*), whose sides are punctuated by hanging valleys from which tributaries plunge in waterfalls to the valley floor (*b*). (Andrew Goudie, University of Oxford)

on the field geologists to come up with a record of past climate change sufficiently accurate to test this hypothesis.

This did not happen until the early 1970s, when an international group of scientists set out to map the Earth's climate over the past 700,000 years. The project, called the Climate Long-Range Investigation, Mapping and Prediction (CLIMAP) project, involved drilling columns of sediment from the sea bed throughout the world's oceans and analysing them chemically for signs of a changing sea temperature. As is the case for polar ice, global temperature changes are recorded in the isotope ratios of compounds in ocean sediments.[2]

The CLIMAP measurements revealed many pronounced changes in climate, and these matched up with the variations predicted by Milankovitch's orbital calculations. In particular, the 100,000-year variation in eccentricity dominates the record: this, it seems, has been the main driving force for the ice ages over this period. Yet climate records extending further back in time – which are more sketchy, since geological records, like historical ones, get more blurred and patchy the further back you go – tell a different tale. From about two million to 700,000 years ago the climate cycles were less pronounced – the ice ages were milder – and they varied predominantly with a period of 40,000 years, which is the beat of obliquity changes. Something seems to have happened about 700,000 years ago to alter the pulse of the Earth's climate to that of the eccentricity cycle, which carried with it the consequence of more severe glaciations.

No one is yet sure what might have caused this switch. One possibility is that the cryosphere provided the trigger. The time of the switch was a period of extensive mountain building as continental plates collided and rumpled up. In particular, the collision of the Asian and Indian plates was creating the Himalayas and the Tibetan plateau. The formation of such a massive region of high ground would have led to an appreciable increase in the Earth's albedo, because of all the snow on the high ground. This albedo effect could have flipped the climate system into a new rhythm.

Looking still further back, the differences in climate are even more striking: before about fifty million years ago, there seems to have been no ice-age cycle at all. In fact, during the mid-Cretaceous period around 100 million years ago, when dinosaurs were still at large, there may have been virtually no cryosphere – the Earth may have been so warm everywhere that no ice sheets existed. It seems likely that the formation of the ice sheets required the right juxtaposition of tectonic plates. They reached something like their present positions, with most of the land mass in the Northern Hemisphere, during the late Tertiary period around fifty million years ago. It may have been only then that ice sheets started to form at the poles. And these ice sheets would have had a positive feedback effect that cooled the planet and led to more glaciation. For as they spread they increase the planetary albedo, decreasing the amount of solar radiation absorbed and producing further cooling. The ice-age cycles of the past several million years could be a creation of the cryosphere itself.

High rise

If ice has engineered such drastic global changes in the past, might it not do so again? There is every reason to expect the seas to recede again in another ice age a few thousand years from now, when the Earth's orbital variations steer us into another of Milankovitch's cold spells. But concerns for the immediate future lie in the other direction. Whether or not the cause is a human-driven excess greenhouse effect, the planet has got indisputably warmer – by about 0.5°C – over the past century. A warmer climate should imply a smaller cryosphere: the ice starts melting. There is enough water in the Greenland and Antarctic ice sheets to raise sea level by a mighty 66 metres, if it were all to melt. The Thames flood barrier and the Dutch dikes would be as effective as a garden fence against such a deluge. Whole Pacific islands would vanish beneath the waves, as would vast tracts of densely populated coastal land lying on flood plains. Most coastlines would change beyond recognition.

But not even the most pessimistic forecasts predict total melting of the ice sheets, even if we do nothing about global warming for at least the next century. On the other hand, it will take nothing like so great a change in sea level to pose a serious threat of inundation to many of the world's lowlands. A rise of a mere five metres would be catastrophic, consigning not just many Pacific islands but also much of Florida, Vietnam, the Netherlands and other coastal regions to a watery grave. Miami, New Orleans and Bangkok will be amongst the cities to share Atlantis's fate.

Is this a real possibility? No one is quite sure; but the best guess right now is that these nightmares are very unlikely to materialize in the twenty-first century. That best guess – an amalgam of state-of-the-art climate modelling hedged around with significant uncertainties – is embodied in the reports of the Intergovernmental Panel on Climate Change (IPCC), produced every two years from discussions amongst the world's leading climate scientists. The 1996 report predicts that global mean sea level will rise between thirteen and ninety-four centimetres over the next hundred years, and that little, if any, of this rise will be due to melting of the ice caps.[3] It is caused instead mostly by the expansion of sea water when it is warmed. This thermal expansion is tiny – a fraction of a per cent – but is enough to make a difference over an entire ocean. Thermal expansion has raised sea level by between two and seven centimetres over the past century – an increase that, while still a cause for concern in some regions, is not the global disaster that became feared in the late 1970s.

At that time, alarms were raised about the possibility that sea-level rise might take an abrupt turn for the worst due to sudden collapse of the West Antarctic Ice Sheet. This sheet is bordered along the continent's coast by ice shelves. But the bedrock on which the ice cap sits is below sea level, so if the ice shelves were to break up as a result of a climate warming of just a few degrees, the exposed land ice might collapse catastrophically once the ocean waters reached it. Disintegration of the West Antarctic Ice Sheet would raise average sea level by five metres.

The west of Antarctica has warmed by 2.5°C over the past fifty years, and there is evidence that some margins of the ice sheets here are cracking up under the strain. In 1995 satellite observations showed that one of the floating shelves of ice that extends over the sea off the Antarctic Peninsula has retreated dramatically since 1966; and there are signs of instability in ice shelves elsewhere. The stability of the grounded ice sheet is another matter, however. The current consensus, in the words of the IPCC report, is that 'the likelihood of a major sea-level rise by the year 2100 due to the collapse of the Western Antarctic Ice Sheet is considered low'.

In fact, the IPCC committee remains uncertain whether the major Antarctic ice sheets will shrink in volume at all over the coming decades, even if climate warming proceeds at its current rate. This is because the cryosphere is not like an ice cube floating in a glass of water; it is part of a complex dynamic system, the hydrological cycle, in which water is constantly on the move. A warmer climate may engender a wetter world, as more water vapour enters the atmosphere and is precipitated back to Earth. Over the poles precipitation occurs as snow, which feeds the ice stores. So a warmer climate might actually cause the ice sheets to thicken owing to greater snowfall. It is very hard to say, from existing records, whether accumulation or melting has dominated in the icy Antarctic wastes over the past century. But climate models tentatively predict that accumulation will dominate slightly in a warmer world. For the Northern Hemisphere ice sheet over Greenland the situation is less clear: melting at the margins might slightly exceed accumulation, but we can't yet be sure. Either way, the chances are that neither ice sheet will experience very much change in the amount of water that it locks away.

A further important contribution to sea-level change comes from the melting of mountain glaciers. Whereas the records and the prognosis for polar ice are both ambiguous, the situation for glaciers is more clear: many, perhaps most, have retreated steadily for the past two centuries (see Fig. 3.2). This retreat is probably a direct result of the rise in global mean temperature. The consequent release of fresh water into the oceans has raised the sea level by between two and six centimetres during the past 100 years.

For all that the current rate of sea-level rise does not obviously signal an impending disaster, it appears to represent a recent acceleration. Over the past century the average sea level globally has risen by 10–25 centimetres, an increase possibly as great as that which occurred over the previous two thousand years. So something has made a big difference during our lifetimes. While we remain uncertain about the reasons for the encroachment of the oceans, it makes sense to hope for the best and prepare for the worst.

4 Oceans in the Sky

Water on other worlds

> The inhabitants of Jupiter must ... it would seem, be cartilaginous and glutinous masses. If life be there, it does not seem in any way likely that the living things can be anything higher in the scale of being than such boneless, watery, pulpy creatures.
>
> William Whewell (1854), *Plurality of Worlds*

> Girdling their globe and stretching from pole to pole, the Martian canal system not only embraces their whole world, but is an organized entity. Each canal joins another, which in turn connects with a third, and so on over the entire surface of the planet.
>
> Percival Lowell (1906), *Mars and Its Canals*

I must be naïve. When the newspapers told me in March 1998 that there would be colonies on the Moon within thirty years, my first thought was 'why would anyone want to go there?' Well, it has its peculiar dusty glamour, I suppose, and I'm told the views are tremendous; but somehow I prefer to have air, warm summers, mild winters, visits to the seaside.

But this flush of astro-colonial enthusiasm was not just a pre-millennial yearning for the optimism of the early space age; it was spurred by the announcement by NASA of the discovery of water on the Moon. We have come to equate water with habitability: where there is water, so can there be life. The discovery of water on other worlds engenders a sense of something like relief – the Universe is not, it seems, so alien after all.

This relief is a release from disillusion. A century ago, scientists accepted almost casually that our sister planets in the solar system teemed with life. It's sobering to discover this blithe confidence in life on other worlds even from the British Astronomer Royal Harold Spencer Jones in 1952:

> ... there are changes from time to time in the Martian surface ... It is difficult to interpret these changes in any other way than by the

> seasonal growth of vegetation ... As the ice-cap melts, the moisture reaches lower latitudes, possibly in the form of rivers or streams, but more probably as rain or dew. With the coming of the moisture, the vegetation commences to grow and the colour of the areas covered with plant growth changes to green.[1]

Such views were reflected in popular accounts of planetary science even into the early 1960s.[2]

But these visions faded as late-twentieth-century astronomy and space exploration revealed that the blueness of Earth is an anomaly in the solar system – that no other planet or satellite has surface oceans of liquid water. Our hopes to find life on neighbouring planets were cruelly trampled, and the solar system suddenly looked like an awfully lonely place. Coincident with that realization, the notion evolved, at least in the popular consciousness, that water is a rare thing in the cosmos.

Its discovery elsewhere therefore continues to be greeted with surprise and excitement. Yet planetary scientists and astronomers know better – they know that water pervades the Universe. We should not be in the slightest bit surprised that water crops up just about wherever we look in the heavens.

The issue, then, is not whether there is water elsewhere, but whether it is liquid – for only in the liquid state does water seem capable of providing the matrix of life. A tour through the solar system reveals a rich panoply of watery environments – some irrevocably icy, some steamy, some more barren than the harshest Arctic tundra, some laced with ancient, dried-up waterways ... and some, perhaps, awash with the precious fluid, tempting us to speculate about the life that might be lurking in these celestial oceans.

Lunar prospecting

Prospecting for water on the Moon sounds like looking for gold in the gutter; but NASA tried it, and struck rich. Actually they found it twice, but most people had forgotten about the first time – and so the beleaguered US space agency got two sets of headlines for the price of one. When NASA announced the findings of the Lunar Prospector mission, launched in January 1998 to orbit the Moon and analyse its surface, scarce mention was made that the same story had appeared just over a year earlier. The Clementine spacecraft, despatched to the Moon in 1994, sent back images in 1996 of a crater on the Moon's south pole that appeared to contain a layer of ice, the size of a small lake and about twenty-five feet thick.

At the south pole there is an immense, roughly circular depression about 2500 kilometres in diameter called the South Pole–Aitken basin, which is thought to be an impact crater formed between 3.9 and 4.3 billion years ago. It is the largest and deepest impact crater yet seen in the solar system, over twelve kilometres deep in places, and the impact that created it must have nearly broken the Moon apart. As it is, the scar is so deep that the entire crust of the Moon may be stripped away there, exposing the mantle beneath. The lunar poles always receive sunlight at a highly oblique angle, so much of the South Pole–Aitken basin is permanently in the shadow of its mountainous rim, and temperatures there may never reach higher than minus 230°C. Clementine was not designed to probe these dark, cold, mysterious lowlands in any detail, but it did conduct a cursory exploration by beaming radio waves down to the surface and picking up their reflection. By analysing how the reflected waves were altered, the project scientists were able to glean clues about the material off which they bounced. In early December of 1996 they revealed that the probe had detected ice deposits.

The later Lunar Prospector provided evidence that water was much more widespread at both of the Moon's poles: the project scientists were no longer talking about an isolated frozen pond, but several hundred million tons of ice (the estimate later ballooned to six billion tons), hidden away in cold, shadowed craters and depressions.

This is no big surprise. If comets and meteorites delivered water to the young Earth, there is every reason to suppose that the same thing happened on the Moon. But because the Moon has next to no atmosphere, sunlight would have evaporated any ice on the exposed surface, and the Moon's weak gravity would have been unable to prevent the vapour from escaping into space. Ice deposits could remain, however, in regions that have never seen the Sun for the past four billion years.

For some planetary scientists, the discovery of water on the Moon is a sign that we must go back there, to the dusty plains where no one has set foot since 1972. A return, says Paul Spudis of the Lunar and Planetary Institute in Houston, Texas, would 'reenergize our space program with a focus'.[3] But what would we be after? 'Knowledge and wealth', according to Spudis. (In which order, though?)

To my mind, if the day comes when we have made our occupation of the Earth so unsustainable that we have to look to so barren a place as the Moon to rescue us, we are in deep trouble. In terms of resources, the Moon doesn't have a great deal to offer. Proponents of lunar settlement argue that the lunar rocks are rich in useful substances: oxygen, silicon, iron, calcium, aluminium, titanium, magnesium. The trouble is, these elements are also abundant in the minerals of Earth – but are expensive to extract. Advocates argue that a lunar colony would only have to be self-sustaining, not profitable – which means that we'd have to ditch the carrot of 'wealth' from the call for Moon travel.

'Ice could be melted to form drinking water, or electrolysed by solar power into hydrogen and oxygen for rocket fuel', said one British newspaper of the Moon's icy lakes. One thing the Moon offers in abundance is solar energy: the daytime side basks in the glare of sunlight, reaching temperatures of 100°C at the equator. Silicon for photovoltaic panels could be mined from the lunar terrain. One can imagine huge solar arrays being erected on the Moon, or in orbit around it, and the energy being used to split water by electrolysis (see page 274) to create oxygen for a habitable environment. Perhaps the power generated could even be beamed back to energy-hungry Earth. This is the one aspect of lunar colonization that might be worth taking seriously, although it is a highly optimistic vision. But why else might we go?

There is a great deal still to be learnt about the Moon in scientific terms, some of which promises to tell us about our own planet's early history. I think it is proper that we should find ways to launch modest missions like Clementine to sniff out such information, both from lunar orbit and with lander probes. Such things are cheap fare in the terms of space exploration as a whole. But a habitable lunar colony is something else again. The idea strikes me as a reflection of our hubris as a species: we haven't really explored until we've gone there in person, regardless of whether that is necessary or productive. You can see this belief in presidential promises to put a person on Mars, in the initiative to build the orbiting manned Space Station. Even many scientists don't want these things, don't look forward to gleaning any intellectual riches from them. They are leftovers from out-moded dreams of conquest (which some would prefer to call 'adventure'), not a means to enhance our understanding of the Universe.

The fact is that, for the foreseeable future, manned spaceflight has no obvious role.[4] There need be no lack of ambition or enquiry in such a statement – it is simply that we have better ways of doing things now. Manned spaceflight is hugely expensive, very dangerous (the two are of course connected) and as redundant as the Cold War that spawned it. To my mind, the unmanned Voyager missions brought us wonders beyond our dreams, enriched our culture and our imaginations. In contrast, after Apollo 11 (a point in the Apollo program reached already at the cost of three lives) we had little to show for our returns to the Moon but a small collection of rock and dust.[5]

Perhaps the most socially important product of these flights was gained by a look backwards – at the blue globe of the world. In comparison with that, it will take more than a few frozen lakes to make the Moon look like an attractive place to live.

Water on Mars: from canals to rivers

During the late 1990s the media fell in love with Mars. In August 1997 NASA's Mars Pathfinder mission deposited its little rover vehicle Sojourner on the red planet's surface, and began sending back to Earth stunning pictures of a kind not seen since the Viking lander missions of the mid-1970s. Public interest in this expedition to our planetary neighbour had already been stirred to a frenzy by the announcement in July 1996 that US scientists had discovered possible fossil evidence for life in a Martian meteorite.

The dream of life on Mars is spectacularly enduring. While this idea is surely encouraged by the similarities with Earth in terms of size and proximity to the Sun, it has long been motivated by one prime factor above all else: water. For whenever these speculations have surfaced as anything like a scientifically based scenario, it is the supposed – and now confirmed – existence of water on Mars that has stimulated them. Without water, most people would have written off Mars long ago as a barren hunk of rock, as dead as the Moon.

Yet legend has it that the whole business began from a trivial linguistic mistake. The Italian astronomer Giovanni Schiaparelli inspected the planet's surface through a telescope in the late nineteenth century, and in 1877 he prepared a map on which he labelled linear geographical markings uncontroversially as 'channels'. But the

Italian word for channels is *canali*, which a hapless translator rendered in English as 'canals'. And canals, of course, are not natural features but artefacts. Inspired by this intriguing thought, the American astronomer Percival Lowell found his own observations of the planet guided by preconceptions, and he somehow managed to convince himself that Mars was laced with these artificial constructions for distributing water. 'On Mars,' he said, 'we see the products of an intelligence. There is a network of irrigation ... Certainly we see hints of beings in advance of us.'[6] It was clear enough even with the telescopes then available that Mars had no oceans like Earth, and so Lowell dreamed up the idea of an ancient civilization that had struggled to sustain itself by irrigating the water-starved planet.

Lowell's scientific contemporaries derided these ideas, and with good reason. The composition of the Martian atmosphere had been probed using the technique of spectroscopy, which looks for the presence of certain chemical compounds by the distinct colours that they pluck out of the spectrum of light as it passes through. These observations gave no indication of water vapour;[7] rather, the atmosphere appeared to be extremely tenuous, and composed primarily of carbon dioxide.

But writers of fiction feel no need to be fettered by facts. Both H.G. Wells and Edgar Rice Burroughs were influenced by Lowell's hypothesis. Wells painted the gloomy picture of a malevolent race of Martians looking enviously at the Earth with plans of conquest – and so set in train a mythology that continues to appeal to the post-Cold War paranoia fantasies of the Western world. Burroughs' Red Planet was more benevolent, peopled by an exotic humanoid civilization apparently lifted largely from the desert legends of Arabia.

It was with some disappointment, then, that the realities of the space missions of the 1960s were received by an expectant public. The fly-by of the Mariner 4 spacecraft in 1965 sent back news of a dead, cold, dry planet less conducive to life than any place on Earth. Later missions and Earth-based observations reinforced this discouraging picture. Temperatures at the height of the Martian summer scarcely top freezing point; on average they fluctuate around minus 60°C at the equator and minus 123°C at the poles. The atmosphere is thin indeed – atmospheric pressure at the planet's surface is barely half a per cent of Earth's. At that pressure, any liquid water would just vaporize, and water ice does not melt if warmed: it hops straight to vapour without passing through the liquid state. The composition of this thin Martian air is not of the sort that would sustain life like ours; it is mostly carbon dioxide and argon, with a water-vapour content of typically a few hundredths of a per cent. The darker expanses of the surface that grow and shrink seasonally are not great swathes of vegetation, as was suggested in 1884 by French astronomer E.L. Trouvelot and was widely believed even into the early 1960s; instead, they are probably the result of arid dust storms that frequently ravage the land. 'So much for Lowell's canals and seasonal plant life,' sighed Bruce Murray, later the director of NASA's planetary-science research centre, the Jet Propulsion Laboratory in Pasadena, California. Popular culture was, however, reluctant to abandon Mars to an eternity of lifelessness. While fictional stories of Martians dwindled in the 1950s and 1960s, writers speculated instead on the human colonization of this barren world. On this theme, Arthur C. Clarke gave us *The Sands of Mars* and Ray Bradbury *The Martian Chronicles*.

Then in 1972, all this changed again when the Mariner 9 spacecraft began to send back images of Mars that told another story altogether. First, there were immense volcanoes, the largest (Mons Olympus) about twenty-six kilometres high and 600 kilometres across, showing that Mars has not always been as geologically senescent as it appears today. But most astonishingly of all, there were river valleys, sinuous and branched like those of Earth (Fig. 4.1).

They are bone dry now, without a doubt; yet the scientists who received these unexpected pictures with shock and wonderment knew of no process that could produce such geological features except the erosive flow of water. There seemed to be no escaping the conclusion that water had once rushed over the Red Planet, perhaps even gathering into lakes and oceans.

According to Bruce Murray,

> By the end of the Mariner 9 mission, in 1972, our view of Mars had completely changed – once again. Lowell's Earth-like Mars was forever gone, but so was the Moon-like Mars portrayed by our first three fly-by missions, Mariners 4, 6 and 7. The Mars revealed by Mariner 9 was not one-dimensional; it was an intriguingly varied planet with a mysterious history. The possibility of early life once more emerged.[8]

The 'rivers of Mars' presented a profound puzzle. The earlier Mariner missions had clearly shown a surface pock-marked with huge craters,

Fig. 4.1 The surface of Mars is engraved with dried-up valley systems, apparently carved by running water. (Michael Carr, US Geological Survey)

like the Moon. Craters, particularly big ones, are nearly always ancient features, since the impacts that caused them were much more prevalent in days of yore than they are now. The principal impactors, the unused debris of the early solar system, were gradually swept up by the planets within the first billion years or so of the solar system's formation. After this time of heavy bombardment, cosmic collisions became considerably more rare and resulted largely from wandering asteroids or comets drawn into a planet's gravitational field. The collision of comet Shoemaker-Levy 9 with Jupiter in 1994 was of this type.

The Earth received the same early bombardment as the Moon and Mars. Yet few impact craters are now visible on Earth, whereas the Moon is covered with them. This is because the terrestrial craters have been wiped away by billions of years of erosion by falling and flowing water, along with the conveyor-belt overturning of the planet's crust. The Moon's pock-marked face tells us that it has never seen a rainy day, has never been washed by the ripples of an ocean.

The persistence of craters on Mars seems to indicate the same thing – that Mars never had long-lived oceans or a hydrological cycle of global extent. Yet there is no avoiding the conclusion that it has known rivers, that running water has carved channels and tributaries into the rock face.

The abrupt, cliff-like valley walls of some of Mars's dry rivers, and the shapes of the tributary heads, suggest that they were not formed in the same way as river valleys on Earth, by the focused flow of run-off from rainfall. Instead, they may have been created by the melting of dirty ice sheets during warmer times, when a denser atmosphere permitted water to flow over long distances before evaporating. Martian river channels of this sort are found only in landscapes that have been undisturbed for around 3.5 to 4 billion years, implying that the conditions that led to their formation have not arisen since.

Another class of Martian river bed, called outflow channels, don't look much like river valleys at all. Their complex, disorderly shape suggests that they were formed by catastrophic floods. Comparable terrain is seen on Earth in regions that were suddenly inundated at the end of an ice age, such as the scablands of Oregon and Washington. Massive floods carve out streamlined islands and deep potholes, whose forms provide some clues about the extent and conditions of the flood. On Mars these outflow channels bear witness to truly awesome events, with discharges of up to half a cubic kilometre of water per second flowing at 170 miles per hour and gouging channels twenty kilometres wide.

Outflow channels are between four and one billion years old, and they may, in Mars's distant past, have provided the planet with temporary lakes and even seas and small oceans. Scientists at the Jet Propulsion Laboratory proposed in 1989 that there was once a huge, shallow ocean – the Oceanus Borealis – in the northern hemisphere, four times the size of the Arctic Ocean. But others contend that the massive flood events would have created little more than vast muddy plains.

Mars possesses other geological features that betray the release of water from substantial deposits of ice below the soil. There are the 'fretted terrains' – large plateaus and isolated, flat-topped mesas, separated by flat-floored valleys. These might have been formed by a combination of running groundwater and the movements of 'rock glaciers', fragmented rock welded together with ice. There are ridges of rubble like the mounds of sediment deposited by melting and retreating glaciers on Earth. And many of the craters on Mars look like muddy splashes, the sort you see when you drop a pebble into a slurry of mud or sand (Fig. 4.2). These may have been caused by the collision of a meteorite into a region of water- or ice-laden soil.

The dusty river beds of Mars are interpreted by comparing them with river valleys on Earth – but this is a bit like interpreting fossils by comparing them with living creatures. It is hard to know how much of what we see is determined by our preconceptions and associations. All the same, taken together these images make it hard to deny that Mars was once wetter and warmer than it is today. The chemistry of the planet's fabric supports this idea. The robotic Viking landers analysed the composition of Martian soil in 1976 and found that it is mostly clay, a mineral formed when volcanic basalt encounters water. And the precious dozen fragments of Mars that have fallen to Earth as meteorites contain minerals that result from broiling volcanic rocks in hot brine, a sign that Mars may once have possessed volcanic hot springs.

An optimistic view of early Mars offers a very different vista to that we see today. There may have been lakes and small oceans, and an atmosphere thick and humid enough to raise the surface temperature appreciably above water's freezing point by greenhouse warming (page 62). This would have set up a hydrological cycle at least on a regional scale, with water evaporating from the surface waters and condensing into clouds. Rain could create river drainage networks, while snow thickened sea ice, ice sheets and glaciers near the poles. All in all, this is not so different a landscape from that envisaged on the early Earth. So why did things change? Where did the water go? To take an anthropocentric view, what went wrong?

Most of the water now on the planet is probably locked up as permanent ice – permafrost – within the soil, and may be destined to remain there for the rest of the planet's existence. In all likelihood, a warmer Mars is now beyond recall.

Only at the poles is the ice visible. The polar ice caps grow and shrink with the seasons: because Mars, like Earth, is tilted on its rotation axis, it too has cycles of alternating seasons in each hemisphere. But the ice that thaws each summer is not water ice – it is frozen

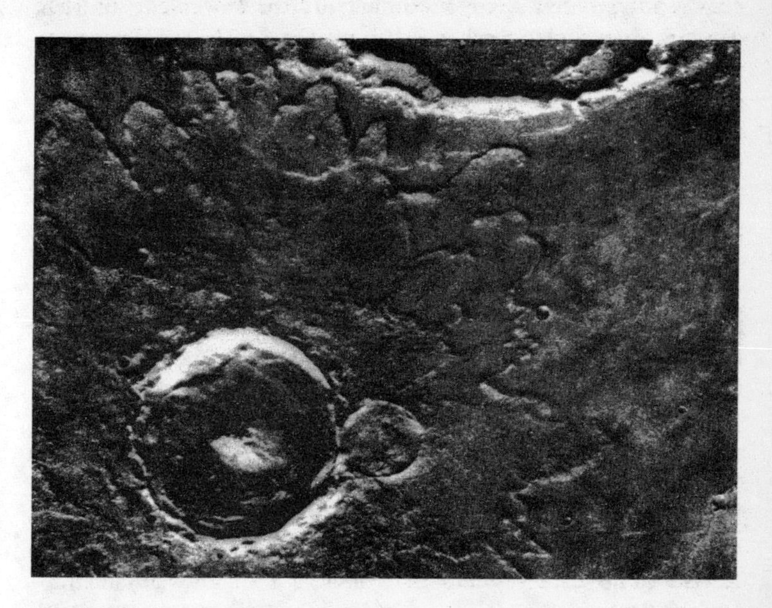

Fig. 4.2 This feature on Mars appears to be a gigantic splash. The impact that formed the 36-kilometre-wide crater probably melted ice below the surface, throwing out splodges of muddy debris. (Michael Carr, US Geological Survey)

carbon dioxide, 'dry ice'. Snow on Mars is carbon dioxide snow, and surely lacks the six-pointed symmetry of earthly snowflakes. There *is* water ice in the polar caps, but it's not known how much. This is one of the questions that NASA's Mars Polar Lander, launched in January 1999, would have answered had it not been lost as it prepared for its planetary descent the following December. Because the water ice stays frozen all year round, however, there is very little water vapour in the Martian atmosphere – never more than the equivalent of 1.3 cubic kilometres of ice, ten times less than that on Earth.

Some – but possibly only a little – of Mars's water has been lost to space: split by sunlight in the upper atmosphere, as it is on Earth, and the fragments then lost from the gravitational field. And some might now reside in the rocks, held between stacked layers of atoms in the clay minerals like the filling of a multi-decker sandwich. Finding water on Mars might be literally like squeezing blood from a rock.

Occasional showers

Mars seems to have slipped into this barren deep freeze only after several warm spells, when water frozen in the soil or the ice caps melted, thickening the atmosphere with water vapour. During the past two billion years, these warm, wet stretches may have lasted no more than a few thousand years each, but earlier still they could have been longer. The young Mars might have spent millions of years as wet as a Monday morning in November, a world mottled blue and red.

To understand why Mars does not now support rain forests, savannahs and, yes, Martians, the key is *size*. Being smaller than Earth, Mars cooled off sooner from its fiery youth. Volcanic activity on the planet seized up, and without the churning of a hot mantle, Mars developed no plate tectonics. In its absence, the early Martian atmosphere rich in carbon dioxide and water vapour was doomed. Water falling and flowing on the rocky surface of the planet induced 'chemical weathering' of the rocks: reactions between the silicate rocks and carbon dioxide which convert the gas into mineral carbonates. Little by little, the carbon dioxide was drawn down from the atmosphere and locked away, and the thinning Martian skies were ever less able to retain the Sun's heat. On Earth this decline is averted because the carbon dioxide removed from the atmosphere by weathering – and now by plants too – is recycled. Dragged down into the Earth's mantle where one tectonic plate plunges beneath another, it is spewed out through the mouths of volcanoes.

Yet Mars was apparently able to free itself, from time to time, to enjoy transient warm spells. These brief thaws might have been caused by changes in the shape of the planet's orbit. Like Earth, Mars cycles around the Sun in an immense ellipse characterized by its eccentricity, inclination and obliquity (see page 71). These orbital characteristics change steadily over time – the orbit pitches, stretches and wobbles. The changes bring about climate variations, just as on Earth they drive the pulse of ice ages. In particular, the obliquity (tilt) of Mars may change by as much as 60° every ten million or so years. This alters the seasonal temperature extremes: summer temperatures could rise above the freezing point of water for several weeks, while the winters would be correspondingly harsher. Even such a brief injection of warmth could have dramatic consequences. During the summer, large amounts of carbon dioxide could evaporate from the permafrost and polar ice caps, and the resulting greenhouse effect could melt water ice, enriching the atmosphere with water vapour. And so the planet would thaw.

Yet as a hydrological cycle kicks into gear, the process of weathering again starts to lock away carbon dioxide in carbonates, and the whole affair literally runs out of steam. It would be like a downpour in a desert oasis: a burst of activity followed by an inexorable return to aridity.

As well as prolonged warm episodes, there have probably been many abrupt and catastrophic events that provoked the dramatic surface flows we see carved into the planet's surface today. Occasional volcanic outbursts would melt the nearby permafrost, sending torrents of water streaming across the plains until they are silenced by evaporation or freezing. And as we've seen, meteorite impacts may melt the ice at the impact site, creating splash patterns and flow channels around the crater.

Anyone home?

After Mariner 9 brought back evidence of water on Mars, there was no question about what the main objective of the Viking lander missions was to be in 1976: to search for life. On Earth life seems to have begun just about as soon as it possibly could. If it is that easy, then why not also on a warm, wet Mars? No one expected to find Lowell's decayed civilizations or Wells's intelligent and malevolent beings; but perhaps there were microbes eking out a precarious living in the frozen Martian soil, just as they survive in the most inhospitable places on Earth. The Viking craft were designed to sniff out signs of biological activity.

But they found nothing – no sign that Martian life has ever existed. Worse, they showed that Mars is an extremely hostile place. Not only is it cold and dry, but the ultraviolet radiation from the Sun sterilizes the soil, creating caustic bleach-like compounds that would burn up any organic matter rapidly. On Earth the ultraviolet light is largely screened out by the ozone layer, which exists only because photosynthesis generates the oxygen from which it is derived.

The Viking missions trampled decisively on the dream of life on Mars. We are alone, they said. For many years Mars lost its appeal, and it was not until the report in 1996 of 'possible evidence for life' in a Martian meteorite[9] that little green men came back in vogue. This report, now regarded as premature, lent a great deal of glamour to the Mars Pathfinder mission of the following year, the first lander mission since the Vikings. Everything Pathfinder revealed was consistent with the picture of a once episodically warm and wet Mars whose chances of harbouring life are now long gone. And yet ... there remain some sober scientists who feel that Mars may still have a chance of hearing once again the lap of waves on a shore, the gurgle of a brook. Though this will never happen of its own accord, perhaps we could *engineer* it: maybe humankind can transform the Red Planet to an Earth-like state, ripe for colonization. This idea has been popularized in science-fiction writer Kim Stanley Robinson's *Mars* trilogy,[10] in which Martian settlers from Earth argue over the rights and wrongs of a 'greening' of Mars, transforming the barren red rock into a life-supporting world.

Now the science is catching up with the fiction, as it always does. In 1991 planetary scientists Christopher McKay, Owen Toon and James Kasting explained how 'the atmosphere and climate of Mars could be altered to allow terrestrial life forms, and possibly human beings, to survive on the surface'.[11] Carl Sagan and others had considered the idea in the 1970s, and James Lovelock had explored it further in the 1980s. But what distinguished the proposals of McKay and colleagues was that they limited themselves 'to technologies that are not far beyond the current art'.

Carbon dioxide in the polar ice caps could be released into the atmosphere by scattering black soot over it to increase its absorption of the Sun's rays, or by using vast mirrors to focus sunlight on the poles. Or small amounts of artificial greenhouse gases, such as CFCs, released into the Martian atmosphere could initiate the thaw. With a thick enough atmosphere, positive feedback would take over: the more carbon dioxide in the atmosphere, the warmer it gets and the more the ice caps evaporate. The rising temperatures would eventually release the water locked up in the Martian permafrost.

McKay and colleagues concluded that making Mars suitable for plant life 'would seem to be feasible' if its frozen reserves of carbon dioxide and water are as large as some estimates suggest. Making a world on which we could walk is another matter. We'd have to rely on the plants to generate enough oxygen, which could take 100,000 years. That's a timescale over which any forecasts for humanity become utterly meaningless. And even then, the atmosphere would remain lethal if it was too rich in carbon dioxide – concentrations more than thirty times those on Earth are fatal.

Inevitably, huge questions remain. Would such a terraformed atmosphere be stable in the long term? And would anyone want to live on a 'synthetic' world? Could we justify the cost and effort, the enormous risk to all concerned? And what ethical obligations do we have in tampering with other planets, even when we are virtually certain that they contain no indigenous life?

We don't really have a map as yet that allows us to navigate such questions. I think it must be factored into the debate that life has not passively colonized a preformed, 'habitable' Earth – it has transformed it utterly, given it a wholly different atmosphere and played an active part in maintaining its climate. And this colonization has been a two-way affair: as the environment has changed under the influence of the biosphere, so evolution has allowed life to adapt itself to those changes. It may be naïve to think that we can replicate our world elsewhere, and then expect it to stay that way. Exploratory studies of 'terraforming' Mars make fascinating reading, but I would be wary of regarding them as blueprints for what is desirable. I am not sure we are mature enough to cope with the thought that we could up roots and leave the planet; for people get notoriously careless about their dwelling when they know, or believe, they are about to leave.

Acid world

Mars is not alone in having a past shaped by water. Deep questions about the evolution of Venus, our nearest neighbour, revolve around its now diminutive reserves of the stuff.

It is hard to imagine two rocky planets less alike than Venus and Mars. Whereas Mars is a world of frozen waste beneath a delicate, rose-tinted atmosphere, Venus is an infernal place that swelters under dense skies full of heavy clouds. Romantic Victorians imagined that these clouds hid steamy primal swamps, perhaps even populated with dinosaurs – for the popular notion of the time was that the sequence of the inner planets reflected the progression of evolution.

Thus Venus was experiencing an era equivalent to that of times past on Earth, while Mars witnessed the tail end of an advanced, dying civilization. In 1918 the Nobel laureate Svante Arrhenius, adopting the hobby popular amongst that illustrious band of speculating about extraterrestrial life, pronounced that 'everything on Venus is dripping wet'. The idea of a warm, wet, richly vegetated Venus persisted even into the 1950s: Isaac Asimov, who took pride in the accuracy of the science in his fiction, wrote of the colonization of this humid planet in *Lucky Starr and the Oceans of Venus* (1954).

But sadly, it was all hogwash. There are no more Venusian dinosaurs than there are Martian princesses. The temperature at the rocky, mountainous surface of Venus is a blistering 500°C or so, and consequently it is bone dry. In the early 1990s, NASA's Magellan spacecraft brought us a sobering picture of this world, once hailed as the 'loveliest of planets': endless rocky terrains, some like the arid plains of Earth and others strange and bizarre. There were volcanic features scattered over ninety per cent of the surface. No swamps, no oceans (not that we had any longer the slightest reason to expect them), nor any sign of such things from the past. The only oceans you'll find on Venus are frozen, and made of rock: vast plains of volcanic lava. Astonishingly, Venus does have features that resemble river valleys, like the 4200-mile Baltis Vallis. But we can be sure these were not carved by flowing water; rather, it seems that they were eroded by lava flows composed of a rock with a low melting temperature.

The indomitable Percival Lowell believed that he could see canal-like features on Venus when he turned his telescope that way in 1896. Clearly, canals were etched deeply into Lowell's mind (and those of his assistants, who saw them too) – for whereas we can at least understand how surface features on Mars might give this impression to an observer already susceptible to the idea of life and civilization on other worlds, Venus offers no such deceptive signs. To the eye, it appears as a featureless yellowish-white ball, since it is wholly enshrouded in thick clouds.

It is because of this veil that the planet's surface remained so mysterious for so long. But it has been known since the 1930s that the clouds are not like the Earthly variety, composed of water. There is scarcely any water in the Venusian atmosphere, which is predominantly carbon dioxide with a dash of nitrogen.

Venus's murky shroud is comprised of droplets of neat sulphuric acid, in a layer about eleven miles thick. On Venus, all rain is acid rain, and much more corrosive than anything we experience in even the most pollution-blighted regions of Earth. But no rain ever reaches the rocky plains: although the droplets grow to raindrop size and sink under gravity through the cloud blanket, the atmosphere is so hot that they evaporate only a little below the cloud base. Rain that evaporates before falling to the ground is common in desert areas on Earth, where it is known as 'virga'.

The picture of a hellish planet was fleshed out by information returned by spacecraft missions to Venus in the 1960s. The first interplanetary spacecraft of any sort, Mariner 2, was dispatched to Venus in 1962, and brought back the news that Venus is an oven. The Soviet craft Venera 4 penetrated below the clouds in 1967 to find a very high atmospheric pressure, many times that on Earth; and its successor Venera 8 touched down on the surface five years later to confirm that the environment was almost water-free.

Venus is closer to the Sun than the Earth, but not enough to explain why it is so much hotter: an Earth-like body in Venus's orbit would be only a few dozen degrees hotter than our planet. In fact, on the basis of the differences in solar heating alone, Venus should be *colder* than Earth, because its brightness gives it a much higher albedo – it reflects much more of the Sun's radiation. But its thick atmosphere of carbon dioxide increases the surface temperature by a full 500°C or so above that which the planet would experience if it had no atmosphere. For not only is Venus's acid-rain problem far in excess of that on Earth – so is its greenhouse effect. Earth and Venus, almost twins at birth, have become strangers to one another.

The carbon on our planet is largely locked up in carbonate rocks such as chalk and limestone, as well as in organic sediments and the coal and oil that forms from their decay. If all this carbon were released into the atmosphere as carbon dioxide, we would have an atmosphere as dense as Venus's – and a comparable surface temperature. This is the greenhouse effect taken to its terrifying conclusion.

So why is so much of Venus's carbon in the atmosphere while Earth's remains in the rock? Water is the key. Greenhouse warming is influenced by both positive feedbacks, which accentuate it, and negative feedbacks, which attenuate it. On the young Venus the positive feedbacks dominated, precipitating the planet into a runaway greenhouse effect that boiled off all its volatile compounds into the atmosphere. Venus may once, very early in its history, have possessed oceans of liquid water. But as volatile substances were exhaled from the planet's interior into the atmosphere by degassing, the greenhouse warming increased the rate of evaporation from the oceans. Water vapour entering the skies contributed to the greenhouse effect, raising temperatures further and creating more evaporation. With nothing to hold this process in check, the entire oceans boiled dry.[12]

Earth seems to have been just far enough from the Sun to avoid this catastrophe – it resides between forty and forty-four million kilometres further out than Venus, which itself orbits at a distance of around 108 million kilometres. Our planet was initially just a few dozen degrees cooler, creating a smaller rate of evaporation. This allowed time for negative feedbacks to come into play to regulate the extent of Earth's natural greenhouse warming to just 30°C or so. Clouds provide one source of negative feedback, cooling the planet by increasing the amount of sunlight reflected back into space. And oceans reduce the amount of atmospheric carbon dioxide by dissolving it. Dissolved carbon dioxide gets converted to carbonate rocks and, once life has evolved, to organic sediments as the carbon gets fixed by photosynthesis. If it were not for this removal, mediated by water and subsequent fixation in carbonate rocks, our planet too would be imperilled by runaway greenhouse warming. Once the oceans of Venus evaporated (if indeed they ever existed), it lost forever its means of salvation from the inferno. Say planetary scientists Jonathan and Cynthia Lunine, 'Once bereft of surface water, the die was cast for Venus.'[13]

If an ocean's worth of water boiled off into Venus's atmosphere, where is it now? Some is tied up in the acid drops of the Venusian clouds, where it has reacted with the sulphur dioxide belched out by Venus's abundant volcanoes to form sulphuric acid. But the rest is largely gone: there are just twenty to one hundred molecules of water in every million molecules of the Venusian atmosphere. Water vapour gets split by sunlight in the upper atmosphere, and the hydrogen atoms produced are light enough to escape the gravity of an Earth- or Venus-sized planet. On Earth this water loss occurs very slowly. But if all Venus's water was in the atmosphere, the loss rate would have been far greater, particularly in view of the more intense solar radiation on Venus and the higher altitude to which the water vapour would have risen. David Grinspoon of the University of Colorado says: 'The more we understand, the more it seems that most differences between these twin planets may boil down to the huge difference in the amount of water.'[14]

Whether Venus really did have ancient oceans is still debated, but the idea receives support from the subtle difference between Venus's remaining water and Earth's. Venus's water contains around 120 times more deuterium (hydrogen-2) than the waters of Earth: fully two per cent of the hydrogen is deuterium. Because it is heavier than hydrogen-1, deuterium is lost more slowly from the planet's gravitational field when it is split off from a water molecule. For this reason, gravity acts like a kind of sieve that selectively retains deuterium over hydrogen. So if the water vapour we see today on Venus is the pitiful remnant of some much vaster ancient reservoir, we'd expect it to be highly enriched in deuterium – just as we observe.

But things are not quite this simple. Judging by the current rate of water loss from the Venusian atmosphere, estimated as something like a swimming pool's capacity a month, it will all be gone in between 100 and 200 million years. That may seem like a long time, but in the life of a planet it is just an instant. So it would be a rather incredible coincidence that human life evolved on Earth just in time to witness the dregs of Venus's water draining away. A hundred million years later and we'd have missed it.

While this coincidence may be just that, it seems more reasonable to suppose that the tiny amount of water vapour that we see on Venus today is not just the tail end of a gradually declining amount, but has somehow been maintained at that level over a long stretch of geological time. So some planetary scientists prefer to believe that Venus's water content is held steady by some source that balances the losses.

Candidate sources are not hard to identify. Volcanoes on Earth recycle water vapour from the deep Earth into the atmosphere, and it is quite possible that they do so on Venus too. And comets, the icemen of the solar system, may come knocking at the door often enough to have topped up Venus's water to a roughly constant level for the past few hundred million years.

Such issues may seem trivial compared with the question, now long settled, of whether Venus has swamps, oceans, dinosaurs, civilizations. But their resolution will tell us much about why our planet now looks so different from its twin, and may even offer insights into the environmental problems now facing Earth itself.

Under ice

Hot or cold, the surfaces of Venus and Mars are more or less dry as dust. But there is no water shortage in the solar system at large: take a trip to Jupiter and you will find it orbited by three moon-sized snowballs.

We have made several trips there already – not in manned spacecraft after the style of Arthur C. Clarke's *2001: A Space Odyssey*, but through the unmanned space missions of the Voyagers in the 1970s and the Galileo Orbiter in the 1990s. That Jupiter has four major moons is something that we can see from Earth, and Galileo himself was the first to do so in 1610.[15] In traditional style, these satellites were named after figures of Greek mythology: Io is the nearest, followed by Europa, Ganymede and Callisto. They range in size from about the same as the Moon in the case of Europa (1563 kilometres radius) and Io to the same as Mercury for Ganymede (2638 kilometres) and Callisto. But the so-called Jovian moons could otherwise hardly present a greater contrast. The outer three have never known anything but bitter ice ages, being covered with a layer of water ice from pole to pole. Io, on the other hand, is like the inner circle of hell, a fiery world of brimstone where immense volcanoes spew molten sulphur and other debris to an altitude of up to fifty kilometres. Although temperatures at Io's equator do well to top minus 150°C, there are localized 'hotspots' where the temperature reaches around 250°C. Io is so hot and volcanic because of the awesome gravitational field of Jupiter: so close is the moon to its giant parent planet that the gravity pulls and tugs at Io's fabric as it orbits, pumping enough energy into the moon to melt much of its interior. This is called tidal heating.

But our quest for water in the solar system takes us out to Io's siblings: Europa, Ganymede and Callisto. The Voyager missions revealed that the latter two are rather similar: heavily disfigured with bright craters and having low average densities, which implies that the dense rocky cores are encased in very thick shells of ice whose surface is so cold that it is as hard and brittle as rock.

The heavy cratering suggests at first glance that not much has changed on these moons for billions of years, give or take the occasional stray meteorite gouging a new blister of bright, fresh ice. Yet Ganymede is crisscrossed by a network of grooves and ridges where the ice has been creased, broken and deformed (Fig. 4.3). These markings are similar to those seen on mountain glaciers, suggesting that the rigid crust has been rumpled by motion below – convective flow in the more ductile icy mantle when the moon was younger and warmer, or maybe even the surging of a sub-surface watery ocean.

And the craters on Ganymede and Callisto are considerably flatter than those seen on the Moon or on rocky planets: they have apparently been smoothed out. Solid ice is plastic; it can flow very slowly through the process known as creep (page 68). The icy crusts of these two moons have gradually relaxed over time to flatten the fingerprints of impacts. Some of the circular features have virtually no topography at all – they remain visible just by virtue of being slightly brighter than the surrounding area. These markings, called palimpsests after the word for old parchments that have been scraped clean and overwritten, are like the ghosts of former craters, all that remains as the wrinkles are soothed by the pull of gravity. One of the most striking of these features is a region on Callisto called Valhalla, a bright area about 300 kilometres across surrounded by a series of concentric ring-like wrinkles extending outwards to about 1500 kilometres (Fig. 4.4).

Compared with these two worlds, Europa tells a strikingly different tale. It is almost wholly free of impact craters, indicating that its surface is probably younger than three billion years old – and maybe *much* younger. The ice is laced with a complex web of dark streaks, which stunned planetary scientists who first saw them beamed back from Voyager 1. Says planetologist Torrence Johnson, 'The aspect of this distant satellite was eerily reminiscent of Percival Lowell's perception of canals crossing ruddy Mars.'[16]

But again, these are no canals, and neither are they like the deep river valleys of Mars. The streaks are shallow, just a hundred metres or so deep, and they resemble the cracks and openings that thread through sea ice at the Earth's poles. To the Voyager scientists, this suggested the exciting possibility that Europa's ice crust might be floating on a sub-surface ocean that is *still liquid*, whose convective motions due to heat in the rocky interior are continually breaking up the ice layer. The continual replenishment and rearrangement of the surface ice would erase most of the early cratering. During the 1980s, several planetary scientists suggested that tidal heating, which produces such dramatic results on Io, could be sufficient on Europa to keep its hidden water fluid.

Scenting the prospect of liquid water on another world, scientists eagerly awaited the results of the Galileo mission in the late 1990s,

Fig. 4.3 The surface of Jupiter's moon Ganymede displays a lacework of ridges – wrinkles in the ice which indicate that the frozen crust might be in sluggish motion like a mountain glacier. (NASA/JPL's Galileo Project; produced by Brown University)

which took close-up photographs of Europa's surface. In February 1997 the orbiting spacecraft made its closest pass, just 363 miles from the surface. The pictures it returned of Europa's striped and fractured surface were spectacular (Fig. 4.5). We saw great creases thrusting through the ice like multi-lane highways about a hundred kilometres wide. We saw huge chunks of grooved ice apparently adrift in a smoother ice 'sea', which could be fitted back together like a jigsaw puzzle. Said Michael Carr from the Galileo imaging team, 'These rafts appear to be floating and may, in fact, be comparable to icebergs here on Earth.'[17] The Galileo team speculated that the rougher terrain

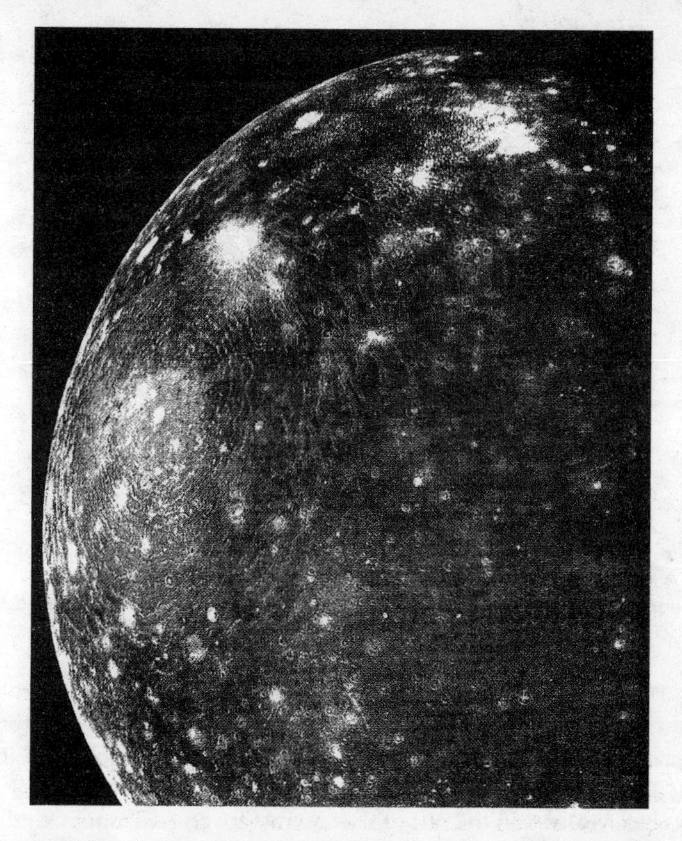

Fig. 4.4 Callisto is coated with a crust of dirty ice, pockmarked with bright craters where meteorite impacts have exposed fresh ice below. The concentric ring structure to the middle left is called Valhalla, and is thought to be the remnant of some huge impact in the distant past. (NASA/JPL)

between the ice blocks might be just a hundred million years old – as recent as last week in geological terms. This would mean that the processes that disrupted the ice plates would still be active today. And in some regions, the network of ridges was disturbed by smooth, roughly circular features, like flat-topped domes. The observers concluded that these might be the caps of columns of water or warm ice rising from below by convection. All of this disturbance at the surface provided a strong indication that there was a fluid below it.

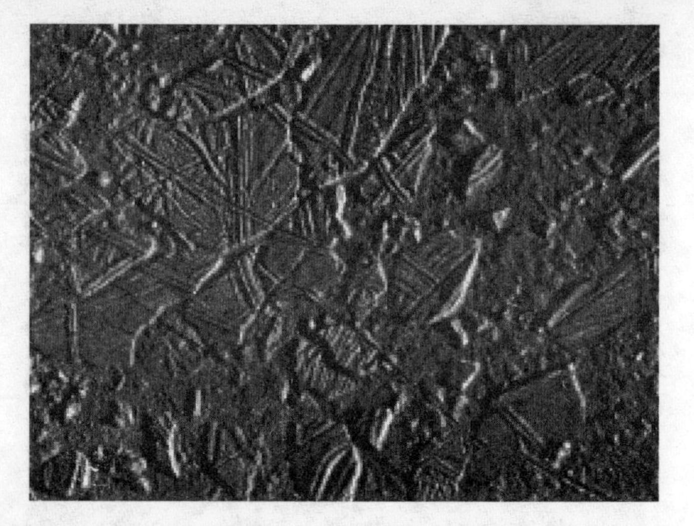

Fig. 4.5 Close up, Europa's surface is spectacular. Huge grooves run through the ice, which in some places is broken into blocky pieces whose features permit reconstruction of the jigsaw. (NASA/JPL's Galileo Project; produced by Arizona State University)

The way that the moon rotates around Jupiter also hints at a subsurface ocean. Jupiter's strong gravity should prevent Europa from rotating on its axis, so that, like our own Moon relative to the Earth, it always shows the same face to its mother planet. But the Galileo probe revealed that the rifts in the icy crust are arrayed as though the ice has been deformed and fractured by rotation – albeit very, very slow rotation, which takes about 10,000 years to complete one revolution. This implies that the icy surface can move independently of the rocky interior, which would be made possible by a liquid ocean lubricating the boundary between ice on the outside and rock below.

The clincher came when the Galileo data revealed that both Europa and Callisto possess their own magnetic fields, which perturb the field of Jupiter in which they are embedded. These fields seem to be generated close to the surfaces of the two moons, and the most likely explanation is that they are produced by eddy currents within an electrically conducting fluid. Because the Galileo observations had already shown that the icy crust of Europa is salty, the corollary is that the moon has a briny (and so electrically conducting) sea below the ice. That Callisto shows the same signature of liquid water is even more of a surprise, since it shows none of the sea-ice-like surface features of Europa.

If there is an ocean on Europa, why not life? The moons of Jupiter are outside the zone of the solar system traditionally considered habitable – the zone, extending from just inside the Earth's orbit to just beyond the orbit of Mars, within which a planet might sustain liquid water on its surface. But Europa, and apparently Callisto too, cheat this expectation by virtue of Jupiter's tremendous size, which provides an energy source independent of the Sun in the form of tidal heating. It seems entirely possible that volcanic activity, like that on Io but hidden below the ice, warms parts of Europa. On Earth, volcanism at deep-sea hydrothermal vents fuels entire ecosystems, which thrive on geothermal energy instead of sunlight (see page 211). Might hidden colonies exist on Europa too?

That is something we can answer only by sending spacecraft to probe beneath the ice cap. NASA now has its sights set on launching such a probe by around the year 2015 to drill hundreds of metres – perhaps several kilometres – through the ice. There are mixed views about what we might find down there. Says oceanographer John Delaney, 'If there is anyplace else in the solar system likely to have life, it is Europa.'[18] But geologist J.A. Jacobs of the University College of Wales is less optimistic in an ode inspired by Lewis Carroll's *Through the Looking-Glass*:

> Europa's crust was dry as dry
> Was underneath it wet?
> You could not see below, because
> The ice above had set,
> You could not see the fish, because
> There were no fish as yet.[19]

One thing seems certain, however – the quest for life elsewhere in the solar system will be far better directed at this icy outpost than at our immediate neighbours. The lesson is simple: prospectors for life should follow the trail towards water.[20]

The outer limits

Some planetary scientists claim that there is one further candidate for hosting nearby extraterrestrial life: Titan, the largest of Saturn's nine moons, 5150 kilometres across. In *The Sirens of Titan*, Kurt Vonnegut did not (characteristically) take the idea particularly seriously:

> The atmosphere of Titan is like the atmosphere outside the back door of an Earthling bakery on a spring morning ... There are three seas on Titan, each the size of Earthling Lake Michigan. The waters of all three are fresh and emerald clear.[21]

Would-be settlers travelling to Titan on the strength of Vonnegut's warm recommendation would be in for a shock. The surface temperature is typically about minus 181°C; and if there are seas, they are composed not of fresh water but of liquid hydrocarbons: methane and ethane.

Titan is a fascinating world, with an atmosphere thicker than the Earth's, beclouded by a smog of complex hydrocarbon particles. But it is hardly a place to anticipate life. All the same, there is no shortage of water there: the planet-sized moon is thought to have an icy surface down to a depth of about a thousand kilometres before you hit rock.

That, however, is nothing unusual. In the outer reaches of the solar system, bodies covered with ice are the norm – with the exception of the gas giants Jupiter and Saturn. Saturn's other moons Mimas, Tethys, Dione, Rhea, Enceladus and Iapetus all seem to be icebound, and the densities of Mimas, Tethys and Rhea are so low that they must be almost totally ice. These worlds are surely frozen solid – no oceans hide beneath their frigid faces. Mimas and Dione, indeed, look much like a rocky, cratered body such as our Moon. Some of these ice worlds show peculiar and unexplained oddities, like the smooth, unmarked face of Enceladus and the dark blotch covering an entire hemisphere of Iapetus, and there remains much to be understood about how their icelands have evolved.

Pluto too, the smallest and most distant of the planets[22] (except when it passes inside Neptune's orbit on its wildly eccentric path), is an iceball; and even Uranus and Neptune, whose outer regions are made up of gaseous hydrogen and helium, contain a substantial inner mantle of water ice mixed with frozen ammonia and methane. And the ice reaches further still: into the Oort cloud, the storehouse of comets. Clearly, water is a major constituent of the solar system's nether reaches.

Why should this be? Was the solar nebula mostly water in this region? No, it is rather that water has the highest melting and condensation temperatures of all of the common small molecules – hydrogen, methane, ammonia, nitrogen, carbon dioxide – in this gaseous disk. For that reason, water would have been the first compound to condense and then freeze around the rocky silicate core of the nascent planetary bodies and their satellites. In the solid or liquid state it would have been less easily lost from the celestial bodies into space. We shall see later that these physical properties are a direct result of water's unique and peculiar molecular structure. It is the fabric of much of the solar system because it is a very odd substance indeed.

The wet Sun

Iceballs in the frigidity of the outer solar system are one thing; far more surprising is the revelation that even in its fiery eye we can see the fingerprint of H_2O. The Sun has steam in its atmosphere.

Well all right, 'steam' is perhaps pushing it – it's not exactly as humid as a sauna up there. Even in the most watery parts of the Sun, a thimble plunged down through the full height of the atmosphere would capture around twenty billion molecules, enough to fill just a thousandth of its volume with liquid water. Yet it's there all the same, revealed by the spectrum of sunlight. The water molecules bite out chunks of light at wavelengths that they can absorb by vibrating or rotating more vigorously.

It was long thought that water couldn't survive in the Sun's furnace. The part of the Sun's atmosphere that glows – the photosphere – seethes at about 5500°C, which is hot enough to split water molecules apart. But in 1995 a group of astronomers focused the McMath-Pierce telescope of the Kitt Peak National Solar Observatory in Tucson, Arizona, on cooler parts of the solar photosphere: sunspots, whose relative darkness betrays lower temperatures. Within the coolest parts of these blemishes on the Sun – the central 'umbrae' – temperatures fall to a modest 3000°C or so.

The spectra of umbrae recorded at Kitt Peak were riddled with little gaps where molecules in the Sun's photosphere had absorbed the light. By comparing the wavelengths of the gaps with laboratory measurements of the spectrum of very hot water, the observers were able to show that it was hot water molecules that had nibbled away at the umbral sunlight.

The discovery of water on the Sun didn't have quite the same impact as the discovery of water on the Moon; even NASA would find it hard to turn it into a call for colonization. But it shows that water finds its way into just about any niche in the Universe that will have it. It has been known since the 1960s that water can be found in the atmospheres of stars cooler than the Sun. Late in their lives, stars like the Sun become red giant stars, which have photospheres cooler than about 3700°C. At these temperatures, water can become the dominant oxygen-containing molecule. And water seems to be the second most abundant molecule, after molecular hydrogen, in star-like objects called brown dwarfs. These are stars that are too small ever to have really ignited. With masses of less than a tenth of that of the Sun, the densities and temperatures generated within them by gravitational collapse of gas are not great enough to spark off hydrogen fusion, and they glimmer dimly at temperatures no greater than 700°C. Because they are so dim, it is hard to estimate how many brown dwarfs exist.

What is so special about the Earth, then, is not that it is a world of water, but that the water is marine blue – we have oceans, not just glassy sheets of bright ice. Perhaps, soon after the solar system was formed, blue worlds were commonplace, until one by one, they turned pearly or ruddy, or became shrouded in bright acid. And then there we were, a lone blue dot, waiting for life to begin.

PART II

Two Hands, Two Feet

5 Open to the Elements

The naming of H_2O

And first, if I would now deal rigidly with my adversary, I might here make a great question of the very way of probation which he and others employ, without the least scruple, to evince that the bodies commonly called mixt are made up of earth, air, water, and fire, which they are pleased also to call elements...

Robert Boyle (1661), *The Sceptical Chymist*

A poor schoolmaster, in rags, introduced himself to a scientific friend with whom I was talking, and announced that he had found out the composition of the sun. 'How was that done?' – 'By consideration of the four elements.' – 'What are they?' – 'Of course, fire, air, earth, and water.' – 'Did you not know that air, earth, and water have long been known to be no elements at all, but compounds?' – 'What do you mean, sir? Who ever heard of such a thing?'

Augustus De Morgan (1872), *A Budget of Paradoxes*

Water has never lost its mystery. After at least two and a half millennia of philosophical and scientific enquiry, the most vital of the world's substances remains surrounded by deep uncertainties. Without too much poetic licence we can reduce these questions to a single bare essential: what exactly *is* water?

The thing no one can ignore about water is that it seems unique. Other stuff of the world is endlessly varied, but of a comparable stamp. Typically opaque, its shapes have some permanence – a leaf, a rock, an animal are comprised of fabrics that do not part before a probing finger. Water is the opposite of all that. It is compliant, mobile, transparent, tasteless. It's not hard to form the impression that, in comparison with the rest of reality, water is somehow unearthly. No matter how much you might be able to convince yourself that all the variegated shades of solid matter are but subtle variations of a single elementary substance (regarded, in different times and places, as earth, metal and wood), you're always going to need a separate category for water, though it appears in contrast to have but a single complexion.

Yet today water is no longer accorded this special status: it is no longer one of a privileged little band of elements. Chemically, it's just a regular old compound – a mixture of more elementary substances, an adulterated element. Yet our interest in it is undiminished. Water draws our sympathies: we *must* go down to the sea again. In part this attraction is surely a cultural inheritance, an echo of the significance that water once had for fragile, formative civilizations or for adventurers and, in their wake, traders before whom water stretched like an invitation to new lands. In part its allure is a biological necessity. But it is also a result of the sheer beauty water has to offer. For scientists too this fascination remains, for all that we now know water's constitution in extraordinary detail.

The journey from water the element to the H_2O molecule is just the prelude to water's mystery. It is a journey that is not about water alone, but about our whole concept of the material world. For water has been a guide towards both modern physics, which tells of the composition of matter and its transformations, and chemistry, which describes the way elements combine. It is also a cipher for much of biology and the sciences of the Earth. By exploring water, the philosophers and scientists of previous generations were exploring the world. William Blake's grain of sand, in which he saw the world reflected, is more properly a raindrop.

The first matter

It's not an easy thing to jump into the minds of our ancestors. Even without a knowledge of chemistry, we do not need much persuading that dull, grey lead will not be transmuted to bright gold no matter how it is boiled, distilled and blended. We do not harbour suspicions that life will be generated spontaneously from sterile dust, that the planets revolve on huge geometrical cages of crystal, that mercury is a miracle cure or the vital ingredient of an immortality potion. Now that such ideas sound like quackery, it is extremely difficult to appreciate what the ancients must have made of water. This is surely the biggest challenge in understanding how water came to have the deep symbolic and emotional significance that obtains even today. Conceiving of what water meant to older civilizations is more than a matter of assuming that they 'knew' less than we do – so that, for instance, they regarded the liquid as an irreducible substance, not as a compound of elemental hydrogen and oxygen. It is a question of realizing that their entire outlook on the physical world was fundamentally different. As French philosopher Gaston Bachelard says, 'The meaning of prescientific research cannot be thoroughly understood until we have formulated a psychology of the seeker.'[1]

It has been suggested that a scientific way of thinking, in which we make deduction and generalization from observation, is innate to humans. We apply inductive reasoning instinctively, making plans for tomorrow on the assumption that the world then will be much as it is today – that there are immutable laws of physics, in other words. This might be ascribed to nothing more than an evolutionary inevitability: if we didn't think like that, the human race would have gone extinct long ago. But in classical Greece, science as we now know it had little part to play in academic debate. Hands-on experience was unseemly; if you wanted to understand how the world worked, you gave the matter some pretty determined thought. If recourse to experiment was made at all, it was not to put ideas to the test but merely to demonstrate their veracity. It is not easy to find a single example of classical Greek philosophical thought that was modified as a consequence of experiment. Naturally, one can't possibly hope to get everything right by thought alone, to find that the Universe falls in line with your ideas of how it ought to be. But the ancient Greeks never seemed to stumble over this difficulty – even Aristotle had little trouble in seeing exactly what he wanted to see. That everyday experience contradicted some of the proclamations of these philosophers did not for a minute disturb them, since crude experience was considered no equal to rational thought.

It is, then, scarcely surprising that the contribution of classical Greek culture to modern scientific thought is most substantial in areas of abstract knowledge, such as geometry and mechanics, and far less so in questions relating to the nature of matter. Yet such questions would surely have arisen in any civilization, since manipulating matter is part of what it means to be civilized. Artisans making metal implements and ceramic vessels could not have failed to wonder about the nature of the transformations they were effecting. In particular, they must have pondered matter's irreducible complexions – its elements.

Although Western culture has come to regard the Aristotelian fourfold categorization of the elements – earth, air, fire, water – as the canonical pre-scientific classification of matter, it was by no means unique: many others have been proposed at various times from the earliest of recorded history to the advent of modern chemistry. Common to nearly all such schemes, however, was water. So was fire: fire and water were indispensable to any culture, so their primacy is not hard to understand. We can see readily enough how this dualism might have been emphasized by the natural associations with the Sun and Moon respectively. Any community dependent on the sea, with a consequent need to anticipate the changing tides, could not for long have failed to notice their relationship with the phases of the Moon, Titania's 'governess of floods' in *A Midsummer Night's Dream*.

The origin of an elemental view of matter – with its implication that all matter can be reduced to a few fundamental substances – is unknown. But it dates back at least as far as Thales of Ionia, who lived from the seventh to the sixth century BC. Thales was a citizen of the Greek colony of Miletus in Ionia, at the western extremity of Asia Minor. He travelled in Egypt, where he learnt of Egyptian cosmology and the cosmogony of the Babylonians. Thales maintained that all of reality was derived from a single elemental substance: water.

This idea was possibly taken from Babylonia, whose mythology held that the entire cosmos was made from water. But despite his broad experience of the world, Thales was reputed to have an indifference to myths, and may instead have derived his inspiration from the evident fact that water was central both to living things and to the geological world. Noting that water can be both vaporized and solidified, Thales concluded that it was sufficient to serve as a basis for all matter. You couldn't make fluids from soil or rock, nor condensed substances from air – only water could generate liquids, vapours and solids. This isn't true, of course, but it is an understandable conclusion given the technologies with which Thales would have been familiar.

In the elemental philosophies of antiquity, transformations of matter are often discussed in terms of changes of *state*, not of composition. We'll comprehend nothing of these ideas if we don't recognize that the distinction had none of the clear meaning that it has today. If we freeze water, the rigid stuff that results is still water – it's simply in its solid state (ice) rather than its liquid form. Yet to the ancient Greeks and their successors, ice is not fundamentally the same stuff as liquid water. It is a substance that has the attributes of earth and metals – hardness, solidity. To Thales, water became a certain kind of earth when it froze, and a certain kind of air when it evaporated. The failure to appreciate distinctions like this meant that the ancient philosophers could not differentiate what water is from what it does – in today's terms, they could not distinguish its chemistry from its physics.

It's only when we recognize this that the old views of matter, including those of alchemical tradition, start to make sense. If you can make earth from water, why not gold from base metals? After all, that's not even converting one 'element' to another (in a pre-scientific framework), but just making one metal from another. Transmutation of the elements is not in itself an irrational idea: we see far stranger transformations all the time. Plants and animals grow, they accrete new mass. Into the oven goes a wet, sloppy dough, and it comes out as a hard-crusted loaf. So long as we judge only by appearances, all kinds of 'elemental' transmutations happen all the time. When salt is added to water, it is apparently transformed into water – it vanishes, and if we were so perspicacious as to weigh the resulting liquid, we'd find that its mass has increased by exactly the mass of the salt. Though there were undoubtedly charlatans in the alchemical tradition who attempted to make their fortune by creating fake gold just about good enough to deceive the jeweller's eye, most had no reason to doubt that a recipe for making from base metals a substance of golden appearance was in fact a recipe for making gold itself.

With this in mind, maybe you can start to appreciate why the idea that water is the root of everything was as good a hypothesis as any in the time of Thales of Ionia.

But Thales's single-element principle was hardly unique. After him followed a succession of philosophers of the 'Ionian school', and it seems that almost every one had a different view of what the Universe was made from. Anaximenes in the sixth century BC gave primacy to air, whereas later Heraclitus proposed that fire was the fundamental element. But Empedocles in the fifty century BC seems to have been the first to make common currency of the classic four-element natural philosophy of Greece. By his time, water, fire and earth were well established as elements, and Empedocles argued the case for air having similar status rather than its being just a transitory stage in the conversion of water to fire.

To Empedocles we owe another facet of the significance of water in physical theory. He showed that it can provide the basis of a primitive timekeeper, a *klepsydra* or water clock. This is a cone-shaped vessel with holes in the base and at the apex, which sinks slowly as it fills with water. The time taken to sink is always the same, and so provides a unit of time. Through water's agency, the elastic days could be segmented into equal measures.

Although Empedocles's elemental scheme is the one familiar in the West today, ancient Chinese tradition has instead five elements: earth, metal, wood, fire and water. This fivefold scheme is thought to have been first advanced by the scholar Tsou Yen around 350 BC. The number five is accorded great significance within the Taoist tradition. There are five directions – north, south, east, west and centre (again, in the West we identify just four) – and five primary colours, yellow, blue, red, white and black. Moreover, the Chinese knew of five planets – Mercury, Venus, Mars, Jupiter and Saturn – and five metals – gold, silver, lead, copper and iron. The number five takes central place in the 'magic square' of the numbers one to nine, whose rows and columns all add up to fifteen – a talisman of deep significance to early Taoism.

Given all these quintets, we shouldn't be surprised that the Chinese sages searched for correspondences between them. So it was that each element was assigned a direction, a colour, a planet and so forth. The qualities of water are: north, black, Mercury, coldness, fluidity. It is associated with *yin*, the passive feminine principle of Taoism, and with the Moon. Within Chinese alchemy, water is the fundamental material of creation, the 'life substance'. The scholar Lao Tzu describes it with reverence in the great Taoist treatise the *Tao Te Ching*:

> The highest good is like water. Water gives life to the ten thousand things and does not strive. It flows in places men reject and so is like the Tao.[2]

Most intriguing is the association of water with the number six. We can only guess at cause and effect here, but to the Chinese this relation found natural expression in the 'flowers of snow' that fell from the sky each winter. Eighteen centuries before the shape of snowflakes was recognized in the West, the sage Han Ying explained that 'Flowers of plants and trees are generally five-pointed, but those of snow, which are called *ying*, are always six-pointed.' Given that it takes rather more than casual observation to deduce this, we cannot dismiss Chinese natural philosophy as so thoroughly based in speculation as that of the Greeks. I will show in Chapter 7 that there are perfectly good reasons to link water to the number six, and that the 'six-petalled flowers' of snow are a direct consequence of this.

Within the language of Chinese alchemy, therefore, we can read at a glance just about all of the symbolism that has come to be associated with water through the ages. Though often revered, it is also a veritable beacon for the stereotyping, and occasionally the denigration, of femininity. In the hierarchical arrangement sometimes depicted in Eastern tradition (as, for example, in the *stupa* form of Buddhist architecture), water lies below masculine fire. It is mysterious, hidden, passive and cold. But at the same time, from it springs all life: it is a 'profoundly maternal' substance, says Gaston Bachelard. It is influenced by the Moon, which in turn symbolizes a whole string of goddesses, from Isis of the Egyptians to Diana of the Romans, Artemis of the Greeks and Qilla of the Incas. Medieval shrines to goddesses were nearly always close to wells, springs, lakes or seas; the Lady of the Lake can be identified with Aphrodite, Greek goddess of love. And in early Christian tradition the baptismal font was described as the womb of Mary, whose name is that of a sea goddess.

I suspect that, as feminist philosophers such as Carolyn Merchant have argued,[3] there is much to be discerned about the way our societies regard the Earth's waters from the way they regard women. Throughout the Middle Ages, Mother Nature was laid bare for study, sometimes in explicitly sexual language, while her waters became regarded as a commodity, ripe for private ownership, to be diverted, exploited and polluted for the benefit of the rising merchant classes.

Pieces of water

Perhaps the most lasting contribution of the Ionian school to our modern understanding of the nature of matter in general comes from its last great philosopher, Anaxagoras of Klazomenae, north of Miletus. To him can be traced the beginnings of an atomic theory, although there is nothing in Anaxagoras's conception of the world that truly corresponds to atoms as we know them. Rather, his contribution was to take a particulate view of matter: he proposed that it consists of an infinite number of 'seeds', each of which contains, to varying degrees, a dash of all that exists. These seeds are eternal, being neither created nor destroyed. Empedocles perpetuated this concept, but with the crucial distinction that the seeds have only four complexions, corresponding to his four elements.

The pre-scientific atomistic picture of matter culminated in the teachings of Leucippus and his pupil Democritus of Abdera, who lived from around 460 to 370 BC. Since he seems to have been the first to refer to the smallest individual fragments of matter as *atomos* – literally, uncuttable – Democritus is often accredited with the inception of atomic theory. Yet more significant were Democritus's ideas about the part of the world that is not atoms: the vacuum or void that separates them. The earlier ideas about a corpuscular character of matter were vague or silent about what lay outside the 'seeds' or particles – that is, the nature of the medium in which they moved. But Democritus assumed that atoms of the four elements reside in a void, within which they are in constant motion. This might seem like a subtle addition to the atomistic picture, but with the introduction of a void the theory becomes for the first time a truly mechanical one, susceptible to ideas about the motion of bodies that are familiar from our everyday experience at larger scales. Atoms become like dancing dust motes in the air. Without the idea of void, the kinetic theory of gases developed during the nineteenth century (see page 153) could not have been formulated.

Democritus's atomic theory was deeply mechanistic. Atoms were physical entities with definite sizes, possessing shapes that explained the properties of the respective elements. Fire atoms were round and immiscible with the atoms of other elements. The shapes of other atoms permitted them to become mixed, entangled and congealed to produce the various substances of the world.

The Greek love of geometry infused this mechanistic view of matter. During the late fifth century BC, Pythagoras founded in Athens a philosophical school based on a mystical approach to mathematics. To the Pythagoreans, numbers were not abstract concepts but had a real material existence. This school, of which Empedocles was initially a disciple, was as much a religious as a philosophical institution, and became ever more concerned with an occult quest towards moral perfection rather than with theories of the material world. The Pythagoreans supposed that the microscopic particles of the four elements of Empedocles possessed, at a microscopic level, mathematically regular geometric shapes separated by a void that was more like air than Democritus's nothingness.

The idea of atoms as regular geometric bodies held a strong appeal for Plato, who lived from around 428 to 348 BC. He neatly combined the numerological tendencies of the Pythagoreans with the materialism of his teacher Socrates and of Democritus, by proposing that the shapes of atoms were harmonious and perfect yet at the same time explained the properties of the elements in a mechanical way. Plato's fire atoms are tetrahedral (Fig. 5.1), their sharp tips allowing them to be the most penetrating of elements. Water atoms are icosahedra, the most rounded of the Platonic solids; air atoms are octahedra. While these three forms are made up of triangular faces, the atoms of earth are cubes, with square faces. Thus the planes of earth atoms cannot be rearranged into the shapes of other kinds of atom, and the solidity and stability of earth is thereby explained. The triangular faces of the other elements can be rearranged, however, and so they can be interconverted. Said Plato of such transmutations:

> We see that what we just now called water, by condensation I suppose becomes stone and earth, and this same element, when melted and dispersed, passes into vapour and air.[4]

Atomists though they were, Socrates and Plato heralded the theory's decline. They did not refute the discrete particulate nature of matter in itself, but were uncomfortable with the notion of void between

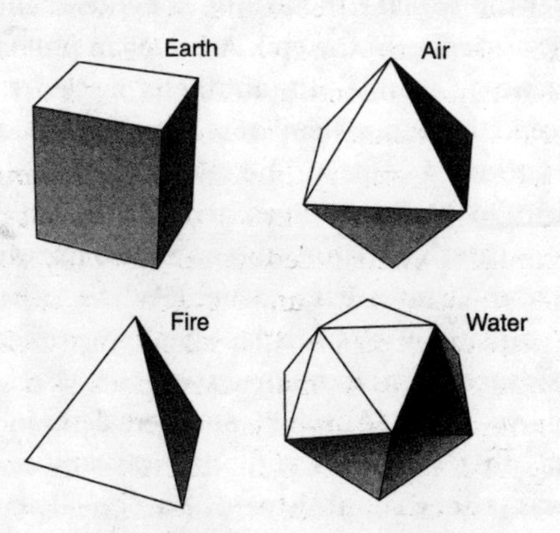

Fig. 5.1 Plato's atoms were particles with perfect geometric forms: a tetrahedron for fire, an icosahedron for water, an octahedron for air and a cube for earth.

the particles. The idea had always been controversial. As Democritus's uncle relates in Gore Vidal's *Creation*, Anaxagoras ('the best of a bad lot')

> ... believes that there is *no nothing*. He believes that all space is filled with something, even if we cannot see it – the wind, for instance.[5]

Anaxagoras was not alone in considering the hypothesis of nothingness an affront to reason; the philosopher Parminedes of the early fifth century BC was especially vocal in his opposition. But it was Aristotle who ensured that atomism did not gain strong support again for nearly two millennia.

A pupil of Plato, Aristotle formulated a scientific programme that was probably the broadest and most ambitious of all those proposed by the Greek philosophers. He was interested in everything, and wanted to explain it all. His writings on physics and biology became dogma for centuries in the West. Throughout the Middle Ages, all new theories had to be consistent with the Aristotelian view of things in both the Islamic and the Christian worlds. This lent a unity to natural philosophy in the two cultures that otherwise found themselves opposed, which surely lubricated the exchange of ideas and learning. But Aristotle's supremacy was also stifling, and eventually proved to be a significant obstacle to the acceptance of the theories of Copernicus, Galileo and Newton which heralded the Enlightenment and much of modern physical science. In the end, Aristotelian philosophy came to stand for resistance to change – the antithesis of science.

Aristotle accepted the four-element scheme of Empedocles, but not Democritus's notion of void. Air, he said, would never permit void, but would rush in to fill it. Even today one can still hear the Aristotelian solecism that 'nature abhors a vacuum', despite the fact that nearly all of the Universe *is* a vacuum. When Aristotelian philosophy became conflated with Christian theology in medieval Europe, later atomists, such as the followers of the Greek philosopher Epicurus (who lived from 341 to 270 BC), were denounced as atheists on the grounds that Aristotle's authority was almost equivalent to God's. Yet it was water that furnished at least one Epicurean with his atomistic beliefs. The Roman poet Titus Lucretius Carus argued in 56 BC that the penetrability of water must arise from the ability of objects, such as fish, to protrude into the void between its atoms and so to separate them:

Therefore that which is humid separates
By minute parts, which no eye penetrates.[6]

Lucretius's poem *De Rerum Natura* narrowly survived the purges of the Church in the Middle Ages to become a crucial influence on the revival of atomism in the seventeenth century.

Aristotle held that all matter was comprised of a single substance, a 'first matter' (*prote hyle* or *prima materia*), on which are impressed the characteristic 'forms' of all different substances. Out of the first matter the four elements arise via the imprint of four qualities, in two pairs of two opposites: hot and cold, moist and dry. Each element represents a combination of two qualities: water, for instance, is moist and cold. In this scheme, elements could be transmuted by inverting the appropriate quality. Water could be transformed to air (moist, hot) by replacing coldness with hotness – in other words, by heating it. Twelve centuries later we can find these ideas guiding the thinking of the Arabic alchemist Jabir ibn Hayyan (latinized to Geber in the West),[7] who describes how water can be distilled ('dried') to remove its moist quality and leave only pure Cold, 'a white and pure substance which ... when it is touched by the smallest degree of moisture, dissolves and is again transformed into water'.[8] We can conjecture that if Geber had tasted his Cold, it would have been suspiciously salty.

But beyond the four mundane elements, Aristotle deemed it necessary to add a fifth: the ether, which played no part in earthly matter but was the perfect 'quintessence' of which the heavens were composed. The ethereal heavens perpetually rotate in a perfect circle, which accounted for the motions of the stars and planets. Gradually metamorphosed into the medium that bears the electromagnetic vibrations of light, the ether was put to rest only at the start of the twentieth century to make way for Albert Einstein's theory of special relativity.

Aristotle's philosophy represents not only the climax but the end of classical Greek philosophy. In the ensuing two centuries after his death in 323 BC, the centre of gravity of Greek thought shifted from Athens to Alexandria, the great city at the mouth of the Nile founded by Aristotle's pupil Alexander the Great. Here classical Greek tradition became infused and enriched by broader influences from Egypt, Asia Minor, Mesopotamia and Persia, and to some extent also from India, which Alexander entered at the farthest extremity of his conquests. Alexandria was the focus of the Hellenistic culture, which flourished until the ascent of Rome at the beginning of the new millennium. It was a culture that owed as much to the oriental world as to Greece. Through these other influences, Hellenistic natural philosophy acquired a more practical flavour, spawning such philosopher-engineers as Archimedes, Ptolemy and Hero. Archimedes put water to practical use by more or less founding the discipline of hydrostatics, and Hero, who lived in Alexandria between about AD 62 and 150, achieved a hydraulic sophistication that was not to be revisited for many centuries. He devised a slot machine for delivering holy water in temples and realized that the evaporation of water by fire could be harnessed to do useful work as the vapour expanded. In other words he foresaw and utilized steam power; and he described, but apparently did not build, a steam engine.

To Hero, the evaporation of water was still as much an elemental transmutation as it had been for Aristotle, rather than simply a change in physical state:

> Water also, when consumed by the action of fire, is transformed into air; for the vapour arising from cauldrons placed upon flames is nothing but the evaporation of liquid passing into air.[9]

And the formation of mud when water met earth was an example of the transformation of water into earth. Yet this Aristotelian outlook did not inhibit a decidedly non-Aristotelian practicality, which is why alchemy blossomed in the Hellenistic culture, and along with it, arts such as metal-working and dyeing. It is to Alexandria, not Athens, that we should look for signs of chemistry's origins, even etymologically: for the Greek *khemia*, meaning 'chemistry', is derived from the Egyptian word for their land, *khami*, meaning 'black' and denoting the dark, fertile soils of the Nile delta. From this cultural melting pot came both an impressive amount of applied science and the oriental mysticism that was to characterize alchemy for centuries.

A pinch of salt

The theoretical framework laid down by the Greeks provided the underpinnings of almost all ideas about the nature of matter for fifteen centuries after the birth of Christ. And so we can, without encountering much of a *conceptual* discontinuity, pick up the story with that most influential, complex and mercurial of characters in the Western Hermetic tradition, Philip Aureolus Theophrastus Bombast von Hohenheim, who called himself Paracelsus.

Born in 1493, Paracelsus provides a bridge between the mysticism and occultism of the Middle Ages and the rationalism of the Enlightenment. To modern eyes he is a mass of contradictions. Though by no means impartial to the obscurantism of much of the alchemical tradition, Paracelsus despised the superstition, self-importance and blind recipe-following of his contemporary physicians and apothecaries. He praised the humility and methodical graft of true 'artists', which is to say, chemists: 'they are sooty and dirty like the smiths and charcoal-burners ... they tend their work at the fire patiently day and night ... and do not highly praise their own remedies'. Yet Paracelsus knew no bounds when it came to self-inflation: his self-styled name was adopted to imply his superiority over, and contempt for, the famed Roman physician Celsius. He was a man of great learning, but that was not always reflected in the language he used in dispute: 'What light do you shed, you doctors of Montpellier, Vienna and Leipzig? About as much light as a Spanish fly in a dysentery stool!'

This Rabelaisian character inherited the four elements of Empedocles and the Aristotelian quartet of principles that related them. But to this backdrop Paracelsus pinned a new layer. While the elements were the ultimate constituents of all matter, its 'immediate' constitution was derived from three 'primary bodies' (*tria prima*): salt, sulphur and mercury. This trinity corresponded respectively to the body, soul and spirit of all things.

The significance ascribed to sulphur and mercury was inherited from long tradition in alchemy, according to which these substances were the fundamental constituents of all metals. Yet 'sulphur' and 'mercury' here do not refer to the elemental substances as we now know them. There is, said the alchemists, a 'subtle' intrinsic sulphur that infuses even pure mercury, and an extrinsic sulphur which corrupts gold into base metals. Paracelsus added salt to this duality to extend the sulphur–mercury theory to all substances – animal, vegetable and mineral.

As the practically inclined investigators of the Hermetic tradition probed further into the transformations of matter, they found it increasingly hard to fit Aristotle's simple, fourfold system of elements to what they saw and experienced in the real world. Yet these proto-scientists were reluctant to abandon the old ideas; and so they consigned them to limbo, accepting that the four elements were at the root of all stuff but then interposing various layers of other 'principles' or 'bodies' responsible for the immediate appearance and interconversions of the ever widening range of substances recognized in nature. Hence Paracelsus's *tria prima*, a scheme that often pays little more than lip service to Aristotle.

In 1586 the Swiss scholar Conrad Gesner showed that there was nothing so sacred about Aristotle's system of elements after all, since no fewer than eight different systems appeared in the ancient literature between the times of Thales and Empedocles. Thereafter the deference to Aristotle began to wither, and other systems of elements started to appear.[10] Girolamo Cardano posited three, dropping the irreducibility of fire; Bernardino Telesio saw just two, 'heat' and 'cold'. And Jean Beguin's system of sulphur, mercury, salt, earth and water, posited in the early seventeenth century, marks the conception of a five-element theory that remained popular throughout that century. But most significant for the present story was the system promulgated by Johann Baptist van Helmont, born in Brussels in 1577. With van Helmont we see a return to the ideas of the Ionian Thales, yet with a decidedly modern flavour. He took there to be but a single element, and that was water.

Wood from water

In his determination to reject old ideas and in his contempt for learned academics, as well as his assurance in his own (often far-fetched) theories and a predilection for the invention of bizarre new words, van Helmont resembles no one so much as Paracelsus. Yet the Paracelsian system of three principles was one of those targeted by van Helmont in his determination to 'break down almost all things that have been delivered by those that went before'. But van Helmont's distinction was his emphasis on quantitative experimentation rather than vague generalization or philosophical speculation.

This is not to say that van Helmont practised experimental science in anything like the modern sense, with its emphasis on theory-testing and its readiness to discard or modify ideas that do not fit quantitatively with the facts. Indeed, many of van Helmont's experiments were devised to meet his thesis, and were made to do so even in the face of the facts. Yet his reliance on quantification, by means of balances and primitive thermometers, as a means of verification was somewhat of an innovation. Nowhere is this better demonstrated than in his celebrated 'willow tree experiment', in which van Helmont purported to show that wood – and by casual extrapolation, all bodies – is made from water.

> All earth, clay, and every body that may be touched, is truly and materially the offspring of water only, and is reduced again into water, by nature and art ... Water always remains whole as it is; or without any dividing of the three beginnings [salt, sulphur and mercury], it is transformed and goes into fruits whither the Seeds do call and withdraw it.[11]

So said van Helmont in 1648, and to demonstrate as much he planted a young willow tree in a pot with two hundred pounds of earth, 'dried in a Furnace, which I moistened with Rainwater'. He covered the vessel with a metal plate to prevent dust from adding its mass to the earth, and over five years he added rain water or distilled water and watched the tree grow. Finally he again dried the earth in the vessel and weighed it and the tree. The soil was lacking but two ounces of the original two hundred pounds, yet the tree had gained 164 pounds – which, van Helmont concluded, 'arose out of water only'. A demonstration as quantitatively precise as this gave his thesis considerable force.[12]

Equally impressive was van Helmont's conversion of 'earth' to 'water' by fusing sand with an alkali to make water glass (sodium silicate), a compound that liquefies on exposure to moisture in the air. Adding acid to this solution reconstituted the silica in the same quantity as the original.

Van Helmont's investigations of evaporation led him to introduce another concept that, however confused from a modern standpoint, was to have important ramifications. The sole exception to his water-centred cosmology was air, which he granted the status of another element – but one that was inert and could not be changed into any other form. This begged the question of what it is that is liberated when water evaporates. Van Helmont identified this tenuous substance as a vapour, distinct from air, which could be demonstrably converted back to liquid water. Many other chemical reactions were also known to generate an air-like substance, which again could not be air itself but which seemed different from a vapour too, as it could not be easily converted back to its previous form. To these exhalations or 'spirits' van Helmont gave the name Gas. The word, thought to be derived from the *Chaos* that Paracelsus used as a general term for an air-like substance, did not catch on at once, while the other bizarre words that the Dutchman coined for different sorts of 'emanations' – Blas and Magnall – are now forgotten. The French chemist Antoine Lavoisier revitalized the word 'gas' (as the French *gaz*) in the eighteenth century, but it did not achieve common usage in England until the nineteenth. The Gas that van Helmont and his contemporaries detected was probably most often carbon dioxide, which is formed in chemical processes such as the dissolution of marble in acid. But it is likely that several other gaseous compounds, such as nitrogen oxides, hydrogen and sulphur dioxide, were also subsumed by van Helmont within this general-purpose term.

You might wonder where, amongst all of this, the idea of atoms could be lurking. The truth is that the chemists of the sixteenth and seventeenth centuries took a rather leisurely attitude to the microstructure of matter. They might casually adopt an atomistic picture when it suited them, but without making much of the critical distinction – highlighted by Democritus – between corpuscles in intimate contact or separated by a void. Many natural philosophers during the Middle Ages adopted the notion of 'minima' as the smallest possible particles of a substance; and while van Helmont spoke in terms of matter being generated from 'seeds', the relation of these to atoms was left unspecified. He did not appear to embrace the notion of void, instead filling up empty space with the emanations Blas or Magnall. Galileo and Francis Bacon in the sixteenth century both accepted a particulate view of matter, as did René Descartes in the seventeenth. But Descartes did not tolerate either of the Democritic principles of void or indivisibility of atoms; rather, he proposed that the particles that make up the Universe are carried along in the whirlpools of an all-pervasive fluid.

Descartes's atomism is three-tiered. The largest particles – the 'first element' – constitute chemical substances, whose shapes, like those of the Platonic atoms, determine their properties. Water particles are long and smooth, so never get entangled. Salt particles are also rod-like, but sharp, and so pierce the tongue to produce a bitter taste. Tangles of irregularly shaped particles form dense substances, such as earth and wood. Between these large bodies are smaller particles of the 'second element', which is the air element or 'ether'; and the fine dust produced from the abrasion of these two sets of particles constitutes the 'third element', fire, whose swiftly moving particles are in intimate contact and leave no room for a vacuum. Although it preserves something of the Aristotelian system of elements, the most striking thing about Descartes's scheme is its emphasis on a mechanical description of phenomena: he accounted for a wide variety of observations in astronomy, meteorology and geology on the basis of the physical interactions of his constantly moving particles. In this, we see the harbinger of the modern theories of chemistry that take their lead from one man, the Anglo-Irishman Robert Boyle.

Boyle lived from 1627 to 1690, and in that time he saw the decline of alchemy as a method of philosophical enquiry and the birth of chemistry. Boyle's *The Skeptical Chymist* (1661) derided the occult and obfuscating tendencies of the alchemists in favour of a rational, mechanical theory of matter, according to which the world could be seen to behave like 'a great piece of clockwork'. Boyle's outlook was atomistic, and he pursued the idea that the size and shape of atoms determine the physical and chemical properties of substances and the affinity of some substances for others. But Boyle also attributed great importance to the motions of these 'corpuscles', and postulated that changes in properties could result from changes in motion. Here we can see the beginnings of a distinction between physical and chemical change.

Yet despite all this, Boyle's concept of elements was still based in an earlier time. His definition of an element as a 'primitive body' from which all other substances are compounded adheres closely to the modern one, but these elements were themselves aggregates of the fundamental particles and so could be transmuted by a rearrangement of these corpuscles. In this sense, nothing about Boyle's much-vaunted characterization of an element was inconsistent with the old ideas that stretched back to Aristotle. Thus, for all his scepticism, Boyle was not opposed to alchemy in itself, but merely to some of its practitioners' mystical, quasi-religious approach to it. He believed in the possibility of the transmutation of gold, and reputedly bequeathed to Isaac Newton a red earthy substance that he believed would effect this transformation.

So while Boyle considered water an element, he did not believe it to be immutable. Indeed, he was content to propound van Helmont's belief that water could be transformed to organic matter:

> I will begin by reminding you of the experiments I not long since related to you concerning the growth of pompions, mint and other vegetables out of fair water. For by these experiments it seems evident that water may be transmuted into all the other elements.[13]

Observing how people and other animals could develop 'heavy stones in their kidneys and bladders, though they feed but upon grass and other vegetables that are perhaps but disguised water', he concluded that even minerals could be congealed from this elemental substance.

Many of Boyle's experiments, which were performed with the quantitative rigour of van Helmont, focused on combustion processes. He observed the flammability of hydrogen gas, an 'impure air' generated by the reaction of iron filings with acid. If indeed Boyle was the first to conduct this experiment, in which hydrogen ignites and explodes with an audible pop, he was unwittingly the first to synthesize water from its elements: hydrogen and oxygen.

Boyle noted that metals gain weight when heated strongly in air, indicating that some substance has been added to them. This result was long known to medieval metallurgists, and in 1630 the French medical doctor Jean Rey made careful measurements of the weight gain. Boyle concluded that the product of combustion was a compound of metal with fire. It seems he was thrown off track by errors in experimental technique: even when he conducted combustion in a sealed glass flask, he reported that the total mass inside the flask increased. In fact it should have remained the same, since matter could neither get in nor get out. Boyle supposed that fire particles were penetrating pores in the glass; but it is likely that his seals were leaky.

In Boyle's work lie the seeds of all that was to come in unravelling the composition of water. He witnessed reactions of both its component elements – hydrogen and oxygen – and he even unknowingly reunited them as water. He focused attention on the combustion processes from which an understanding of water's make-up was ultimately to emerge, and he conducted, albeit with only partial success, the kind of quantitative experiments that were essential for this to happen. Yet Boyle's studies seemed at face value to be much more concerned with two of the other Aristotelian elements, air and fire, in whose 'hotness' it was hard to imagine that there was any connection with cold, moist water. From such studies was water's secret finally to emerge, but not before one of the most curious detours in chemistry's history.

Burning backwards

In Johann Becher we find a clear reminder of how thoroughly Boyle's contemporaries were rooted in the alchemical tradition of the Middle Ages, with its acceptance of magic and mystery. Becher was a Dutchman from Speyer who claimed in his book *Foolish Wisdom and Wise Folly* to have encountered stones that conferred invisibility and geese that grew on trees. In 1673 he offered to provide Prince Herman of Baden with gold on an industrial scale, transmuted from Dutch sand, to fund his war with France. Yet Becher believed that this transmutation could be effected by rational, chemical means rather than by occult powers, and he set in train a theory of matter and its transformations that was to be replaced only by Antoine Lavoisier, founder of chemistry as we know it today.

Becher's preliminary demonstrations of the transmutation of sand to gold proved persuasive to a commission from the Dutch government and the Mayor of Amsterdam. But he was forced to flee from an intrigue that threatened his life before he could turn this process to its intended use. The theoretical basis of his work was fluid to say the least – at different times he adopted schemes of two, three, four and five elements. He was finally forced to conclude that one earthy element was not enough to account for all of the solid substances observed, and so he proposed no fewer than three: *terra fluida* or *mercurialis* (fluid earth), *terra pinguis* (fatty earth) and *terra lapida* or *lapis fusilis* (fusible stone/earth). Together with air and water, these made up Becher's five-element scheme. The three earths are in fact nothing more than Paracelsus's mercury, sulphur and salt, thinly disguised by renaming; but by this sleight of hand Becher shook off some of the metaphysical baggage that adhered to Paracelsus's philosophy.

The conceptual origin of fatty earth can be traced directly to Paracelsus, who wrote 'the life of metals is a secret fatness, which they have received from sulphur'. This *terra pinguis*, said Becher, is lost from matter when it is burned, an idea that harks back to Plato. 'Metals contain an inflammable principle which by the action of fire goes off into the air', was how Becher put it. Notice how this idea unites three of the four ancient elements: earth, fire and air. Water becomes sidelined, and remained so for much of the next century as

Becher's ideas took on a new form that was to infuse the whole of chemistry.

The influence of Becher's theory was due in large part to his disciple Georg Ernst Stahl, who was able to transform it from a rehash of old alchemical notions into a theory that looked recognizably like modern chemistry. Stahl renamed *terra pinguis* 'phlogiston', a term derived from the Greek for 'to set on fire'. The phlogiston theory is regarded now with much ambivalence – some see it as 'the first great unifying principle in chemistry', others as a dead end that hindered the development of modern chemistry for at least a century. Either way, it is not hard to understand the tenacity of the idea, for the postulation of phlogiston accounted for many of the facts as they were then known, and could be fitted to others discovered subsequently.

According to the phlogiston theory, when a metal is heated in air, phlogiston is given off and the residue is 'dephlogisticated'. The residue can be restored to metal ('rephlogisticated') by heating it with a substance rich in phlogiston, such as charcoal. Phlogiston is also given off by living creatures. It is absorbed from the air by plants, from whence it can be restored to animals. Boyle's observation that combustion will not take place in a vacuum could be explained by supposing that air is required as a mechanical medium for carrying off the phlogiston, just as it bears the tang of the sea to our nostrils. And by demonstrating a series of chemical transformations that evinced the passage of phlogiston between different substances, Stahl was even able to show that the theory had predictive power.

The reason it all looked so persuasive is that phlogiston turns out not to be a substance or an element that passes back and forth in chemical transformations, but more or less precisely the inverse of that. You could say that phlogiston is the 'absence of oxygen'. When a metal is burned, it takes up oxygen rather than giving out phlogiston, forming an oxide. When the metal oxide is heated with charcoal, the oxygen is removed, not the phlogiston put back, to restore the pure metal. This is how iron is smelted from its oxide ore. The reason combustion of a metal does not happen in a vacuum is because there is no oxygen in the evacuated vessel to combine with the metal, whereas Stahl believed it was because the phlogiston cannot get out. Plants give out oxygen, they don't take in phlogiston. Stahl and the chemists of the eighteenth century who succeeded him were unfortunate enough to hit on the 'negative' of a theory – the existence of oxygen – that in the end proved to be the right one.

But the argument worked so well in reverse that it took a century to recognize the inversion.

So powerful, indeed, did the phlogiston theory prove to be that even give-away clues to its falsity were explained away by turning other well-established concepts on their head. As Jean Rey had shown, metals don't lose weight when burned – they gain it. Where does this extra weight come from, if phlogiston is being lost? Rey himself supposed correctly that the metals were combining with a part of the air; Boyle proposed correctly that they were combining with fire particles. But Becher asserted triumphantly that phlogiston 'may sometimes weigh less than nothing'. Then, 'something minus another thing that weighs less than nothing, weighs more'. No other known substance had a negative weight, but no matter – look at phlogiston's successes! Besides, some other substances, such as wood, *do* lose weight when burned. Yet the weight gain of combusted metals continued to trouble chemists, who were unsure whether they should regard phlogiston as weightless, imbued with buoyancy, or something else again.

This confusion is largely due to a lack of clarity about the concept of a gas. Van Helmont had established the idea of a gas as an emanation, different from air and chemically inert, which was given off during certain chemical processes; but this concept remained undeveloped, and most chemists at the beginning of the eighteenth century believed that gases were merely air in different states of purity. Thus when in the middle of the century Joseph Black, a Scottish medical student, studied carbon dioxide formed by the action of acid or heat on magnesium carbonate or limestone (calcium carbonate), he regarded it as 'fixed air': a form of air that could be 'fixed' back into solid form by passing it over magnesia (magnesium oxide) or lime (calcium oxide). That a substance as subtle as 'air' could exist in the form of a hard rock like limestone astonished Black's contemporaries.

And when in 1772 Black's student Daniel Rutherford observed that some gas remained after burning materials in a sealed container of air, he called it 'mephitic air'. This component of air, which we now know as nitrogen, had been recognized by Boyle's assistant Robert Hooke in 1665 and by the Englishman John Mayow in 1674.[14] But neither they nor Rutherford knew quite what to do with the observation, although Rutherford at least quantified it, deducing that 'mephitic air' comprised about four fifths of 'common air'.

The turning point in our understanding of these various forms of air came with the work of Joseph Priestley, an Englishman born in 1733 who trained as a Presbyterian minister and took up chemistry as an untrained amateur only in his mid-thirties. But because Priestley was deeply convinced of the phlogiston theory, it was left to others to mould his observations into a form consistent with modern chemistry. To Priestley, Rutherford's 'mephitic air' was simply common air saturated in phlogiston, so that it could take up no more: therefore it could not support combustion. Again this is the inverse of the oxygen-based picture that we now see: mephitic air is air devoid of oxygen, not saturated in phlogiston.

When in 1774 Priestley discovered pure oxygen gas, again he saw it in negative, as 'dephlogisticated air'. With this discovery, Priestley stood at the edge of chemistry's modern vista, but he was looking a different way.

The way Priestley made oxygen gas was to heat mercury oxide so that it released the oxygen and left behind pure mercury. There was nothing new about this experiment: Robert Boyle had certainly performed it in the seventeenth century, noting that the red substance was transformed to the 'quicksilver' of mercury; and it is virtually certain, given that the oxide was known at least as far back as Geber's time, that many earlier alchemists had done the same. But Priestley was less interested in the silvery residue than in the 'air' liberated by the reaction. He found that it promoted combustion even more vigorously than ordinary air: a candle held in a bottle of the gas burned more brightly, a piece of glowing charcoal became incandescent and sparkled with fire. This, to Priestley, indicated that the 'air' was especially depleted in phlogiston, and so even more eager to take up this substance from a burning object. By 1775 he was studying the effect of this gas on respiration, noting that dephlogisticated air supported the breathing of a mouse long after another unfortunate creature had exhausted the same volume of normal air and died of asphyxiation. This encouraged Priestley to try breathing the gas himself, and he noted that 'my breath felt peculiarly light and easy for some time afterward'. 'Who can tell', he went on to speculate, 'but that in time this pure air may become a fashionable article in luxury', although 'hitherto only my mice and myself have had the privilege of breathing it'.[15] But he also saw dangers in this: 'For as a candle burns out much faster in this air than in common air, so we may live out too fast.'

While Priestley was excitedly telling others of his discovery of this new gas, he remained blithely unaware that his observations were not new. Several years earlier, the Swedish pharmacist Carl Wilhelm Scheele had found and collected the very same gas by heating a variety of salts, such as potassium nitrate, mercury oxide, silver carbonate and saltpetre. Because of its combustion-enhancing properties, Scheele called it 'fire air', and he measured accurately the proportions of the two major constituents of common air, pronouncing it one part 'fire air' to four parts 'spent air' (nitrogen). But Scheele did not publish his discovery until 1777, by which time Priestley's work was well known.

As if this does not complicate matters enough, into the fray stepped the man who was finally to overturn the phlogiston theory and set chemistry on its modern rails. Antoine Laurent Lavoisier was a suave and ambitious Frenchman who trained in law and spent his early years as a tax collector, but took up chemistry in his early twenties. By the age of twenty-five he had achieved a sufficient reputation to become elected as a member of the French Academy of Sciences, and immediately thereafter he succeeded in disproving the old alchemical idea that water could be transmuted to earth by prolonged heating. This idea no doubt stemmed from the observation that evaporated water invariably leaves a residue of salts. Lavoisier showed that distilled water heated in a sealed glass container for 101 days acquired no solid particles beyond a few specks whose origin could, by careful weighing, be ascribed to the walls of the container. Another old idea about 'elemental' water was laid to rest.

By the 1770s Lavoisier's reputation was solidly established, and when Priestley visited Paris with his patron Lord Shelburne in 1774, they dined with Lavoisier and other distinguished French scientists. At this gathering Priestley described his discovery of the new 'dephlogisticated air'. March 1775 saw Lavoisier announcing much the same result: that mercury oxide gives off a gas when heated that resembled common air. But on seeing this report, Priestley pointed out that the gas is not simply like common air – it had a 'goodness' that supported combustion even more avidly. Lavoisier confirmed that this was so. In April of that year he presented a paper to the French Academy on the combustion of metals, in which he explained that the gas released by heating mercury oxide is taken up in equal measure by mercury when it is heated. Clearly this gas was Priestley's dephlogisticated air, but Lavoisier made no mention of

Priestley's work in his presentation – nor did he mention that of Scheele, or of his fellow Frenchman Pierre Bayen, who had also described the effect of heat on mercury oxide in 1774.

Where Lavoisier's report went beyond the work of his predecessors, however, is that he no longer made any reference to phlogiston. Lavoisier saw clearly at last that the phlogiston theory made no sense of the facts: it proposed that a substance was expelled when mercury was heated, and taken up when the oxide was heated, whereas quite evidently the reverse was true. Though he did not coin the new name until two years later, Lavoisier had conceptually replaced phlogiston with oxygen, and ushered in the era of modern chemistry.

The water-former, misconstrued

Hydrogen, meanwhile, had made its appearance already, but suffered a case of mistaken identity. Carl Scheele discovered and investigated a great many other substances in the course of his short but productive career. He discovered chlorine gas and called it 'dephlogisticated marine acid'; and in 1770 he collected hydrogen gas from the action of an acid on iron and zinc. This 'inflammable air' was, as we have seen, known already to Boyle, and apparently also to van Helmont and Paracelsus before him. Scheele concluded that it might be pure phlogiston, given its propensity to burn vigorously.

This same conclusion was proposed four years earlier by one of the most significant, and probably one of the oddest, players in this tale. Born in 1731, Henry Cavendish was an eccentric millionaire and grandson of a duke, a self-financed natural philosopher whose social peculiarities did not prevent him from becoming a distinguished member of the Royal Society in London. He was by all accounts reclusive to the point of misanthropy, dressing in shabby and outdated clothes, never marrying, shunning conversation and women and disliking all towns but London. He seems to have been driven by a curiosity about nature, which he pursued methodically to the exclusion of any curiosity about his fellow people. Cavendish seemed to care very little for the high esteem in which he came to be held; indeed, he seems to have been positively embarrassed by it.

Cavendish was an avowed proponent of the phlogiston theory, so that when he found that various acids would liberate an identical, extremely light gas from iron, zinc and tin, and that this gas burned readily in air, it was only natural for him to suppose, like Scheele, that he had isolated pure phlogiston. This he reported in a paper on 'factious airs' in 1766. That phlogiston was a substance like any other, which could be isolated and weighed, was marvelled at by some of his contemporaries. Meanwhile, others found that the gas offered fancy tricks: the Frenchman Pilatre de Rozier breathed fire by inhaling and igniting it, and was lucky to survive the explosion that resulted when he tried the same thing with hydrogen mixed with common air. And the physicist Jacques Charles of Paris demonstrated the first hydrogen-balloon flight to enthralled spectators in 1783.

From H to *Eau:* making water

You can see that the story is now becoming a tangled one. All of a sudden, so many pieces were falling into place that the same discoveries were being made independently and almost simultaneously, and issues of priority become increasingly hard to disentangle. The history is made all the more convoluted by the informal exchange of ideas and observations between several key players on both sides of the English Channel, as well as by differing attitudes towards urgency of publication.[16] The fact was that chemists throughout Europe were on the verge of a tremendous breakthrough in understanding, which was sensed only vaguely in 1775 but which was all but completed by the turn of the century. And for all that the phlogiston theory was always destined to perish, it helped to accelerate this revolution in chemistry by providing a theoretical framework which not only made sense of several different observations but also helped to guide genuinely predictive research.

By 1775 chemists had isolated both elemental components of water, oxygen and hydrogen; yet the fact that these two were pure elements, not 'impure airs', was still to be established. And the idea that water itself might not be elemental was still deeply challenging, if not quite the affront to dogma that it would once have been. The ambivalence of contemporary opinion on this matter is reflected in Priestley's words:

> There are, I believe, very few maxims in philosophy that have laid firmer hold upon the mind, than that...atmospherical air...is a simple elementary substance, indestructible, and unalterable, at least as much so as water is supposed to be.[17]

Intentionally or not, Priestley's choice of words here represents a masterful example of keeping your options open.

And well he might have done so, for Priestley knew of an experiment performed in 1774 by his colleague John Warltire in which common air and 'inflammable air' (hydrogen) were mixed in a copper flask and ignited with an electric spark. After the mixture exploded, Warltire found that the gases had lost weight and that a 'dew' had formed on the walls of the vessel. In the same year, the Frenchman Pierre Joseph Macquer, a colleague of Lavoisier's, also burned hydrogen in air and observed that the flame produced no smoke. Instead of a sooty deposit on a white porcelain plate placed above the flame, Macquer saw that 'it was wetted by drops of a liquid like water, which indeed appeared to be nothing else but pure water'. James Watt, a university technician who worked under Joseph Black, also performed this experiment, and Priestley repeated Warltire's procedure in 1781 – both with the same results. The implication was staring them all in the face: hydrogen combines with oxygen in the air to make water. Here, perhaps, is the truth that lurks in Honoré de Balzac's cryptic comment 'Water is a burned body.'[18]

But the appearance of water in these experiments was at the time regarded as unremarkable. After all, condensation of water from air is commonplace, and can be seen on any laboratory window on a cold day. The phlogistonian Priestley suggested later that the observations could be explained as the release of water from common air due to its uptake of phlogiston from the 'inflammable air': 'common air deposits its moisture by phlogistication', he said.

While Priestley and his contemporaries Cavendish, Black and Watt continued to juggle with phlogiston, Lavoisier was busy burying that concept. Since his discovery in 1773 that air is 'fixed' in metals during combustion, he had become convinced that phlogiston had to go. Priestley's dephlogisticated air was not an impure compound but a substance in its own right, and apparently an elemental one. Lavoisier spoke initially of this substance as 'pure air' or 'true air', saying 'it appears to have been proven that the air we breathe contains only one-quarter true air'; but in 1777 he gave it a new name. Noting that the burned residues of certain substances, such as sulphur and phosphorus, form acids when mixed with water, he concluded that the 'pure air' responsible for combustion is a component of all acids. That all acidity derives from a single substance was a common idea at the time, and 'pure air' seemed to be the ideal candidate. So Lavoisier called it *oxygène*, 'acid former'. When some of his peers pointed out later that certain acids, notably hydrochloric acid, contain no 'airy' component (it is composed of hydrogen and chlorine), he retorted that they simply did not yet know its composition well enough. By invoking oxygen as the principle of combustion, Lavoisier had a theory that could supplant phlogiston. He stated boldly that his oxygen theory 'directly contradicts Stahl's theory and those of famous men who have followed him'.

In 1781 Henry Cavendish, extending the studies of Warltire, Macquer, Watt and Priestley on the burning of dephlogisticated air, used electric sparks to ignite common air and hydrogen (which he regarded as phlogiston) in sealed glass vessels. In this way, Cavendish synthesized considerable quantities of water from its elements. He noted that when common air was used, only about one-fifth of it (the oxygen component) 'lost its elasticity' – that is, became bound up in the water; the mephitic or spent air (the nitrogen component) remained. In his report to the Royal Society in 1784, Cavendish said:

> ... it appeared that when inflammable and common air are exploded in a proper proportion, almost all the inflammable air, and nearly one-fifth of the common air, lose their elasticity, and are condensed into dew. And by this experiment it appears that this dew is plain water, and consequently that almost all the inflammable air, and about one-fifth of the common air, are turned into pure water.[19]

Pure water? Well, not quite. At first Cavendish found the water to be slightly acidic, a result of the formation of small amounts of nitric oxide in the explosion which dissolved in the water to form nitric acid. While tracking down the source of this acidity, which involved no small prowess in analytical technique, Cavendish was delayed in publishing his results. By nature he was unconcerned with the acclaim that publication might bring and more bothered about elucidating all that was happening in his experiments.

Cavendish went on to provide a further clue to the composition of water. When he used Priestley's 'dephlogisticated air' instead of common air, he found that a given volume of this air would always unite with twice that volume of inflammable air: in modern terms, one volume of oxygen to two of hydrogen. But the significance of this was not to be appreciated for some time to come.

Had the alchemists or the natural philosophers of Boyle's time performed the same experiment – and the means and materials were surely available to them, except that they had less sophistication in collecting gases – they would surely have seen here evidence for the transmutation of impure but elemental air into elemental water. By the 1780s there was no room for these primitive concepts in the expanding theoretical web of chemistry. Yet Cavendish's own interpretations were still steeped in phlogiston. It remained for Antoine Lavoisier to clarify that what was really happening here was the reaction of two elements to form a compound – water.

In June 1783, while Cavendish's results had not yet been made public knowledge, his assistant Charles Blagden visited Paris, and attended a demonstration by Lavoisier and his colleague Pierre Laplace. Having persuaded himself that oxygen was the begetter of all acids, Lavoisier concluded that an acid should result when oxygen was combined with inflammable air (hydrogen). He conducted experiments to this end in 1781–1782, but found no acid. In the demonstration of 1783, he and Laplace fed oxygen and hydrogen into a glass sphere and ignited them to show that they produced pure water whose weight roughly equalled that of the two gases. Blagden told the French scientists that Cavendish, and also Watt, had made just the same observation, to which they replied that this, of course, was not news, since Priestley had already shown as much.

But none had framed the conclusion with such directness as Lavoisier:

> It is difficult to refuse to recognize that in this experiment, water is made artificially and from scratch...[20]

And he went further, closing a crucial link in the argument. If water was the compound product of a reaction between oxygen and inflammable air, it should be possible to split water back into these two gases. Indeed Lavoisier considered it a general principle in chemistry that one should not be satisfied to conclude that a substance was a compound of two or more elements until it had been both synthesized from these component parts and broken down into them. In late 1783 he showed that water could be decomposed by the rusting of iron filings immersed in the liquid. The iron takes up the oxygen, and the hydrogen gas escapes.

Thus one is led still more nearly inevitably to conclude that water is

> not a simple substance at all, not properly called an element, as had always been thought.[21]

A subsequent demonstration in 1784 was more forceful: Lavoisier split water into hydrogen and rust by passing steam through a red-hot gun barrel.

Finally accepting that oxygen gave only pure water, and not an acid, when combined with inflammable air, Lavoisier named the latter *hydrogène*, water-former. That water was indeed not an element but a compound of oxygen and hydrogen became widely accepted in France, but elsewhere the habits of two millennia were not easily given up. William Ford Stevenson in England denounced Lavoisier's 'deception', saying,

> This arch-magician so far imposed upon our credulity as to persuade us that water, the most powerful natural antiphlogistic we possess, is a compound of two gases, one of which surpasses all other substances in its inflammability.[22]

Cavendish railed at the idea too, and so, inevitably, did Priestley. But Joseph Black took Lavoisier's side, and James Watt likewise, while simultaneously making strenuous attempts to claim the discovery of water's synthesis as his own.

The final part of the proof fell to two Englishmen, William Nicholson and Anthony Carlisle, who discovered while experimenting with electrical conduction that they could split water into its pure elements. In 1800 the Italian Alessandro Volta sent to Sir Joseph Banks, the president of the Royal Society in London, a letter describing a battery made from stacked disks of zinc alternating with silver. Just a month later, Nicholson and Carlisle were looking at the transmission of electricity from a 'Voltaic pile' through water. They connected brass wires to the two terminals of the pile and dipped them in water. From the negative wire they observed a gas bubbling off. The gas was hydrogen, exploding when ignited in air. The other wire became tarnished and eventually black, which led the researchers to conclude that here the oxygen from the water had 'fixed itself' with the metal to form an oxide. When they used platinum wires instead of brass, the oxygen no longer reacted with the metal; instead they saw oxygen gas bubbling out. The volume of hydrogen generated was twice the volume of oxygen, which accorded with Cavendish's measurements of the volumes consumed when the two gases were ignited to form water. Nicholson and Carlisle had achieved the electrolysis – 'splitting with electricity' – of water.[23]

Lavoisier, meanwhile, executed a masterful strategy for infusing his ideas throughout all of chemical practice: he set about reforming the language. With his countrymen Guyton de Morveau, Claude Louis Berthollet and Antoine François Fourcroy, he devised a scheme for renaming the elements as they were then known, and in the process obliterating phlogiston from the chemist's lexicon. It was a clever move, for the system that they proposed implicitly endorsed Lavoisier's oxygen theory of combustion. Combusted metals, previously called calxes, were now metal oxides. Mephitic air was 'azot' or 'azotic gas' (from the Greek for 'without life'), making it easier to see it as an element rather than as an impure air. In 1790 Chaptal promoted the name 'nitrogen' (nitre-former), which Lavoisier toyed with but rejected. It became accepted in the English language, though *azote* persists today in France.[24]

The new scheme was described in *Méthode de nomenclature chimique* (1787), and Lavoisier summarized his ideas in *Traité élémentaire de chimie* (1789) – a book that stands in relation to the development of chemistry such as Newton's *Principia* does to physics. Its influence spread quickly, and most chemists accepted Lavoisier's chemistry as evidently the best framework for rationalizing the transformations of matter. It was well that he did not delay in providing an account of his theories, for the French Revolution was by then in full swing, and five years later, at the height of the Reign of Terror, Lavoisier was denounced for his role as a collector of taxes. Instrumental in bringing charges was Jean Marat, a poor chemist whose ideas on fire Lavoisier had disparaged. Despite his eminence and pleas for his value to science and to France, Coffinhal, the president of the Tribunal, proclaimed that 'the Republic has no use for savants', and Lavoisier was sent to the guillotine in May 1794.

The pieces put together

So water is a compound of hydrogen and oxygen. But what exactly does that mean? John Dalton, a Quaker and school teacher from Manchester, provided the answer at the beginning of the nineteenth century. While the idea of atoms had been around since Leucippus, it had never acquired a form that was anything less than vague. In 1704 Isaac Newton had given it a somewhat more mechanical basis, calling atoms 'solid, massy, hard, impenetrable, movable particles' and explaining that a gas is made up of 'mutually repulsive particles', the forces between which are 'reciprocally proportional to the distances between their centres' – suggesting that atomic motions in gases were as acquiescent to his new mechanics as the tracks of the planets around the Sun. But Dalton was to unite such a picture with Lavoisier's new chemistry by providing a graphic image of what it is that atoms do when they combine, as hydrogen and oxygen do, to form a compound like water.

Dalton's contribution was really rather simple, but of tremendous conceptual significance. He drew atoms. Recall that Democritus and Plato had their own concept of what atoms looked like: those of different elements had different shapes, which accounted for their properties. But this simplistic idea had already been eroded away by Newton's day, and Dalton conceived of atoms as spheres whose differences were of size only. That is to say, he believed that atoms of the same element were all identical in size and weight, but distinguished in these respects from those of different elements. He elected to represent these atoms as circular symbols, shaded or elaborated to show their differences.

The union of atoms into compounds could be represented by the bringing together of two of these atomic symbols to make one 'compound atom', which we would now call a molecule. So the molecules that Dalton drew contained atoms of different elements in fixed ratios – an explanation at the microscopic scale of the Law of Fixed Proportions proposed in 1799 by chemist Joseph Louis Proust: 'a compound is a substance to which Nature assigns fixed ratios'.

The remaining question was: what are these proportions? How many atoms of each element make up a molecule? The raw data that Dalton had at his disposal was not quite accurate enough to enable him to answer this unambiguously. Several chemists before him had performed careful measurements of the proportions (by weight) of different elements in compounds. Notably, the German chemist Jeremias Richter's book *Foundations of Stoichiometry or Art of Measuring the Chemical Elements* (1792) contained a great deal of data, albeit hazily presented and interpreted, about the elemental proportions of many compounds. So what Dalton had available was the information that, say, water results from the combination of about eight parts by weight of oxygen with one part by weight of hydrogen. But it is not possible to deduce from this how many *atoms* of each element are involved in these unions – that is, the ratio of oxygen to hydrogen atoms in a molecule of water – unless one knows the weight of an oxygen atom relative to a hydrogen atom.

So what Dalton had to do was to work the other way around. He would assume a chemical formula – the atomic ratios in a given molecule – and calculate from that the relative weights of the atoms. Once the weights had been obtained in this way from a few simple molecules, they could be used to deduce the atomic composition of other compounds. Not surprisingly, Dalton's guesses about composition turned out to be sometimes right and sometimes wrong. His default assumption was that molecules were binary, consisting of just two atoms. Thus he supposed that 'carbonic oxide', the gas that we now know as carbon monoxide, has molecules comprised of one atom of carbon and one of oxygen – which is true enough. But he assumed the same for water, regarding it as a binary compound of oxygen and hydrogen in a 1:1 ratio. The true proportion is, of course, 1:2. The error had been made before: in 1789 the Irish chemist William Higgins published a pamphlet propounding much the same atomic theory as Dalton's, in which he stated that:

> Water is composed of molecules formed by the union of a single particle of oxygen to a single ultimate particle of hydrogen.[25]

Help was at hand. The Swedish chemist Jons Jakob Berzelius improved on many of Dalton's atomic weights in 1814, and by 1826 he had corrected oxygen's atomic weight to sixteen, not seven (as Dalton had supposed in 1808) or eight (as Dalton would have deduced if the data he used had been more accurate). Although not universally accepted at first, Berzelius's weights made it clear that a molecule of water must contain two hydrogen atoms and one oxygen. All the same, erroneous formulas invoking one atom each of hydrogen and oxygen, or even two each, could be found in chemistry textbooks at least up to the 1860s.[26]

Berzelius introduced a new way of depicting the chemical formulae of compounds. Up to and including Dalton's atomic theory, systems of representation of the elements had been almost always symbolic, from the arcane astrological symbols employed by the alchemists to Dalton's embellished circles. These symbols left non-initiates utterly in the dark – there was no way of deducing the identity of an element from its symbol alone. Berzelius saw the benefits of a mnemonic system, in which one or two letters derived from the full names served as abbreviations. Thus hydrogen became H, oxygen O, nitrogen N, while some other symbols found their origins in the (then still familiar) Latin forms of the names: Au for gold (*aurum*), Ag for silver (*argentum*), Cu for copper (*cuprum*), Fe for iron (*ferrum*). And after dabbling for a while with systems of dots and bars to represent compounds such as oxides, Berzelius settled on denoting them by sequences of the respective element symbols, with superscripts indicating the number of each atom in a molecule of the compound if there was more than one. So water became H^2O.

The sole modification of this scheme today is that the superscripts are rendered as subscripts, an adaptation proposed later in the nineteenth century by German chemists Justus von Liebig and J.C. Poggendorff.[27] Thus we arrive at last at the well-known representation of the water molecule: H_2O.

The picture of the water molecule as a union of two atoms of hydrogen with one of oxygen is a long way from Democritus's idea of an element comprised of rounded, slippery, indivisible particles. And yet this is just our starting point. For there is no clue here as to why water is so special, why it is the quintessential liquid and solvent, the stuff of snowflakes and glaciers and the essence of life on Earth. Two thousand years and more of natural philosophy bring us to the beginning of our journey.

6 Between Heaven and Earth

Why water is the weirdest liquid

> ... we live in the hope and the faith that, by the advance of molecular physics, we shall by-and-by be able to see our way as clearly from the constituents of water to the properties of water, as we are now able to deduce the operations of a watch from the form of its parts and the manner in which they are put together.
>
> T.H. Huxley (1869)

> The full area of ignorance is not mapped: we are at present only exploring its fringes.
>
> J.D. Bernal

The wintry scene depicted by Pieter Bruegel in *Hunters in the Snow* (1565) was a common sight in the late Middle Ages. Ponds, lakes and rivers froze over quite regularly in Northern Europe, during an unusually cold period called the Little Ice Age that lasted from the mid-sixteenth to mid-nineteenth century. London's River Thames became the temporary, slippery stage for fairs and markets, its sluggish current hidden by a thick sheet of ice.

Since water informs all our preconceptions about what liquids are like, it is scarcely surprising that none of Bruegel's carousers would have been puzzled by the tendency of lakes and rivers to grow a frozen skin in winter time. Yet this turns out to be a very strange thing for a liquid to do.

The vast majority of substances are more dense when solidified than when liquid. This is just as you might expect, if you think about what is happening to the individual atoms and molecules as the substance melts. In a crystalline solid, the components are all packed together in a regular, orderly fashion; in a liquid, this orderliness is lost and the component atoms or molecules move about relatively freely. Will a room hold more people if they agree to line up in an orderly, static manner or if they insist on milling around? Surely the former – people on the move are likely to take up more space, on average, than those who have consented to stand still in regimented positions. So then, a crystalline solid would be expected to contain more atoms in a given volume – that is, to have a greater density – than the corresponding liquid. And this is precisely what we find: liquids tend to shrink and become denser, sometimes by as much as ten per cent, when they freeze.

But unlike almost every other liquid, when water freezes it *expands* and becomes *less* dense. If this were not the case, the world would be a very different place. The *Titanic* would never have sunk, for ice would not float on water and there would be no icebergs. We would not suffer floods from burst water pipes in the springtime thaw: for it is the expansion of the water as it freezes to ice that causes the rupture. Stone bridges and buildings face the same problem: water seeps into tiny cracks and, on freezing in cold weather, it expands and wedges the crack wider, eventually weakening the structure. Rock faces are reduced to scree slopes over the course of many centuries of such freeze–thaw cycles.

This inverted behaviour upon freezing is just one of the many anomalous properties that water displays. Only when we compare water with other, less familiar liquids do we start to appreciate just how profoundly odd it is. None of the intuition we have developed from our experience of water can be relied on for predicting and understanding the behaviour of liquids more generally. It would be like trying to make deductions about English life by studying the residents of Buckingham Palace. There's no shortage of information – they're probably the most written-about family in the country – but you could scarcely have chosen a less representative household. Says Felix Franks, a rare expert on the physics and chemistry of water, 'of all the known liquids, water is probably the most studied and least understood'.

The anomalous liquid

First, a confession. I have taken liberties in opening this chapter with Bruegel's scene and proceeding to the expansion-on-freezing of water as if to imply that the one explains the other. The reality is a little more complex, and it highlights an even more bizarre and mysterious example of water's anomalous nature. It's true that a sheet of ice floating on liquid water would sink like granite if, like most substances, the solid was denser than the liquid. Yet that is not the full story, since it is a puzzle even why ice forms first on the surface of a pond, let alone why it is then insufficiently dense to sink.

Let's consider what a freezing pond might be like if water were a 'normal' liquid. As it cools in the depths of winter, heat is radiated from the surface and the surface water gets colder. In any other liquid, this means it would become denser.

This is because most substances expand as they warm up, and shrink as they cool. An engine seizes when a hot piston expands by a tiny fraction and jams in its chamber. By putting the top on a jar of newly made jam while it is still hot, we achieve a vacuum seal owing to the contraction of the air in the top of the jar as it cools. Thermal expansion has an obvious explanation at the atomic scale: hotter atoms jiggle about more exuberantly, and so on average occupy more space.

Therefore the density of a substance increases as it cools, because the same number of atoms occupy a smaller amount of space. This is *not* the same issue as densification on freezing, although it sounds rather similar. In that case the substance changes its state – from a liquid to a solid – and so the arrangement of the atoms changes completely, from disorderly and mobile to regular and fixed. During normal thermal expansion, meanwhile, the state of the substance does not change – a piece of iron at room temperature has the same crystal structure as one at 100°C. It's just that each atom effectively takes up more space. Think again of our bizarrely regimented party: if the guests remain static but start to gesticulate, they'll have to shuffle slightly further away from their neighbours if they are not going to knock one another.

This applies equally to liquids, even though their atoms or molecules are disorderly and mobile – they'll occupy more space when higher temperatures cause them to jiggle more, and so the liquid will expand on heating and contract (densify) on cooling. As the 'normal' liquid at the top of our hypothetical pond cools, it gets more dense and it sinks. The pond reaches a state in which it gets gradually colder the deeper we go.

This means that the liquid at the bottom of the pond will reach freezing point first, and the pond will freeze from the bottom up. If that's how things were with water, Bruegel might never have had the opportunity to paint his skating scene, because the ice would be at the bottom, not the top. Only if temperatures plunged far enough to freeze a lake solid would one be able to take to it on skates – and even the Little Ice Age was not cold enough to solidify totally most of the lakes and rivers of Northern Europe.

But evidently water is different, because rivers, lakes and ponds freeze from the top down. The reason for this has been known for at least three hundred years: water is densest not when it is coldest, at zero degrees centigrade, but at four degrees above this. As the temperature increases from freezing point, at first the density *increases*. Only above 4°C does the density behave 'normally', declining with increasing temperature. So water close to freezing point can happily ride on top of water a few degrees warmer, because it is less dense.

This might seem a trivial matter. OK, there is a little eccentricity close to freezing point, but once you're beyond that, water looks pretty normal, right? Wrong. This minor deviance is a vital clue that true strangeness lies deep in water's character. Once you start to tug at the puzzle of the density anomaly, all the neat ideas that have been developed to describe other liquids start to become frayed, and then to fall apart. If you prefer explanations that are simple and tidy, you had best stick with the one offered by the good Reverend Dr Brewer in his *Guide to the Scientific Knowledge of Things Familiar* (1876):

> Q. When does WATER begin to EXPAND from cold?
>
> A. When it is reduced to 40 degrees [fahrenheit, equal to 4°C]. It is wisely ordained by God that water shall be an exception to a very general rule, it contracts till it is reduced to 40 degrees, and then it expands till it freezes.[1]

The repercussions of this unusual variation in density are global and awesome. The North Pole and large areas of the Southern Ocean are covered all the year round with sea ice – floating ice sheets that grow and shrink in a seasonal rhythm. But what if the polar oceans froze instead from the bottom up? Then only the surface layers would thaw in summer. We've seen earlier that the bottom waters of the polar oceans play a crucial role in transporting heat around the globe, since they participate in a vast conveyor-belt circulation that brings warm tropical waters to cooler regions. If the ocean bottom waters were frozen, this circulation could not take place, with the result that northern countries would be much colder. In that case, perhaps Bruegel would get his frozen-solid lakes after all; but the scenic and leisure opportunities they offer would be small consolation for a perpetually Arctic climate.

The effect of ocean circulation on global climate depends also on another of water's most pronounced anomalies. It takes more heat to raise the temperature of water than to warm up most other substances – liquid or solid! – by the same amount. The temperature rise induced by a given amount of heat is determined by a material's *heat capacity*. Just as it takes a lot of water to raise the water level of a capacious tank, so it takes a lot of heat to raise the temperature of a substance with a high heat capacity. Because of water's large heat capacity, we have to use more energy to boil a kettle – to raise the water's temperature to boiling point – than we would if it were a 'regular' kind of liquid. On the other hand, there are benefits too: hot water cools very slowly, since it must lose a lot of heat for its temperature to fall significantly. So our hot water tanks and our baths stay hot for a long time.

Water's large heat capacity means that warm ocean currents can carry a phenomenal amount of heat. The Gulf Stream, which ultimately keeps Northern Europe warmer than Labrador (at the same latitude) by carrying heat from tropical South America northwards across the Atlantic Ocean, bears with it every day twice as much heat as would be produced by burning all of the coal mined globally in a year.

Some of water's anomalous properties are nothing like as dramatic, but are equally perplexing all the same. For example, cold water gets more runny when it is squeezed, whereas most liquids become more viscous under pressure. Water's other peculiarities are chemical, not physical: it will dissolve almost anything, and is chemically reactive to a degree that makes it highly corrosive (and yet how good it is for us!). Water is not necessarily unique, or the most extreme example, in displaying any one of these anomalies, but their accumulation in a single substance makes it stand out as a decidedly eccentric representative of the liquid state.

Another fine state

To suggest that water is a most unusual liquid might seem absurd – rather like saying that bread is a most unusual food. Surely water is the most common, most ubiquitous of all liquids? If I am to convince you otherwise, I shall have to say a little about liquids in general: I need to show the rest of the family before you can see why water is the black sheep. The more one comes to know this family, the more water stands out as an oddball. For this reason, many scientists interested in liquids have shied away from water, diverting their attention towards far less familiar substances – for the simple reason that one stands at least an outside chance of making sense of them.

It's a well-kept secret that liquids have posed one of the great scientific challenges of the twentieth century. I think that we are told more these days about DNA and quarks, which in all probability we will never see, than about the liquid state, which we cannot fail to encounter every day. Tom Stoppard expressed it well in his play *Arcadia*:

> The ordinary-sized stuff which is our lives, the things people write poetry about – clouds – daffodils – waterfalls – and what happens in a cup of coffee when the cream goes in – these things are full of mystery, as mysterious to us as the heavens were to the Greeks. We're better at predicting events at the edge of the galaxy or inside the nucleus of an atom than whether it'll rain on auntie's garden party three Sundays from now ... We can't even predict the next drip from a dripping tap when it gets irregular.[2]

Stoppard was talking about chaos theory; but how apt it is that his words are saturated with watery stuff.

Journeys in phase space

I mentioned earlier that Buddhist shrines commonly display a tower-shaped structure called a *stupa*, in which the five ancient elements are represented as shapes stacked one atop the other. Earth, a square form, provides the sturdy base; the tenuous elements fire, air and ether are arrayed at the top. Between them is water, awarded a circular form – sandwiched, like the oceans, between the heavens and the earth.

In physics too, water is the intermediary between the airy and the ponderous, between gas and solid. If you want to turn steam to ice by cooling it, you first have to go through the liquid state – the steam condenses to liquid water, which then freezes. These abrupt changes of physical state – condensation and evaporation, freezing and melting – are called *phase transitions*.

The three states – solid, liquid, gas – are like different countries on a map, and to travel between them you have to cross borders with precise geographical locations (Fig. 6.1). The geographical coordinates of this world are defined not by longitude and latitude, but by temperature and pressure. As you travel from, say, low to high temperatures, you might cross first the boundary between solid and liquid and then that between liquid and gas. These are the national borders between the kingdoms of Solidland, Liquidland and Gasland: kingdoms in a so-called *phase diagram*, which depicts the stability regions of the different states in a two-dimensional *phase space* – a Phase Land – of temperature and pressure.

The temperature at which you cross a border depends on the pressure – just as walking eastwards from Tanzania to Kenya along different lines of latitude takes you across the border at different longitudes. At the ambient atmospheric pressure at sea level (a pressure of 'one atmosphere'), liquid water freezes at exactly 0°C, and boils at 100°C. But you can boil an egg in cooler water on a mountain top – because the atmospheric pressure is lower up there, the boiling point of water is also lower. In Tibet, people never have to sip their tea: up on the high plateau it boils far enough below 100°C that one can gulp it back at once.

It's tempting to imagine that a liquid evaporates to a gas when it gets hotter. But at the border of Liquidland and Gasland, the liquid changes to gas *at the same temperature*. A millimetre (so to speak) one way and you're in Liquidland; a millimetre the other and you've moved into Gasland. The change in state is driven not by a change in temperature, but by an intake of heat. When it evaporates, the liquid soaks up heat without changing its temperature. This sounds odd – we usually imagine that heating entails an increase in temperature. But during one of these phase transitions, that isn't so: the substance alters its heat content, and consequently the arrangement of its particles, without altering its temperature. The heat absorbed by a gas as it evaporates from a liquid is called latent heat – it is 'latent' because it seems hidden, unaccompanied by a rise in temperature. Latent heat is released again when a gas condenses to a liquid, and that is why steam at 100°C scalds more seriously than boiling water – the steam dumps an additional load of latent heat when it condenses on skin.

We can think of the latent heat of a phase transition as a sort of toll that must be incurred when a border is crossed. But it is a strange kind of toll, because you pay it (by releasing the heat) only when you cross the border in one direction (say, from Gasland to Liquidland). If you cross in the other direction, the same amount of money is instead given to you at the border. This peculiar system has a

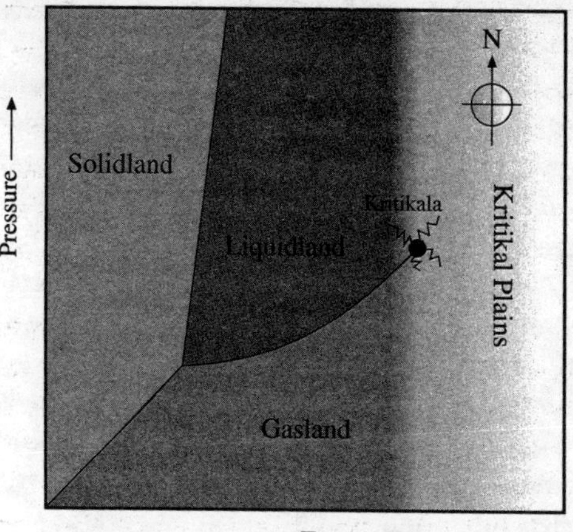

Fig. 6.1 The phase diagram of a substance is like a map, showing the borders between the three states of solid, liquid and gas. Borders define the conditions of temperature and pressure at which phase transitions such as melting and freezing occur. The border of Liquidland and Gasland stops abruptly at the city of Kritikala, which is seismically unstable. In a real phase diagram this corresponds to the critical point. East of Kritikala there is no boundary, only the continuous Kritikal Plains from south to north – no distinction, in other words, between liquid and gas.

rationale, however: Gasland is a richer and more expensive country than Liquidland, so the authorities figure that a Gaslander coming to Liquidland can afford to relinquish some of her money, whereas a poor Liquidlander venturing into wealthy Gasland will need the extra cash he is given at the border.

Gasland and Solidland stretch all the way to zero temperature and pressure, whereas Liquidland does not. In other words, there is a point at which the melting and boiling lines converge, below which only the gas and solid are stable. This point is called a triple point, because at precisely this temperature and pressure all three states – liquid, solid and gas – can coexist. Below the triple point, a gas will condense straight to a solid as it is cooled; and a solid will evaporate ('sublime') to a gas on heating. In this sense, we could say that all substances must possess solid and gaseous states, but a liquid is *contingent* – wedged in between the other two like the flowing ocean between the land and the sky.

Continuity

There's something strange going on at high temperatures too. The boundary of Liquidland and Solidland extends to the 'north' – to ever higher pressures – for as far as one can follow. But the boundary between Liquidland and Gasland does not. Instead, it stops at a point marked with a dot.

The dot is called a critical point, and beyond it there is no longer any distinction between a gas-like and a liquid-like state. In this 'supercritical' region there is just a single fluid state whose density can range from gas-like to liquid-like. That the liquid–gas boundary has a limited extent was discovered in 1822 by the French nobleman Charles Cagniard de la Tour, who heated liquid ether, alcohol and water in sealed tubes and found that they became gas-like *without evaporating* at high temperature and pressure.

John Herschel, son of the famous astronomer William, deduced from this in 1830 that 'the solid, liquid and aeriform states of bodies are merely stages in a progress of gradual transition from one extreme to the other' which need not necessarily be separated by 'sudden or violent lines of demarcation' (by which he meant phase transitions). He was over-generalizing here, however, since the transition from liquid to solid has no critical point. In the 1860s, the chemist Thomas Andrews at Queen's College in Belfast showed that carbon dioxide could be guided along the circuitous tour from Gasland to Liquidland around the far side of the critical point, and concluded that 'the ordinary gaseous and liquid states are ... only widely separated forms of the same condition of matter'. Above the 'critical temperature', he said, the question of whether a fluid is liquid or gas 'does not admit of a positive reply'.[3]

Eastwards of the critical point in Phase Land, there are no longer two neighbouring countries but only a single realm that extends from north to south. I'm not aware of any regions in the real world where such a situation can be found; but we can imagine what it might imply. As you go northeastwards in Gasland towards the city of Kritikala at the border's end, the language and customs change gradually: the dialects share more in common with those of Liquidland across the border, where the same applies. Kritikala itself is a melting pot: it is quite impossible to say whether its inhabitants are really Gaslanders or Liquidlanders, because they share characteristics and traits of both in equal measure.[4] And while this remains true further eastwards, on the Kritikal Plains, you'll find if you take a trip from the south (east of Gasland) to the north (northeast of Liquidland) that the language and traditions change slowly. There is no abrupt change in speech like that at the border westwards of Kritikala, but nonetheless by the time you've journeyed to the far north you realize that the people here are distinctly akin to the regular Liquidlanders further west. One more thing: Kritikala is not a safe place to be. It is a city prone to earthquakes of all sizes, and no one tends to linger there for long. The ground tremors can be felt as you approach the city from any direction.

What does the map of Water World look like? Superficially, it is unremarkable (Fig. 6.2) except for one thing: the phase boundary between the liquid and the solid slopes *backwards*, from north-north-west to south-south-east. For reasons that need not detain us, this connotes the peculiarity that liquid water is denser than ice. The reversed slope of this boundary looks deceptively like a minor aberration on water's behalf; but we will see that herein lies a vital clue to all of water's oddness.

On the brink of order and chaos

I imagine it may come as a surprise to find that theoretical physicists still study things like liquids or melting and freezing. It sounds a little bit like using a supercomputer to add up the grocery bill. We are encouraged to believe that theoretical physics is primarily a tool for studying life's Big Questions: the origin of the Universe, the meaning of time, the fundamental constitution of subatomic matter. But a huge amount of modern physics is about phase transitions and critical points, which are of mind-bogglingly broad relevance – to superconductors, to the early Universe and to particle physics as much as to the humble old liquid state. Physicists have been curiously reluctant to admit this, which is why liquids have never quite received their due. Noble reasons, then, for asking: what is a liquid really?

Let me be upfront – even regular, well-behaved liquids (let alone a rebel like water) are not an easy family to come to terms with. So we had best begin with the other two familiar states of matter, which are altogether more approachable: solids and gases. By 'state of matter', what I mean is a characteristic, generic disposition of the component parts. In some substances, like metals, these parts are individual atoms; in others, like water, the basic units are molecules. To keep

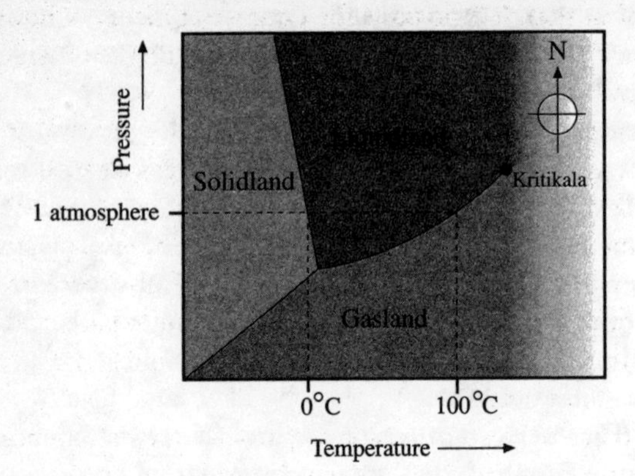

Fig. 6.2 The Phase Land of water – Water World – is subtlely different from that of most substances: the boundary of Solidland and Liquidland slopes north-north-west, not north-north-east. This is a direct consequence of the fact that water has the unusual property of getting less dense when it freezes.

things simple, I'll just call these individual, discrete components 'particles'. In a gas, the particles dance in virtual isolation from one another: scaled up to human size, they'd each typically gyrate over several hundred metres at a time before encountering another dancer. For gases, the number of dancers in the ballroom depends only on the pressure and temperature, not on the chemical composition – that is, whether the gas is oxygen, hydrogen or whatever. The particles in crystalline solids, meanwhile, are densely regimented: regularly arranged at positions on a geometric grid, the crystal lattice. Strong forces between neighbours hold each individual to its place.[5] The dance of solids is no dance at all, but a frozen tableau.

So gases are diffuse, while solids are densely packed. Liquids lie in between the two – denser than gases but, except for eccentrics like water, less dense than solids. The particles feel one another's presence, yet not so much as to pin them to the spot. In liquids the dance floor is a crowded place, with much unruly jostling and colliding.

A description of liquid structure can at best provide only a picture of what each particle's environment is like *on average* over time. This statistical picture is the best we can hope for, short of a detailed account of the trajectories of every single molecule over all time. That's not a model, it's a narrative – and an unbelievably detailed and tedious one at that. It is no basis for a universal theory of liquids, just as you won't arrive at a meaningful analysis of English literature simply by reiterating all its books.

While the chaotic rambling of particles in a liquid means that on average it looks as unstructured and featureless as a gas, there is some orderliness to be found if you look in the right way. Suppose we were to alight in some microscopic craft, as in the movie *Fantastic Voyage,* on a particular particle and take snapshots of what we see around us. We'd observe other particles tumbling about in such apparent disarray that you'd think a superposition of snapshots would generate just a uniform blur. But it doesn't.

Instead, we'd see a greater density of particles on the surface of several concentric spherical shells, of which our craft is at the centre (Fig. 6.3). The enhancement in density is quite pronounced in the first shell, but the second is typically much more blurred, and the third may be barely perceptible in many liquids. Each particle carries these shells with it as it moves – not in the sense that it carries along a well-defined party of neighbouring particles, but that on average it is more likely that the particle will possess neighbours around the surfaces of the shells than sandwiched between them. This, then, is the structure of a liquid – constantly changing but not quite random all the same.

Where does the structure come from, if the particles are simply jiggling at random? All molecules are attracted to one another by forces that are much weaker than those that bind the atoms together in a molecule. A liquid is therefore somewhat like a box of fridge magnets being shaken about in the back of a lorry – the molecules are constantly making and breaking unions with their neighbours.

So we might be tempted to imagine that each particle in a liquid tends to gather a coterie around it due to these weak forces of attraction, which, although not strong enough to bind the material into some rigid arrangement, nevertheless suffice to maintain a degree of local average structure. There is some truth in this picture – the attractive forces between the particles of a real liquid undoubtedly play a role in creating the shell structure. But surprisingly, even a hypothetical liquid in which there are no attractive forces between the components at all has this local structure. That's to say, hard, non-attracting 'billiard ball' particles still manage to gather their shells of neighbours in the liquid state. The reason for this is that in a liquid the close proximity of particles forces them to pack together

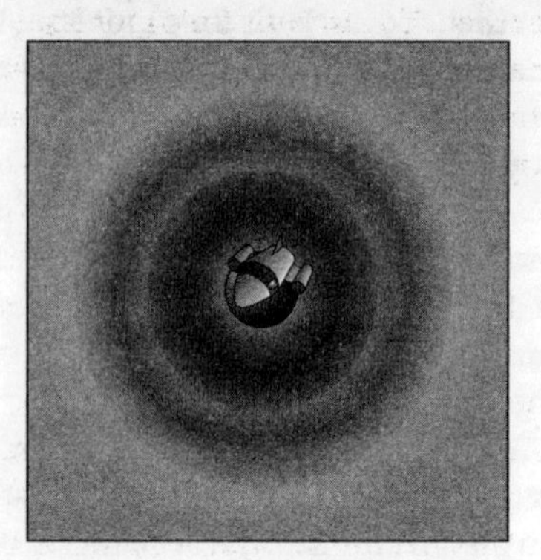

Fig. 6.3 The structure of a liquid can be discerned by measuring the average density of particles around any individual – for example, by superimposing many snapshots. Amidst all the apparently disorderly jostling, a definite average structure emerges in which neighbours are more likely to be found in a series of concentric shells centred on the chosen individual, here augmented by the tiny craft that takes the 'soundings'. Beyond the second or third shells any deviations from a uniform blur become impossible to discern.

with some degree of local order, simply so they can all fit within the available space. There is not, if you like, enough room in the crowd to allow each particle to prance around totally at random. By unconscious mutual consent, each particle tends to keep at an optimal distance from its neighbours, and on average this gives rise to the shell structure.

So the localized order in a real liquid is the result of a combination of two influences: the weak attractive forces, and the packing constraints imposed by the repulsive forces that stop particles from overlapping. In simple liquids it is the *repulsive* forces that dominate. In 1962 British physicist J. Desmond Bernal expressed it thus:

> ... the key word in the structure of liquids is the one which Humpty Dumpty used in *Alice Through the Looking-Glass*[*sic*], 'impenetrability'.[6]

Simply liquid

Being so much conceptually easier to picture, solids and gases were fairly well understood at the molecular level some time before any theory of liquids existed. The turning point for liquids came in 1873, when the Dutchman Johannes Diderik van der Waals realized that it is the forces of attraction and repulsion that distinguish a liquid from a gas. The forces are *there* in both states of matter – but in a gas they don't have much effect, since the particles are too far apart, on the whole, to feel them. Van der Waals knew that liquids and gases are intimately related, because of the possibility of converting one to the other 'smoothly' – without the abrupt phase transition that always separates a liquid or a gas from a solid.

Rudolf Clausius in Germany and James Clerk Maxwell in England had the theory of gases almost sewn up by the 1870s; the German Ludwig Boltzmann applied the finishing touches during that decade. By assuming no more than that a gas consists of trillions of particles dancing about independently and at random, with a vigour that increases when the gas gets hotter, the 'kinetic theory of gases' accounted for the relationship between a gas's density, pressure and temperature. When Van der Waals embellished this theory to allow for the forces of attraction[7] and repulsion between the particles, out popped the liquid state. Van der Waals's new theory predicted that, when cooled or squeezed, gases will condense to a denser fluid state – the liquid. Moreover, it implied that this condensation transition vanishes at sufficiently high temperatures and pressures – in other words, that the phase boundary between liquid and gas ends at a critical point, so that there is *continuity* between the two states.

The theory was eventually hailed as a tremendous success, and it won van der Waals the Nobel prize for physics in 1910 – proof, if proof be needed, that physics does not have to be exotic to be important. But some eminent scientists around the turn of the century remained uncomfortable that the theory rested on untested assumptions, the most profound of which was that the liquid had no structure. The way in which van der Waals had taken account of the attractive forces between particles was to assume that each molecule experiences a uniform average attractive force due to its interactions with all the others. Because of these attractions, the particles are pulled together like stars in a galaxy, and so the outwards pressure that they exert on the walls of a container is less than it would be if there were no attractive forces at all. Boltzmann realized that this way of treating the attractive forces is valid only if the force exerted by each particle acts over much larger distances than the average separation between them. Only then does each particle 'feel' enough others to experience a uniform mean attraction. The crude estimates of the range of the attractive force available in Boltzmann's time suggested that there was no guarantee that this was the case. Today an assumption like this is called a mean-field approximation, meaning that the individual components of the system are assumed to experience a uniform mean field of attraction due to all the others.

Going beyond Van der Waals's mean-field approximation, to develop a theory of liquids that remains true to the microscopic structure I discussed earlier, has taken the best part of the twentieth century. The modern theory of liquids contains a mathematical sophistication that rivals anything found in other fields of physics. Indeed, some of the theoretical ideas that liquid-state theory has generated have proved immensely valuable for studying other problems, such as the structure of subatomic particles. I'm going to do this tremendous body of work grave disservice by dispensing with it in a paragraph. The vast majority of the theoretical work on liquids is concerned with so-called simple liquids, in which the particles are approximated as spheres that attract one another with a force that, like gravity between planets, gets progressively weaker with increasing separation. Once the mathematical dependence of the force on separation is specified, it is now generally possible to calculate what the structure of the liquid is like – to deduce where and how pronounced the density peaks are in a picture like Fig. 6.3. We can generally calculate under what conditions such a liquid freezes, where its critical point lies, and how the liquid behaves as it is brought close to this point.

It is all, I have to say, rather impressive – except for one catch.

Sadly, very few liquids are anything like 'simple'. And this could hardly be more true for the most important liquid of all: water. When it comes to water, the refined tools of the theory of simple liquids are often woefully inadequate. It becomes like chiselling with a screwdriver or banging in nails with a mallet – you can just about get the job done, but it doesn't always look very convincing or elegant. The nails bend and, likely as not, snap. Since most of the other liquids that we encounter in daily life, in technology and industry – milk, fruit juice, blood, beer – are nothing but tainted water, full of dissolved or suspended substances, you can appreciate that this is a problem. Why, then, is water so difficult?

Why water is crooked

Water is bent – that's the bluntest way to put it. Its molecules, composites of an oxygen atom and two hydrogen atoms, have a characteristic kink at an angle of 104.5° (Fig. 6.4*a*). This is water's *molecular structure* – every molecule of water in the streams and oceans of the world, in raindrops and in our bloodstream, even in distant Europa's oceans if they truly exist, looks like this.[8]

Why 104.5°? The real answer resides within the quantum-mechanical description of atoms and their bond-forming habits, which was developed in the early part of the twentieth century. But it can be paraphrased, without losing anything important, as follows. The true shape of water is not a 'V' but a tetrahedron, Plato's regular four-cornered solid.[9] It just looks kinked because we are seeing in a diagram like Fig. 6.4*a* only two of the corners. The other two are occupied not by atoms but by pairs of electrons, called lone pairs. These are electrons from the oxygen atom's complement, which don't partake in the bonding between the molecule's atoms but which nevertheless have to go somewhere. They pair up, as electrons in atoms are wont to do, and take up residence about as far from each other and from the hydrogen atoms as they can get. The tetrahedral arrangement affords the greatest distance between each

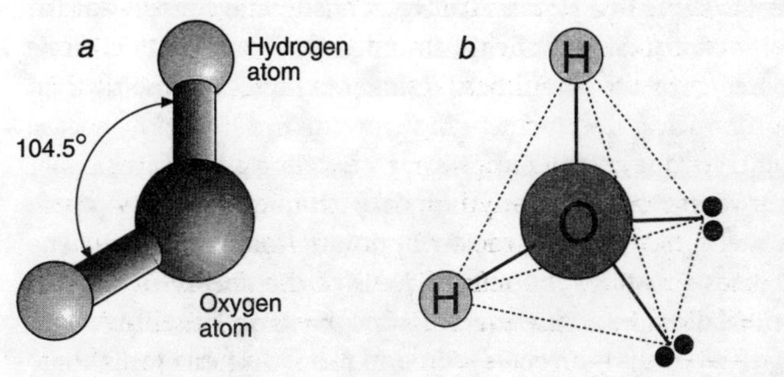

Fig. 6.4 The water molecule is bent, with the two bonds between oxygen and hydrogen splayed at an angle of 104.5° (*a*). To understand the structure of liquid water, we must also take into account the two 'lone pairs' of electrons on the oxygen atom. The hydrogen atoms and the lone pairs sit more or less at the corners of a tetrahedron (*b*). At the molecular scale, the structure of water is imprinted with this tetrahedral geometry.

of these four entities – the two hydrogen atoms and the two lone pairs (Fig. 6.4*b*). If it was a *perfectly* tetrahedral arrangement, the angle of the kink would be 109.5°; the difference between that and 104.5° is the consequence of the slightly stronger aversion of the lone pairs for each other than for the hydrogen atoms, so that the hydrogens are pinched together.

Hydrogen glue

If you had to guess how water molecules might behave by comparison with other, apparently similar small molecules, you'd anticipate that it would boil at temperatures below 0°C. That's what methane does, a substance whose molecules are comprised of carbon atoms surrounded by a tetrahedron of hydrogens. It's what hydrogen sulphide does too, whose molecules have the same kinked shape as H_2O. If we judge a molecule by appearances, the oceans should be up there in the atmosphere, as they are on Venus.

But water has anomalously high melting and boiling points. Somehow its molecules acquire an extra cohesion which prevents vaporization. In the 1930s the American chemist Linus Pauling explained how the water molecule gets this stickiness.

If a non-scientist knows just one chemist, it's a fair bet that it is Linus Pauling. (Sorry, but I refuse to count Margaret Thatcher.) Pauling is the Einstein of modern chemistry, and lived almost from one end of the century to the other. He is the only scientist to have been awarded two Nobel prizes in different fields: the chemistry prize in 1954 came for his contributions to the understanding of chemical bonding, and his efforts for nuclear disarmament brought him the 1963 Nobel peace prize. Trained in the technique of X-ray crystallography, a means of finding out where the atoms sit in crystals, Pauling went on to apply quantum mechanics to the question of how atoms form bonds. The fruits of his insights were collected in *The Nature of the Chemical Bond* (1939), one of the most influential scientific books of the century.

Pauling showed that the chemical bond formed when atoms share electrons is not necessarily an equitable arrangement. Some atoms, he said, hold on more tightly to the bonding electrons than others. In its greed for electrons, oxygen is surpassed only by fluorine. So in the water molecule, oxygen hogs the electrons like a selfish lover stealing most of the duvet. As a result the oxygen atom acquires a negative charge, and the hydrogens are left positively charged.

Pauling proposed that these charges in the water molecule give rise to an electrical force of attraction between neighbouring molecules, in which the hydrogen atoms of one molecule point towards the oxygen atoms of another. This attraction can be regarded as a kind of chemical bond, about ten times stronger than the van der Waals forces that hold 'regular' liquids together – but ten times weaker than the bonds that link hydrogen and oxygen atoms into discrete molecules. This is called a hydrogen bond.

The idea that some kind of weak interaction exists between a hydrogen atom in one molecule and the oxygen atom in another did not in fact originate with Pauling – it was first proposed in 1920 by Maurice Huggins, an undergraduate student of American chemist Gilbert Lewis, to account for an aspect of bonding in organic chemistry. Lewis immediately saw the value of the proposition, and in 1923 he coined the term 'hydrogen bond' for such an interaction. Two of Lewis's colleagues, Wendell Latimer and Worth Rodebush, realized that Huggins' idea might provide the key to understanding some of the strange properties of water.

The hypothesis of the hydrogen bond contradicted the accepted chemical wisdom of the time, which dictated that hydrogen can form a bond with one and only one other atom. For this reason, some chemists were most reluctant to consider the idea: in 1926 the British chemist Henry Armstrong ridiculed the notion that hydrogen could act as a 'bigamist'. Pauling, however, helped to provide a theoretical basis for the proposal by showing how this bigamy might arise from purely electrostatic effects, thus relinquishing any need for a single pair of bonding electrons to be shared between two bonds.[10]

The hydrogen atom in a hydrogen bond does not just stick indiscriminately to the oxygen of another molecule; being positively charged, it goes where the electrons are. So the hydrogen bond is really a bond between a hydrogen atom and a lone pair. This means that a water molecule can form four hydrogen bonds: the molecule's two hydrogens form two bonds with neighbouring oxygens, while the molecule's two lone pairs interact with neighbouring hydrogens. A water molecule can be regarded as having 'hooks' for ensnaring other water molecules at each of the four corners of its tetrahedron.

The best way to understand this is to impersonate a water molecule.[11] Your hands are hydrogen atoms, your ankles are the lone pairs of electrons on oxygen. Stand legs apart (if you can get an angle of around 109° between them, good for you – but don't push it). Twist 90° at the waist, stretch out your arms – and you're H_2O.

The way that water molecules join up has just one rule: hands can grasp ankles, but nothing else. That grasp is a hydrogen bond. In everything further that I shall say about how water behaves at the molecular scale, I'd like you to be thinking of this picture (Fig. 6.5). If it helps, get up and twist.

Water isn't alone in forming hydrogen bonds. Ammonia and hydrogen fluoride molecules also have what it takes to be sticky: lone pairs and electron-depleted hydrogen atoms. So they too have unusually high melting and boiling points.

Hydrogen bonds help to glue biological molecules into their distinctive shapes. In DNA, they provide the zipper that holds the famous double helix together. They secure the delicate folds of protein molecules, the workhorses of the cell and the structural fabric of our soft tissues.

The infinite network

The hydrogen bond, then, is what sets water apart from other liquids. But it doesn't immediately explain why water is *so* odd – why it is denser than ice, why it is central to life, why it has such a capacity to absorb heat and so forth. Can't we think of water as being just like any other liquid, except more strongly bound?

Not a bit of it. I stated earlier that the structure of 'normal' liquids takes rather little heed of the forces of *attraction* between molecules, hinging instead on the packing constraints imposed by the *repulsive* forces. In water, this is no longer the case – it is the attractive forces, the hydrogen bonds, that play the largest role in the way the molecules are arranged. And these attractive forces introduce pronounced preferences for the positions and orientations of neighbouring molecules: each oxygen atom sits at the vertex of a tetrahedral network of bonds. In this respect, water is far less disordered, far more highly structured, than most liquids. It is more akin to a crystal than to a gas.

Iced water

With this realization, water becomes altogether unworldly – a

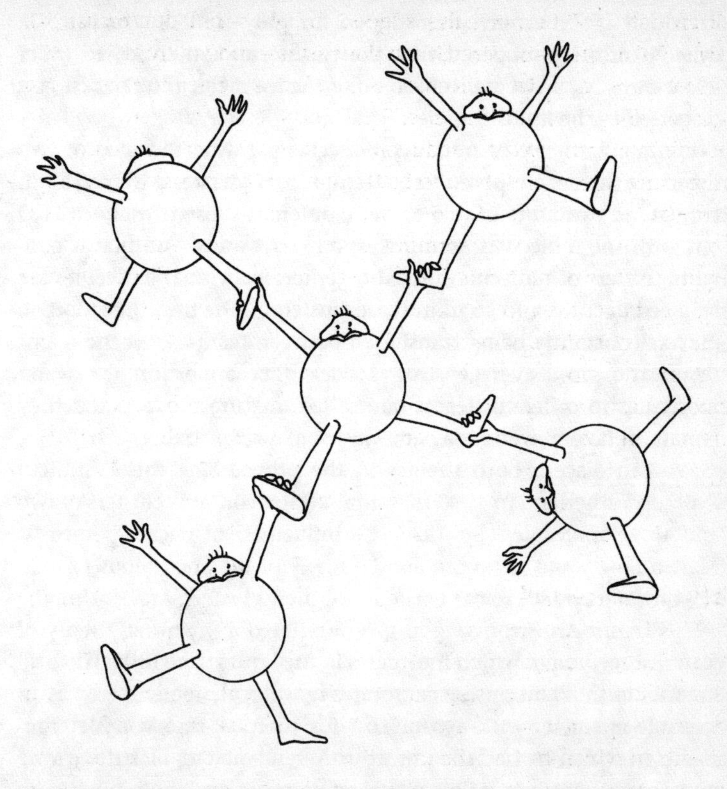

Fig. 6.5 Water molecules – with hands representing hydrogens and feet representing lone pairs of electrons – perform a dance that involves grabbing neighbours by the ankles. These clasps, due to hydrogen bonding, lead to a tetrahedral arrangement of neighbours around each molecule. This is the central motif of the structure of water, and the key to all its anomalous properties.

substance seemingly poised, like the rivers and forests of J.G. Ballard's *The Crystal World*,[12] on the verge of coalescing into prismatic solidity. The spectre of crystallinity has haunted attempts to understand water just as it haunted Ballard's protagonists – even, in fact, before we had any inkling of its uniquely structured matrix.

In 1892, the German physicist Wilhelm Röntgen[13] suggested that the density anomaly of water – the fact that it is denser at 4°C than at freezing point – might be explained by assuming that cold water is a mixture of tiny crystalline, ice-like fragments and a fluid component in which mobile molecules adopt a roughly ice-like structure. In other words, he supposed that the transformation of ice to water involves only *partial* melting: some of the ice melts and the rest gets dispersed in the fluid.[14]

Röntgen's theory of microscopic icebergs sowed the seed of two recurring themes in subsequent attempts to understand water: that it consists of a mixture of two or more different substances, both H_2O but with the molecules arranged in different ways; and that it contains clusters of molecules linked together. Individual molecules are believed to come and go from these clusters all the time, like football players constantly being transferred between teams – but the teams retain the same average sizes. Models that apportion the water molecules to different 'teams' are called mixture models, and they remain in favour even today amongst some water researchers.

We can discern both themes in the proposal by the Australian William Sutherland in 1900 that liquid water contains clusters of two and three molecules. By 1922 the influential physical chemist T. Martin Lowry was happy to ascribe many of the anomalous properties of water and ice to the existence of these clusters. And during the 1920s Henry Armstrong vigorously promoted a 'hydrone' theory of water, according to which molecules in the liquid were linked into all manner of small unions. Armstrong is now largely remembered as an irascible character with a penchant for entering into voluble arguments in which he had the unfortunate tendency to back the wrong horse. Another target of his mistaken diatribes was the ionic theory of Svante Arrhenius, which explains what happens to salts when they dissolve – its adherents, said Armstrong, were 'people without knowledge of the laboratory arts'. He even took against the word 'scientist', saying 'The real men, those who do things – bakers, butchers, builders, boxers, grocers, even green-grocers – all have names ending in *er*.'[15] Armstrong's hydrone theory was no more valid than his erroneous views on hydrogen bonding, which ironically provides a means for his hydrone clusters to form. But a serious, if reactionary, scientist lay behind the bluster,[16] and he was instrumental in improving the state of chemical education in Britain, transforming it from rote learning to an experiential, investigative process.

A popular mixture model of water from the 1950s invoked a brew of no fewer than five different teams of molecules. This is not just an arbitrary proliferation; there are good reasons for setting the number at five. As each molecule can form up to four hydrogen bonds in total, it seemed reasonable to suppose that the liquid contains molecules attached to one, two, three and four others, as well as some molecules with no partners at all. That sounds all very well – but deducing the size of each team is complicated by the fact that the hydrogen bonds are not made and broken independently of one another. Just as in the football league the chances that players will want to stay in a particular team may depend on whether or not others have just left or joined, so too do hydrogen bonds behave cooperatively: the breaking of one bond alters the chance that another will be made or broken nearby. This aspect of hydrogen bonding makes the whole question of water's structure immensely more complicated, since it introduces an interdependence amongst participants like that which makes the stock market so volatile.

One size fits all

A very different approach to the problem of water's structure is to suppose that the liquid is uniform on average, and to focus on the average structural features. In other words, one does not assume that there are several different and distinct classes of water molecule but that there is a continuous gradation of the kinds of environments in which water molecules might find themselves. Then the central questions are: what does a 'typical' molecule experience, and how varied are the deviations from this average environment? In the same way, a survey of national earnings would aim to tell you what the average wage is and how rapidly the numbers tail off above and below this average.

This is the 'uniformist' approach, and its most celebrated example is the model proposed by Desmond Bernal and Ralph Fowler in 1933. They reasoned that, rather than regarding hydrogen bonds as make-or-break affairs – so that either two molecules are hooked together via a hydrogen bond or they are wholly free of each other – one could postulate a varying degree of 'bondedness'. In particular, they pictured a more elastic bond which can be bent to greater or lesser degrees, with the more highly bent bonds being less stable and so more easily broken. In terms of our spread-eagled water dancers grasping one another's ankles, a bent hydrogen bond means that the leg and the arm that grasps it do not lie along the same straight line. Bernal and Fowler said: what if melting of ice isn't really about bonds breaking at all, but about their becoming increasingly floppy? Of course, *some* bonds must break totally, because the characteristic feature of a liquid is that the molecules can move around; otherwise, the substance would be rubbery. But Bernal and Fowler suggested that the most well-defined average property of the liquid might be the degree to which it is distorted from the crystalline ice structure by hydrogen-bond bending.

Again, we see the orderly structure of the solid state – ice – being taken here as the starting point for modelling water's liquid structure, in contrast to van der Waals's assumption that the random structure of a gas is the appropriate base from which to comprehend liquids. It's no surprise, then, that some of the key figures in studies of water's structure are scientists whose interests lay more with the crystalline states of matter, such as Linus Pauling and Desmond Bernal, rather than the giants of traditional liquid-state theory.

Each water dancer in the crystal lattice of ice grasps, or is held by, four other molecules arrayed tetrahedrally around it (Fig. 6.6). You can see that this creates a rather open network with lots of space between the molecules. The molecules could be packed together considerably more closely – but if they were, they'd no longer be in the right positions to seize one another via unbent hydrogen bonds. So somewhat paradoxically, while the hydrogen bonds hold the crystal together and convey strength and rigidity, they are also responsible for its unusually low density. When ice melts to liquid water, this regular network of hydrogen bonding is disrupted and so there is less constraint on the molecules to maintain the open structure – the void space partially collapses. The central question of water's liquid structure is in what manner, and to what extent, this collapse occurs.

In 1946 the Russian scientist O. Samoilov provided a new perspective on this crystal-based approach to water. Like Bernal and Fowler, he considered how the structure changes as ice melts. But Samoilov blended this view with the mixture-model ideal by assuming that melting creates different populations of molecules: some that remain in place in the ice lattice, and others that break loose and move about in the empty spaces (the 'interstices') between them. This is the basis of so-called 'interstitial' models of water.

It was common in Samoilov's era for ideas developed in the Soviet Union to remain unknown to researchers in the West, since Soviet scientists tended to publish in Soviet journals and, with the Cold War just beginning to gather pace, the flow of scientific information

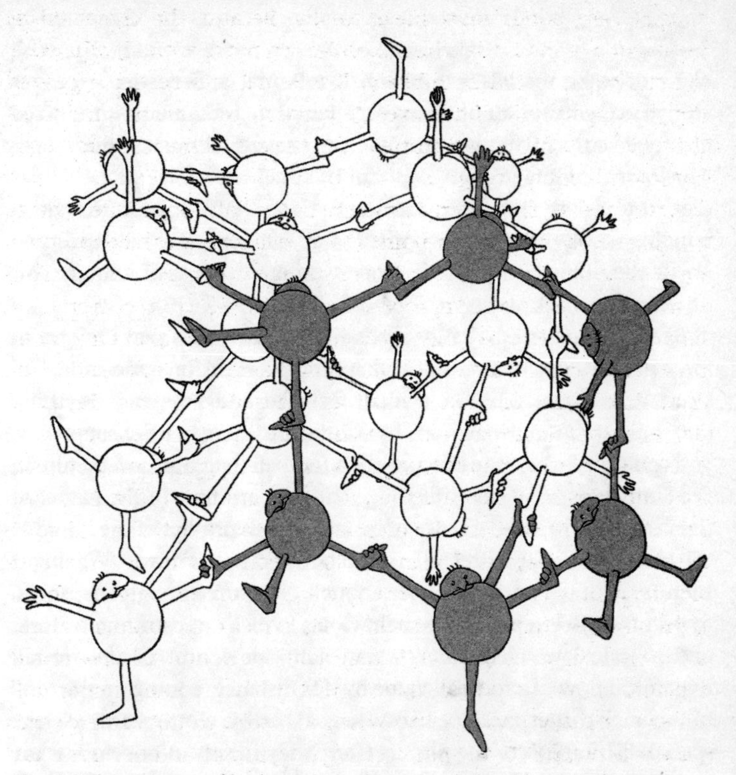

Fig. 6.6 The structure of ice is no dance but an orderly tableau. Each water molecule is hydrogen-bonded to four others in an arrangement that displays hexagonal symmetry. Here I have highlighted one of the hexagonal rings in grey.

between the West and the Soviet Union was impeded. This explains why interstitial models similar to Samoilov's were put forward by two groups of Western scientists in the 1950s and 1960s in apparent ignorance of the Soviet scientist's idea. Linus Pauling proposed another related model in 1959 which took as its starting point the crystal structure not of ice but of another kind of open framework that water molecules may form, called a clathrate. Only when Samoilov's book on aqueous solutions was published in 1957 and translated into German and English some years later did the precedence of his ideas become appreciated.

Virtual water

None of these models – uniformist, mixture or interstitial – seems able to account for all of the experimental measurements conducted to probe water's structure. Yet like the proverbial blind Indian sages developing an image of an elephant by touch alone, researchers have tended to assert that *their* model is the one that fits, even though it remains wholly inconsistent with what someone else has discerned about the object under study from a different point of enquiry. No one can step back and see the whole elephant.

The fundamental difficulty with all these models is that they try to force a static description on a changing scene. Any static model of structure is doomed by the evident fact that liquid water is anything but static – the molecules are in constant motion. In the same way, you couldn't deduce the outcome of a football match from a picture in the Sunday papers – less still gauge the ebb and flow of the game. At the very least, we need a series of instantaneous profiles of the liquid at the molecular scale; better still would be a molecular movie. We might then hope, if we watch for long enough or peruse enough snapshots, to begin to discern patterns of behaviour, to pick out common features and to calculate averages. You can gain some hint of the overall dynamical flow of a football game by, for instance, adding up the total number of corner kicks or shots at goal, or by plotting the average spatial distribution of the players (were they mostly at one end or the other?) or the most common geometries of passes.

We don't have the technologies yet for tracking huge numbers of water molecules as they play out their tactics in the real liquid. Although it is now possible to take frame-by-frame snapshots of molecular processes over the incredibly short timescales – perhaps a few trillionths of a second – during which they occur, it may never be feasible to monitor directly and simultaneously the environments and trajectories of thousands of individual molecules in a complex and messy system like a liquid.

But computer technology gives us the next best thing. Today's computers are gateways to synthetic worlds: we can use their digital might to construct all manner of alternative realities. Scientific 'experiments' can now be conducted on a computer – we can journey into the molecular world by simulating it, just as we can simulate a motor race or an aircraft flight.

It works like this. The computer is programmed to calculate the paths that a whole boxful of molecules will take if they are allowed to wander around freely under the influence of each other's fields of attraction and repulsion. It's like simulating a snooker game – you specify the speeds of the balls, and use Newton's laws of motion to determine how these alter when balls collide. In a computer simulation one can control everything to be just as one likes it – something rarely possible in the lab. The temperature, density, volume and pressure of the substance can be chosen and fixed with complete accuracy. There are no impurities unless they are wanted. And the shapes and force fields of the molecules can be designed to order. The only limitation is the amount of computing power available. Sadly, this can often be a serious limitation. The earliest computer simulation of a fluid, conducted in 1953, modelled 224 particles – and then only let them move on a flat plane, not in three-dimensional space. Simulations today can be conducted for millions of particles. But that still corresponds to only a minuscule volume of liquid: a billion trillion such volumes would just about give you a raindrop.

For water, there's an additional complication: it is very tough to make a good computer model of an H_2O molecule, because the way a hydrogen bond forms as two water molecules approach one another is so complex. The crucial difficulty is that hydrogen bonds form cooperatively, not independently.

So computer models of liquid water are best regarded as cartoons rather than movies depicting the 'real world'. One of the earliest of these cartoon depictions of water was sketched in the 1950s by Danish scientist Niels Bjerrum, who modelled the molecule as a four-armed tetrahedral creature with electric charges at the ends of each arm. Two of the charges are positive, representing the hydrogen atoms, and the other two are negative, representing the lone pairs. This creature is of course just like one of our dancers, with a preference for linking up with others in the tetrahedral arrangement.

Well, cartoons have come a long way since the 1950s, and likewise computer simulations of liquid water today are impressively adept at capturing many of its key attributes. They can, for example, reproduce anomalies such as the density maximum at 4°C and the high heat capacity. That these unusual traits emerge spontaneously – that is, without being imposed at the outset – from a collection of 'model' water molecules can give us considerable faith that the model is a good one, and hence that the molecular dance revealed by the computer bears some relation to that being conducted in the real liquid.

So what does simulated water look like? One thing is for sure: it doesn't much resemble ice after all. Forget crystallinity; the network is a jumble, like a collapsed climbing frame (Fig. 6.7). What's more, computer simulations lay to rest the notion, inherent in interstitial models, that there are two classes of molecule, 'ice-like' in the framework and 'free' in the interstices. If that were the case, the majority of molecules would either have four hydrogen bonds on average, or none. Quite the opposite is found in simulations: most molecules have two or three bonds, and a relatively small proportion has four or none.

Clearly there is a significant rearrangement of hydrogen bonds when ice melts. In particular, whereas hydrogen bonding in ice links each water molecule into rings of six (see Fig. 6.6), the most common ring structure in liquid water contains five molecules, not six. And while each vertex of the ice network is tetrahedral, the confluence of two arms and two legs, in water some molecules are left dangling – broken struts on a mangled frame. And a few of the molecules appear

Fig. 6.7 Computer models of liquid water indicate that its hydrogen-bonded network is random and disorderly, and extends throughout the entire collection of molecules like a crazy climbing frame. Here the network is depicted as a lattice of struts representing the hydrogen bonds between molecules. The oxygen atoms of the molecules are located at the junctions of struts. (Stephen Harrington and Gene Stanley, Boston University)

to break the rules of the dance by forming five hydrogen bonds instead of four, cheekily grasping two ankles in one hand.

The picture of water structure that emerges from computer simulations, therefore, is neither ice-like nor like a fully random vapour. The molecules form a continuous, disordered and dynamic network of hydrogen bonds in which each molecule is linked with up to four (or very rarely, five) others. Because the hydrogen-bonded network is floppier, more distorted and more defective than that in ice, the molecules are able to occupy some of ice's interstitial space, making the liquid denser than the solid. The degree of distortion becomes greater as the temperature is raised from the melting point of 0°C, and so the liquid goes on getting denser until it reaches a maximum of 4°C. Above this temperature, enough hydrogen bonds have been broken that the liquid starts to behave normally (or at least, more normally), becoming less dense with increasing temperature because the thermal jiggling of the molecules tends to keep its neighbours at an increasing distance.

I'd like to be able to say that you can experience all this for yourself by taking part in a genuine dance of water, according to the rules mentioned earlier, and seeing what kind of fantastic Twister-style contortions you and your partners generate. But you'd need to do it in three dimensions, and short of trying it out in the deep sea or in zero gravity I don't see how that is possible. And there's one further consideration that might dim your enthusiasm: to be entirely accurate, the dancers would have to fall apart from time to time, an arm coming adrift from the rest of the body and floating off attached to someone else's ankle. For about one in every 500 million molecules in a glass of tap water is fragmented into a positively charged hydrogen ion and a negatively charged hydroxide ion, which contains the remaining oxygen and hydrogen atom.

Adding an acid to water increases the number of spare arms – hydrogen ions – while an alkali (more generally, a substance that chemists call a base) like caustic soda increases the number of one-armed bodies, the hydroxide ions. The rule here is that if fragments are added by acids or alkalis, correspondingly fewer water molecules fall apart. You might think that the chances that *you*'d be the one to lose an arm would be negligible – but the fact is that every dancer will take a turn at joining, if only fleetingly, the one in 500 million. Do the dance for long enough, and your time will come.

Because it is so rare, this fragmentation has little significance for water's structure; but it is crucial to water's chemistry. For instance, it allows water to act as an acid or a base by providing a source of hydrogen or hydroxide ions. And water's structure does have an important influence on this chemical behaviour. Hydrogen ions can be shunted around the hydrogen-bonded network much more quickly than they can travel by drifting through any other liquid. If an extra 'arm' gets attached to one end of a chain of linked water molecules, all the handholds can be rearranged sequentially along the chain until another arm falls off the other end – as though the arm has been magically transported from one end to the other. This process was anticipated in 1806 by the French chemist C.J.T. Grotthuss in the course of his attempts to explain how water conducts electricity.

The secret

This, then, is the current molecular-scale picture of the strange stuff called water: a liquid that has a high degree of internal structure brought about by the 'stickiness' that can bind the H_2O molecules into a dynamic, ever-changing labyrinth. Water's strangeness resides almost wholly in its hydrogen-bonded character. While it is not by any means the only molecule that can form hydrogen bonds, no other has just the right shape to allow the network to extend throughout all space: the crucial twist of the hips to impersonate water opens up the third dimension. The hydrogen bonds impose structural constraints that are most unusual for a liquid, and these in turn affect the physical properties such as density, heat capacity and heat conductance – as well as the way that water accommodates dissolved molecules, an issue discussed in Chapter 9. We've seen already why these eccentricities are not just arcane footnotes to the palimpsest of science, but have consequences of global and indeed universal significance.

7 Cold Truths

Water below freezing point

The world of ice and of eternal snow, as unfolded to us on the summits of the neighbouring Alpine chain, so stern, so solitary, so dangerous, it may be, has yet its own peculiar charm.

Hermann von Helmholtz (1865)

I have hydrophobia. But I know something about ice.

Peter Høeg, *Miss Smilla's Feeling for Snow*

'Now suppose,' chortled Dr Breed, enjoying himself, 'that there were many ways in which water could crystallize, could freeze. Suppose that the sort of ice we skate upon and put into highballs – what we might call ice-one – is only one of several types of ice. Suppose water always froze as ice-one on Earth because it had never had a seed to teach it how to form ice-two, ice-three, ice-four...? And suppose,' he rapped on his desk with his old hand again, 'that there were one form, which we will call ice-nine – a crystal as hard as this desk – with a melting point of, let us say, one hundred degrees Fahrenheit, or better still, a melting point of one hundred and thirty degrees.'

With me so far? Dr Breed clearly means to speak in degrees centigrade, not fahrenheit; but let's continue. What if you drop this ice-nine into a puddle?

'The puddle would freeze?' I guessed.
'And the muck around the puddle?'
'It would freeze.'
'And all the puddles in the frozen muck?'
'They would freeze?'

Yes. But –

'There is no such stuff?' I asked.

> 'No, no, no, no,' said Dr Breed, losing patience with me again...
> 'If the streams flowing through the swamp froze as ice-nine, what about the rivers and lakes the streams fed?'
> 'They'd freeze. But there is no such thing as ice-nine.'
> 'And the oceans the frozen rivers fed?'
> 'They'd freeze, of course,' he snapped...
> 'And the springs feeding the frozen lakes and streams, and all the water underground feeding the springs?'
> 'They'd freeze, damn it!' he cried.

Except that –

> Dr Breed was mistaken about at least one thing: there *was* such a thing as ice-nine.[1]

My copy of Kurt Vonnegut's *Cat's Cradle* has 'science fiction' emblazoned across the cover, but the narrator is right: there *is* such a thing as ice-nine. In Vonnegut's book, a sliver of it is dropped into the ocean, and it seeds the transformation of all the world's liquid water into this exotic form of ice. The Earth is converted to a 'blue-white pearl', the seas frozen forever.

Dr Breed, however, was also wrong about another thing. Ice-nine does not melt at one hundred or so degrees centigrade: it cannot even exist above minus one hundred. Relieved? Let me introduce you to ice-seven. This is an ice that *does* melt at water's normal boiling temperature, and it too is real stuff, not Vonnegut's invention or anyone else's. It is nearly twice as dense as normal ice, ice-one, and hard as a rock. But I am not losing any sleep over it, and I will tell you why.

The ice we know

Water brings life; but ice is a metaphor for a certain kind of death. Not the fast, flickering fury of Hell's fire but the sluggish, groaning emptiness that waits at the edge of the world. In Norse mythology, the icy Fimbul winter presages the apocalypse of Ragnarok, just as a nuclear winter threatens to complete the lethal work of a nuclear holocaust. Amongst ice, life is not consumed; it just stops.

Beyond recorded time, before the murky beginnings of human

civilization, lurks a fearsome ancestral memory of an ice age, hedged with glaciers that extended to Hamburg, Birmingham and New York. Imaginable time, the last few thousand years, runs into mythical, geological Deep Time across a frozen landscape. When it snows in London, we don't know whether to be delighted or dismayed: children play and people die of hypothermia. In places where the ice never departs, it becomes a building material, a means of survival; and even if tales of the Inuit people's classifications of ice are perhaps exaggerated, they nevertheless have a thing or two to teach the glaciologist about snow and ice.

When water freezes, it reveals a beauty hitherto hidden. Windows are laced with its delicate traceries, trees are dusted with sparkling powder, gables are adorned with fangs. The whiteness of Antarctica is revered for a primal purity that we ascribe to no tawny desert. But the beauty is that of the Ice Queen, cold, lifeless, sterile and without mercy:

> For this powder was not made of tiny grains of stone; but of myriads of tiniest drops of water, which in freezing had darted together in symmetrical precision – parts, then, of the same inorganic substance which was the source of protoplasm, of plant life, of the human body. And among these myriads of enchanting little stars, in their hidden splendour that was too small for man's naked eye to see, there was not one like unto another; an endless inventiveness governed the development and unthinkable differentiation of one and the same basic scheme, the equilateral, equiangled hexagon. Yet each, in itself – this was the uncanny, the anti-organic, the life-denying character of them all – each of them was absolutely symmetrical, icily regular in form. They were too regular, as substance adapted to life never was to this degree – the living principle shuddered at this perfect precision, found it deathly, the very marrow of death.[2]

So says Hans Castorp of snow in Thomas Mann's *The Magic Mountain*, and maybe he speaks for us all in implying that water has somehow betrayed us when it forms a crystal – as though its former liveliness and bounty were just a wicked trick. And indeed we'll see that life encounters severe difficulties in the fact that its vital fluid exists in an environment that can regularly cool below the liquid's freezing point. But we have also seen already how ice is not wholly so inimical, for its remarkable property of expanding on freezing operates to the advantage of aquatic life, keeping lakes liquid beneath a solid cap. With ice, what meets the eye can be spectacular enough; but there is more to it than that.

The six-petalled flowers

We've glimpsed ice-one already in the previous chapter, but now it's time to take a closer look. For it is indeed true that ice has many relatives – thirteen at the last count (though at least one remains rather elusive) – and that is something of a record for a solid material. To understand where all these relatives come from, we need to fix our gaze first of all on good old ice-one, the kind that comes in Martinis and glaciers.

When water freezes in our refrigerators, in winter puddles and in the Southern Ocean around Antarctica, in snowflakes and on frosted branches, it forms a crystal with the structure shown in Fig. 6.6. Each water molecule is joined by hydrogen bonds to four others – two hands grasping ankles, two ankles grasped by another's hands. The network of linked molecules has plenty of empty space, giving ice a density less than that of liquid water.

The ice lattice has the same symmetry as a hexagon: you can rotate it by a sixth of a full revolution, and it looks just the same as before. This symmetry at the atomic scale exerts a strong influence on the shapes that the crystals acquire. Here's Vonnegut again, as Dr Breed talks about ice-nine:

> That old man with spotted hands invited me to think of the several ways in which cannon balls might be stacked on a court-house lawn, of the several ways in which oranges might be packed into a crate.[3]

Those stacking patterns create flat faces – crystal facets – with square and triangular shapes. The arrangement of atoms in a crystal determines the shapes and orientations of the faces. Table salt has cubic crystals because of the cube-like arrangement of its constituent ions.

The sixfold symmetry of the ice crystal lattice therefore gives ice crystals the same symmetry. But how! While most crystals are compact, blocky objects, ice seems to use its sixfold symmetry as a licence for geometrical excess (Fig. 7.1). The patterns formed by ice crystals are some of the richest in all of nature.

Amongst crystals, snowflakes are a rather special case. They form in the upper atmosphere as water vapour freezes out onto some tiny 'seed' particle, perhaps a speck of dust. This is a very different process from the freezing of a beaker of liquid – snow crystals typically grow at a temperature well below water's normal freezing point by accumulating water molecules on their ramified arms, like Byzantine icons accreting dust from the air.

If molten metals are cooled very quickly below their freezing point, they too may form branched fingers reminiscent of the snowflake's arms (Fig. 7.2). The formation of these structures is called dendritic growth, and the complex science behind their ornamented, Christmas-tree-like appearance became fully understood only during the 1980s. Dendritic crystals grow as needle-like fingers rather than in the blocky, faceted form that we normally associate with crystals because the flat face of the solid is unstable: small bumps that form by chance blossom quickly into long peninsulas. This is called a growth instability, and something rather similar governs the complex, branching shape of a crack in a brittle material. The

Fig. 7.1 Snowflakes are crystals of ice that bear witness to nature's inexhaustible pattern-forming capability. (After Bentley and Humphreys, 1931)

side-branches that adorn the dendrite's needle-like tip are a consequence of the same instability, but modulated by the underlying symmetry of the crystal lattice which makes some growth directions 'special'. It is this specialness of certain directions that compels the arms of a snowflake to branch and branch again at a diverging angle of 60°, the characteristic angle of hexagonal symmetry. Last winter I found ice dendrites as wide as my little finger, like gigantic snowflake arms, on the underside of the ice cap on my water butt.

Fig. 7.2 When grown very rapidly, crystals commonly form dendrites: snub-nosed needles decorated with side branches, like individual snowflake arms. Here I show dendrites formed by sudden solidification of a molten metal. (Lynn Boatner, Oak Ridge National Laboratory)

This sixfold symmetry of snowflakes was, as I indicated earlier, known to Chinese scholars at least as early as the first few centuries BC, and found an echo in the affinity between the 'element' water and the number six in Chinese alchemy: 'since Six is the true number of water, when water congeals into flowers they must be six-pointed', said the scholar T'ang Chin. Another Chinese scholar, Han Ying, astutely remarked in 135 BC that in this regard these tiny 'flowers of snow' differed from the flowers that grow in the ground, which are commonly five-petalled.[4]

The sixfold symmetry of snow crystals seems to have been common knowledge in China throughout the first millennium AD, but it was not recognized in the West until the seventeenth century. Western references to 'star-shaped' forms do appear as early as the thirteenth century in the writings of the Scandinavian Albertus Magnus, and an engraving from *The History of the Nordic Peoples* (1555) by Olaf Mansson shows not only six-pointed stars but an array of snowflake shapes that betrays an over-active imagination: bells, arrows, crescents, hands. It was not until 1610, however, that we find a clear statement in the Western literature of the six-ness of snow. In an essay 'On the Six-Cornered Snowflake',[5] astronomer Johannes Kepler asked, 'Why always six-sided?' Coming at the dawn of the scientific age, this seems to have been the first attempt to understand the fantastic shapes in mechanical rather than symbolic terms. Though lacking any clear concept of atoms, Kepler speculated that the symmetry of snowflakes derives from the packing of their constituent particles, which he supposed to be globules of condensed moisture. Ultimately this insight has a sound basis; but Kepler's proposal fails totally to explain why snowflakes are highly branched rather than the dense, faceted forms that generally result from packing spheres together.

Kepler's enquiries heralded a flurry of investigations into nature's ice flowers. René Descartes drew several detailed sketches of hexagonal forms in 1637 (Fig. 7.3*a*), and Erasmus Bartholin noted their branched character in 1660 (Fig. 7.3*b*) – something that is also evident in the sketches of the Italian astronomer Giovanni Domenico Cassini from 1692 (Fig. 7.3*c*). The development of the microscope in the early seventeenth century gave natural philosophers a new and powerful third eye with which to examine nature's tiny secrets. Robert Hooke made sketches of his microscopic observations of snowflakes in 1665, while in 1675 Friedrich Martens, a ship's barber on a voyage between Spitzbergen and Greenland, noted how their shapes changed with the prevailing meteorological conditions.

In the late nineteenth century the technique of taking photographs through the microscope – microphotography – alleviated the need to record observations as sketches, and so dispensed with reliance on the artistic and observational skill of the experimenter. Photomicrographs of snow crystals aided attempts at classifying their different forms in the 1890s and early 1900s; but the classic collection of images of snow's inexhaustible diversity is that of the American William Bentley, who with the assistance of W.J. Humphreys assembled at least 6000 photomicrographs over a period of almost fifty years. Their book *Snow Crystals* (1931) remains unrivalled today. Bentley's hunger for new specimens of 'beautiful snow' was insatiable, but their deathly precision, as Hans Castorp has it, was indeed his downfall. In 1935 his gathering was brought to an untimely end when, after wandering through a blizzard, he contracted a fatal bout of pneumonia.

To the biologist Thomas Huxley in 1869, these fantastic patterns were 'frosty imitations of the most complex forms of vegetable foliage'. This very comparison, indeed, prompted Huxley to believe that vitalism – the common idea of the time that life arises from some mysterious and non-physical force – was a redundant hypothesis:

> We do not assume that a something called 'aquosity' entered into and took possession of the oxide of hydrogen as soon as it was formed, and then guided the aqueous particles to their places in the facets of the crystal, or amongst the leaflets of the hoar-frost.[6]

On the contrary, said Huxley, we should be able to understand these forms from the basic physics of molecular motion and interaction.

Huxley's wish is now fulfilled as far as the symmetry of the branching pattern is concerned. But the full glory of a snowflake's shape remains a matter of intense debate. No other substance is known to crystallize in quite so many different ways. Our most familiar image of a snowflake is that of a flat star – and why the ice crystal should confine itself to a plane, when all three dimensions are available, is still something of a mystery. But not all snowflakes adopt this form. As Martens the barber first noted, and as Miss Smilla would surely have appreciated, the type of snow depends on the weather. This dependence was verified by the Frenchman M. Guettard in 1762, who proposed that the air temperature determines the shape of a snowflake. In 1820, Arctic explorer William Scoresby arranged these shapes in a formal classification scheme which included, in addition to the six-pointed stars, such forms as needle-like hexagonal prismatic columns – the shape of the shafts of pencils and Biro pens. Observations in the early twentieth century showed that prisms become more common than plates when the air temperature is below minus 20°C.

In the 1930s, Ukichiro Nakaya at Hokkaido University in Japan developed techniques for growing snowflakes artificially in the laboratory, allowing him to explore systematically how shape and form vary with the conditions of growth. These studies showed that, for all their variety, the 'six-petalled flowers' are just one breed of a whole menagerie of forms. Nakaya and his coworkers grew the ice crystals on a rabbit's hair suspended within a stream of moist air, in a chamber placed within a room that could be cooled to minus 30°C. The main influence on the crystal shape is the temperature, although varying the humidity of the air can also affect the shape. At around 0 to –3°C, the snowflakes are 'plates' – flat flakes of ice with hexagonal shapes. Between about –3 and –5°C, they take on a needle-like form instead. At lower temperatures down to about –22 to –25°C, plates are formed if the air is not too moist, whereas dendritic stars and other complex forms appear in more moist air. If it is colder still, prismatic crystals appear. Much the same relationships have been observed in the real atmosphere. A feeling for snow is, it seems, in large measure a feeling for coldness and humidity.

A continuing mystery about dendritic snowflakes is why all six of their branches seem to be more or less identical (see Fig. 7.1). The theory of dendritic growth explains why the side branches will develop at certain angles, but it contains no guarantee that they will all appear at equivalent places on different branches, or will grow to the same dimensions; indeed, these branching events are expected to happen at random. Yet snowflakes can present astonishing examples of coordination, as if each branch knows what the other is doing. One hypothesis is that vibrations of the crystal lattice bounce back and forth through the crystal like standing waves in an organ pipe, providing a degree of coordination and communication in the growth process. Another is that the apparent similarity of the arms is illusory, a result of the spatial constraints imposed

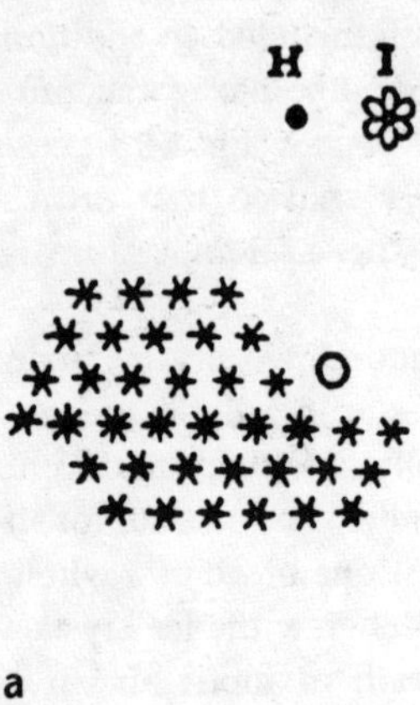
H
I
K
E
F
O
R
Q
Z
M

a

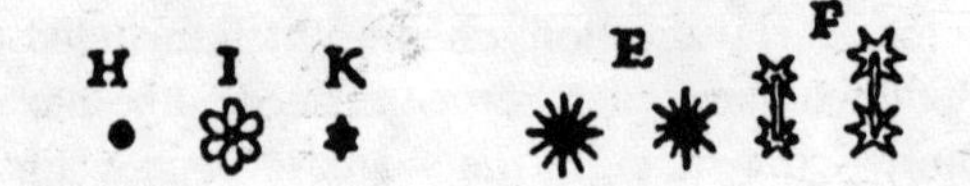

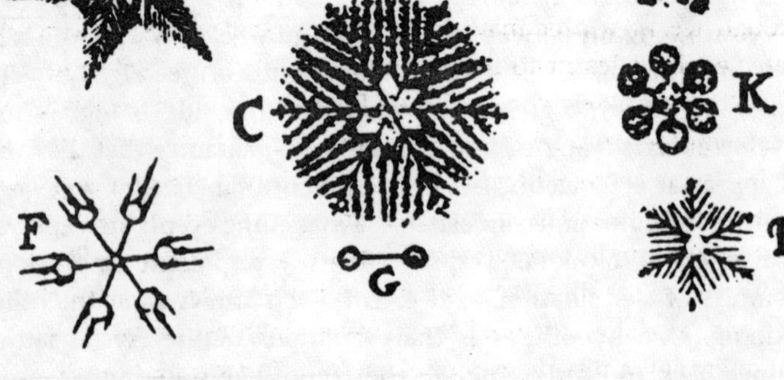
fig. I.
Pag. 19.
B
A
E
H
C
K
F
G
D

b

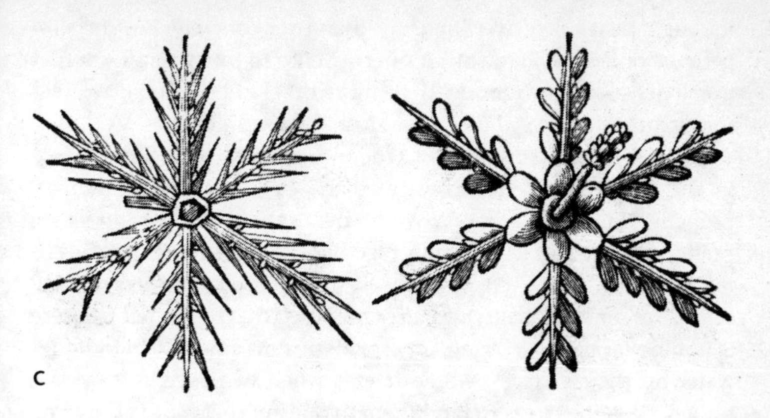

Fig. 7.3 Early impressions of snowflakes in the seventeenth century get progressively more detailed – from René Descartes (1637; *a*) to Erasmus Bartholin (1660; *b*) to Giovanni Domenico Cassini (1692; *c*) – due in part to the concomitant development of the microscope.

because all the branches grow close together at more or less the same rate. But for the present, the secret of the snowflakes endures.

Slide rules

Though ice in winter may bring chaos and death to treacherous roads, we have learnt to make the most of its slippery nature. One thing that has barely changed since Bruegel's day is the way we take to frozen ponds and lakes for recreation, and we'll positively pray for the big freeze as the skiing holiday or the winter Olympics approach.

The slipperiness of ice is ingrained in our experience. Yet again we must take a step back and make comparisons with other substances before we can see that water is doing something strange here. For ice is a solid, a crystal, is it not? Over how many other solids, then, can you travel downhill on a pair of planks at fifty miles an hour?

When one solid slides over another, the movement is inhibited by the rubbing of the two surfaces, manifested in the friction that opposes the motion. Some surfaces are slidier than others: it's easier to drag a heavy box over linoleum than over concrete. But the slipperiness of ice is remarkable: it offers twenty to thirty times less frictional resistance to motion than most other solids. Small wonder that so many cars wind up bonnet-first in snowdrifts.

Michael Faraday was drawn to the strange surface properties of ice in the 1840s, when he investigated the curiously complementary property of ice's stickiness. When brought into contact near freezing point, two pieces of ice will stick together. This is why it is possible to fashion snowballs from powdery ice (have you ever tried making sand-balls, or sugar-balls?). Faraday proposed that, if the temperature is not too far below freezing, ice crystals possess a thin layer of liquid water on their surfaces, which freezes when two such surfaces are brought together so as to bind them to one another. But at the same time another distinguished British scientist, James Thomson, proposed a rival theory for why ice sticks to itself. According to Thomson, a melted layer on the surface of ice is created by the very act of pushing the pieces together. In other words, the pressure lowers ice's melting point.

Legend has it that these two giants of nineteenth-century science battled for some years over the issue of how snowballs are formed; but the correspondence on the matter between Faraday and Thomson's brother William, also an eminent physicist, is nothing but good-natured: 'Ever my dear sir, yours truly' and so forth. It seems possible that the dispute was engineered by the Irish scientist John Tyndall, who seized on Faraday's idea in support of his own theory of how glaciers form from compacted snow – at the expense of Thomson. Said Tyndall loftily, 'I am far from denying the operation under proper circumstances of the *vera causa* to which Mr. Thomson refers, but I do not think it explains the facts.'[7]

Yet despite Tyndall's protestations, it was Thomson's explanation that got the upper hand. And in 1901 the British engineer Osborne Reynolds realized that if Thomson was right, his idea explained not only why ice sticks to ice but why our skis do not. The pressure of their surfaces against ice or snow would melt the surface and enable the skis to slide over a thin film of liquid water. In this way, ice would be self-lubricating.

But Thomson's theory can't be the full story either. It sounds all well and good until you put in some numbers. You can ski or skate quite merrily on ice at minus 10°C, but a quick calculation – which F.P. Bowden and T.P. Hughes of Cambridge University performed in

1939 – shows that to melt the top layer of ice this cold requires phenomenal pressures, far in excess of those generated under the feet of even the stoutest skier. So Bowden and Hughes proposed a new theory of skiing, in which local melting beneath the skis is the result of the heat produced by friction as they slide over the surface. This frictional heating is a familiar effect: it's why tyres burn when a vehicle skids on a dry road, and why you can warm up your limbs by rubbing them. Frictional heating makes skiing a self-sustaining process: once in motion, the skis generate enough heat to maintain the lubrication beneath. Bowden and Hughes verified their proposal in ingenious experiments with skis made of different materials: brass and a hard plastic called ebonite. Brass conducts heat more efficiently than ebonite, and so produces less frictional melting of ice because the frictional heat is carried away more rapidly from the surface by conduction through the metal. For this reason, the frictional force is greater for brass skis than for ebonite – they slide less well. Clearly, this is the sort of thing that ski manufacturers need to know. The fact that friction melts the surface of ice at all, however, is also in part a consequence of one of ice's peculiarities: it is a very poor conductor of heat, and so helps to ensure that the frictional heat at the interface is not quickly dissipated.

Yet Thomson was not wholly wrong: pressure *can* melt ice even when sliding is not occurring, provided that the pressure is large and the ice not too cold. This is why a heavily weighted wire draped over a block of ice will gradually cut through it like a cheese-cutter's wire. The incision freezes up again behind the wire, so that the wire seems to pass through the solid ice as if by magic. Because all of the weight is concentrated at the tiny area of contact between the wire and the ice surface, the pressure there is very high. And it now appears that even Faraday was at least partly right about ice too: a very thin film of liquid water does seem to exist at the surface of ice just below its melting point. Even though this surface layer of water may be very thin – several tens of millionths of a millimetre thick at a few degrees below freezing point – it can nevertheless still act as a lubricating layer.

Change of scene

When the Dance of Water, in which spread-eagled astronauts

floating in space grasp one another by the ankles, is arrested into the static, orderly Tableau of Ice, those outstretched arms and legs keep a good distance between the dancers. The ice crystal contains a lot of empty space.

This makes ice a most unusual solid. In most solid materials the molecular or atomic constituents are packed densely, to make the most of the attractive interactions between them. This means that there is little scope for rearrangements, and all that happens if the solid is squeezed is that the atoms get a little closer together. When the pressures get really big, however, a crystal may change its structure abruptly to some totally new atomic arrangement. This is a phase transition, taking place between two alternative solid forms of the same substance. For example, graphite – a form of pure carbon – is transformed to diamond, a denser form of carbon, when squeezed to around 200,000 times atmospheric pressure at high temperatures. This is how industrial diamonds are manufactured from graphite-like carbon. The difference between the crystal structures of graphite and diamond is profound: diamond possesses carbon atoms linked in a tetrahedral arrangement of bonds, whereas graphite contains sheets of carbon atoms linked into hexagonal rings, like chicken wire. You can see by comparing hard, brilliant diamond with soft, black graphite that the arrangement of the atoms, rather than the identity of the atoms themselves, can have a strong influence on a solid's properties.

Typically, solids may exhibit perhaps two or three different crystal structures at different pressures. But ice is much more versatile – so far, twelve different crystal structures have been identified when ice is compressed. When ice-I is squeezed, there is so much space to take up the strain that it faces many more options than other solids of how to arrange its component parts. And the hydrogen bonds that hold the crystal lattice together are weak enough to be bent, stretched and shortened substantially under pressure. Each phase transition in ice corresponds to a change of scene, a shift of the performers into a different tableau.

These transformations of the structure of ice augment the map of water's different states. Water World becomes a veritable patchwork of little nation states at the northern extremity of the border between Solidland and Liquidland (see Fig. 6.2). Because the pressures required for these phase transitions are so high, and the temperatures are often very low, it is a stiff challenge for experimenters to map out the terrain in any detail, and some boundaries are known only approximately. There is still a lot of mapping to do out there, although the conditions are cold and harsh, and deter explorers.

That ice-I under pressure rearranges to different structural forms was discovered in 1900 by G. Tammann, who squeezed ice to pressures of three and a half thousand atmospheres and saw it jolt twice into new states: ice-II and ice-III. But much of what we now know of the ice family stems from the work of Percy Bridgman, who pioneered high-pressure research at Harvard University during the first half of the twentieth century. In those early days, high-pressure work was a hazardous business: the risk of explosion of vessels under tremendous compression claimed several lives. Bridgman was a reclusive and complex character who committed suicide in 1961 following a diagnosis of bone cancer. But the genius of his experimental technique spoke for itself, and it won him the Nobel prize for physics in 1946. He revolutionized the field by devising new ways both to achieve high pressures and to measure them. In 1910 he constructed an apparatus that could achieve compressions of over 20,000 atmospheres, equivalent to the mass of 40,000 elephants concentrated over one square metre. Bridgman used this instrument in 1911 to map a large part of the phase diagram of ice, identifying five different structural forms – including the two that Tammann had seen earlier.

One of these states, ice-VI, seemed to remain solid up to at least 80°C. It became known as 'hot ice', and some excited business entrepreneurs wondered if there might be money in it. Bridgman received a letter from the manager of a fruit-shipping company who sensed a market opportunity:

> I have read a brief note in the newspaper regarding your production of a solidified form of hot water. Might we inquire if in your opinion this process has commercial possibilities for use in transit in a manner similar to ice to protect fruit and vegetables . . . against cold instead of heat?[8]

But sadly there was no future in it, since ice-VI remains stable only while a pressure of at least six and a half thousand atmospheres is piled on. It melts immediately if the pressure is released.

Ice-VI transforms to ice-VII at around 22,000 atmospheres, and ice-VII is the *really* hot ice: you can heat it to well above 100°C without melting it, provided that the pressure is stepped up too. So could a seed of ice-VII ever solidify the oceans, like Kurt Vonnegut's ice-nine? No, we're quite safe from that fate: again, release the pressure and ice-VII melts.[9]

All these forms of ice contain the tetrahedral bonding arrangement of ice-I, but in various distorted or reorganized forms. In general the network of hydrogen bonds gets bent, like a garden trellis contorting under the weight of a honeysuckle bush, so as to fill up some of the empty space and give the crystal a greater density than ice-I. In ice-V, for example, some water molecules have to bring together their outstretched arms or legs to make angles of just 84°, while others splay out to an uncomfortable 135°. So, unlike the Tableau of Ice-I, the players in the Ice-V Tableau are no longer identical but have several different parts.

But the very-high-pressure phases of ice – ice-VI, VII and VIII – have a density almost twice as great as ice-I – there's twice as many performers packed into the same volume of space. No amount of bending or splaying will fill this much empty space. Instead, these tableaux are dramatically innovative. The performers link hands and ankles into not one but *two* networks, each threading its way through the other so that they are intimately entwined without being linked by any hand-holds.

To visualize what this means, I'll set aside for a moment the spread-eagled tetrahedral dancers and consider a simpler formation, in which dancers with their feet firmly on the ground can forge links to just two neighbours each, joining hands with arms outstretched to make an angle of 120°. In the crystallized tableau, they make up groups of six, joined in rings (Fig. 7.4*a*). The instruction that all arms be outstretched and at angles of 120° keeps the individuals at arm's length, if they are to remain linked together. So we end up with rings with empty space in the middle, analogous to the 'wasted space' in ice-I.

If the area of the dance floor is reduced, say by a caretaker positioning chairs around the edge, the performers have to move closer together. This is like squeezing a crystal. But there's not a lot of scope for getting closer, if the arms are to remain outstretched and the hand-clasps unbroken. Evidently, some of the hand-holds will get bent.

If the area of the floor space is drastically reduced, there is an option that allows the dancers to keep to their rings while reducing the amount of space they occupy almost by half: the rings can

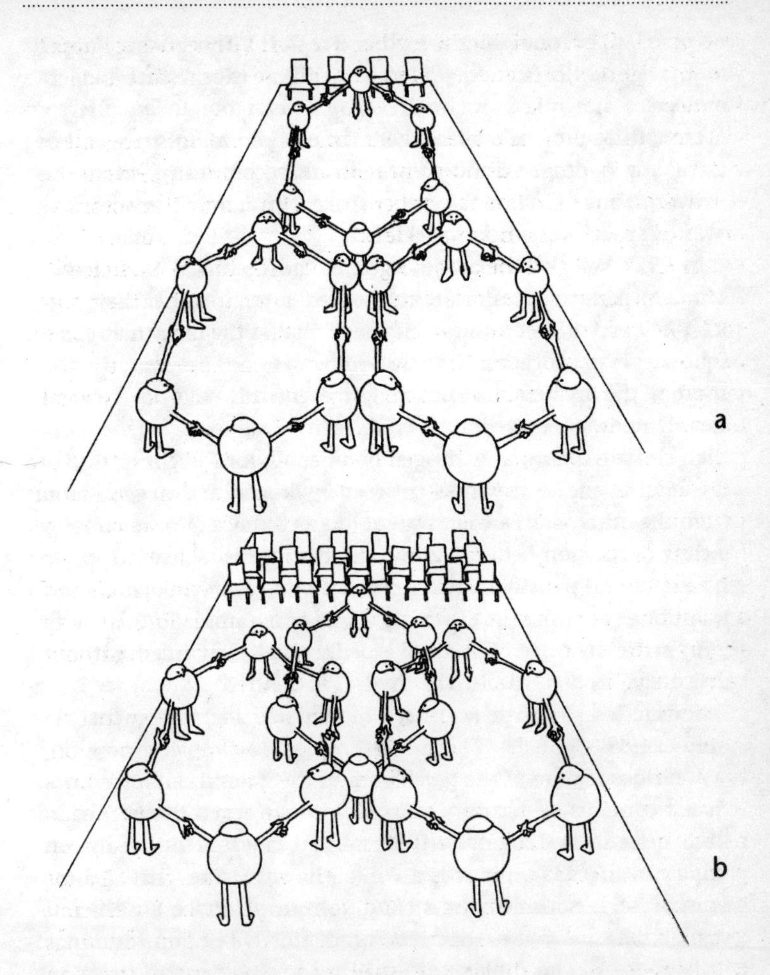

Fig. 7.4 How ices get denser at very high pressures. I represent the three-dimensional structure of ice here with a two-dimensional analogue in which dancers join hands to form six-membered rings. The structure in *a* is the analogue of ice-I. When the dancers are crowded into a smaller volume by chairs placed around the periphery, the rings edge closer together. But they can't come very much closer before they start to overlap. A more efficient use of space is to rearrange the positions of the rings drastically so that they interlock (*b*) – this enables the available space to be filled with roughly twice as many dancers. Notice that it causes the dancers to have some near neighbours to which they are not linked. The interlocked structure is the analogue of ice-VI, ice-VII and ice-VIII.

become looped, one inside the other (Fig. 7.4*b*). This is what happens to the ice lattice in ice-VI, VII and VIII – except that here the molecules are linked not in hexagonal rings but in an interconnected, three-dimensional network. To economize on space without sacrificing hydrogen bonds, the network becomes two intertwined networks, one filling up the empty space within the other so that the 'wasted space' is virtually halved.

In 1972 Wilfried Holzapfel, then at the Technical University of Munich, performed calculations to predict what would happen to ice-VII at very high pressures. He found that as the oxygen atoms get squeezed closer together and the hydrogen bonds between them get shorter, the hydrogen atoms might eventually take up residence exactly midway between the two oxygens. This arrangement cannot then contain a normal hydrogen bond at all, for there is nothing to distinguish one of the links between hydrogen and oxygen atoms from the other: both are equivalent. It's as though there is no longer a clear distinction between arms and legs in this tableau. Hydrogen here is bound equally to two oxygen atoms, even though chemists traditionally think of it as strictly a one-bond atom. So this kind of 'symmetric' hydrogen bond is decidedly odd – chemists call it, more delicately, 'non-classical'. The 'hydrogen-centred' state of ice corresponds to ice-X, and its unusual structure was seen for the first time only in 1998.

A curious feature of the patchwork of Ice States in Water World is that it contains no territory marked 'ice-four', even though there is ice-five, ice-six and so on. Ice-IV does exist – it was identified by Percy Bridgman in 1935 – but it is a will-o'-the-wisp, a tentative, ghostly form of ice. It is formed under conditions in which ice-V is the more stable form, and so it cannot persist indefinitely but transforms gradually to ice-V. The Tableau of Ice-IV is too wobbly, you might say, and its performers slowly shift their positions to adopt the more steady Ice-V arrangement. This means that ice-IV is hard to study experimentally, and its structure was not determined until 1981. States of matter that are only provisionally stable are said to be *metastable*, and they can be compared with the state of mountain lakes. All that water high up in the mountain would be more stable – it would have less gravitational energy – if it were to flow down the mountainside to sea level. But it does not do that spontaneously because it does not have the energy needed to get over the edge of the lake. True, the energy would be more than recouped by the water flowing down the mountainside; but nonetheless the water has to get the energy from somewhere in the first place before it can start flowing. So the water in the mountain lake remains in a less stable state unless the energy can be found to overcome the barrier to its relocation. Similarly, metastable states persist only so long as an energy barrier prevents their conversion to the more stable state.

New states are still being found out in the frozen north of Water World. In 1998 a twelfth structure of ice – ice-XII – was reported by researchers at University College, London, and the University of Göttingen. Like ice-IV, this phase appears at pressures and temperatures for which ice-V is usually observed, and it too is metastable. It would not be at all surprising if still more forms of ice are discovered as it becomes possible to extend the search down to lower temperatures and out to higher pressures. There is no telling what lies out there.

Chilled glasses

All of the ice tableaux we've seen so far are well disciplined: they have crystalline, orderly structures. But the performers can be tricked into adopting dishevelled displays too, by freezing water so suddenly that wild dancers are stopped right in their tracks before they have a chance to take up regimented positions. In 1936 E.F. Burton and W.F. Oliver in Toronto made disorderly ice by condensing water vapour onto a copper pipe cooled below minus 140°C. Disordered solids are said to be glassy or amorphous.

If amorphous ice is warmed up above –140°C, its molecules can just about start to move again. The result is an extremely peculiar form of liquid water, which is about as viscous as asphalt. In such frigid conditions this viscous water exists in a highly precarious state: once the temperature exceeds –120°C the molecules have enough mobility to rearrange themselves into the crystalline ice lattice, and the stuff freezes to regular old ice-I.

It's no surprise that water can form a glass: most other crystal-forming substances can do it if melted and then cooled rapidly.[10] But water has a *second* glassy form too. In 1984, a group of researchers at the National Research Council of Canada in Ottawa found that when ice-I is squeezed to 10,000 atmospheres at a temperature of –196°C, it 'melts' not into a liquid but into an amorphous ice with a greater density than that of Burton and Oliver's glassy ice. The greater density indicates that the water molecules in this material are packed together more closely: the structure, while still disorderly, is different. This denser form is called high-density amorphous ice, or HDA, while the less-dense form is low-density amorphous ice, LDA.

The molecular structure of HDA looks like that of liquid water compressed to around 1000 atmospheres, which is about the pressure experienced by sea water at the bottom of the deepest ocean trench, the Marianas Trench in the Pacific Ocean. But in HDA this structure is no longer a dynamic one: it is frozen, a tableau of the compressed liquid.

Glasses are not truly stable phases – like ice-IV and ice-XII, they are only provisionally stable, and the only thing that prevents them from transforming into a crystalline form is that the molecules are so cold they can barely move.

Into no man's land

If it is news to you that water can freeze (to ice-VII) at 100°C, here is another revelation: water can be cooled at least 38°C *below zero* without freezing to ice. Liquid water below its freezing point is said to be *supercooled*.

Supercooled water is an interloper: in the phase map of Water World it is a liquid in Solidland, a liquid that's strayed over the border. This means that supercooled water is another metastable state, persisting in conditions that favour a different state. A metastable supercooled liquid can be compared to Belgian Flanders. You can cross the border from France and not notice it: on the Belgian side everyone still speaks French, and all the town names and street signs are in that language. The customs are typically French. You could say – although the Belgians might not approve – that there is a metastable region of France extending into Belgium, so that a Frenchman can wander in this region and not be perceived as a foreigner. So too in Water World can a Liquidlander wander into Solidland and not be forced to adopt the Solid way of life.

Supercooled water probably sounds like exotic stuff, but you see it often enough: many clouds, including altocumulus, altostratus, cirrocumulus, cirrostratus and cirrus, are composed at least partly of supercooled water. Air temperatures at altitudes of greater than two kilometres are usually low enough to freeze water; and yet clouds consisting of tiny water droplets can be found up to altitudes of around thirteen kilometres. The water droplets in these clouds are typically between about 5 and 200 micrometres across. Just as it gets harder to muster a company sports team the smaller your workforce gets, so it gets more and more difficult to grow ice crystals in progressively tinier droplets of water. The chances of enough water molecules coming together at random in the right orientation to form a 'nucleus' of ice become ever smaller. Cloud droplets generally have to be colder than about –15°C before they'll freeze, and even then a particle of mineral dust is usually needed to 'seed' the ice (see page 58). Liquid droplets supercooled to as much as –40°C have been observed in some high-altitude cirrus clouds. Because the way in which clouds absorb and reflect the Sun's radiation depends critically on whether the particles are ice or liquid water, the possibility of supercooling has important consequences for the climate of the planet.

Seed particles such as dust are the first places where freezing happens when a liquid is cooled below its freezing point. So to supercool water, you need to be assiduous in removing such impurities. Even then, supercooled water is poised on the brink of freezing: below freezing point, there is a chance that a nucleus will form and ice will start to grow even without a seed. Even a slight agitation of the liquid might trigger crystallization, as American chemist Irving Langmuir related in 1943:

> The formation of crystals on cooling a liquid involves the formation of nuclei or crystallization centres . . . The spontaneous formation of these nuclei often depends upon chance. At a camp at Lake George, in winter, I have often found that a pail of water is unfrozen in the morning after being in a room far below freezing point, but it suddenly turns to slush upon being lifted from the floor.[11]

The lower the temperature, the more likely it is that ice will nucleate. Below about –38°C it becomes practically impossible to stop a beaker of supercooled water freezing by any means.[12]

This is a great shame. For if we could keep water from freezing until just a few degrees lower, we might see the supercooled liquid do something very strange indeed. In the 1970s, American physical chemist Austen Angell and his coworkers began to unearth evidence that at around –45°C it meets with catastrophe. It now appears that this dramatic behaviour could hold profound insights into water's fundamental character – as I guess breakdowns often do.

Catastrophe

The drastic behaviour of supercooled water shows up in its so-called 'response functions', which are measures of its response to change. The response function of a substance is rather like the squishiness of a rubber bung. It is easier to squeeze the bung when it is warm than when cold. Physicists would say that the response function called the compressibility – loosely speaking, the amount by which the bung shrinks for a particular degree of squeezing – increases with temperature in this case. Other response functions include such things as the heat capacity (the change in temperature occasioned by an input of heat) and the thermal expansivity (the change in volume caused by a change in temperature). In effect, they characterize a material's *sensitivity to change.*

Typically, the response functions of a liquid decrease in a gradual, dignified manner as the temperature is lowered. But supercooled water is much more excitable. As it is cooled towards the experimental limit of –38°C the response functions begin to increase dramatically. If the trend continued down to still lower temperatures, said Angell, it seemed that the responses might blow up out of control at –45°C. An unfettered increase in the heat capacity would mean that you could pour as much heat into supercooled water as you like, and it wouldn't get a degree warmer. For the compressibility, it means that the liquid becomes absurdly soft and squishy: the smallest increase in pressure makes it shrink to nothing. If this 'catastrophe temperature' really exists, it corresponds to a point at which supercooled water is hypersensitive to change. But frustratingly, we can't test whether this is really so, because nature draws a veil over this unseemly behaviour once we reach –38°C. It's reminiscent of the event horizons that hide from view the crazy, singular behaviour at the eye of a black hole, something that astrophysicists call 'cosmic censorship'.

Two waters, no ice

Liquid water therefore has a curious, metastable no man's land in which strange things seem to happen; but we can't get there (Fig. 7.5). In which case, you might say, why should we care? Well, it now appears that Angell's 'catastrophe' could be the McGuffin of water's story, like Frodo Baggins's ring in *The Lord of the Rings* – a seemingly trivial curiosity that in fact holds tremendous import for the broader picture. To explain this behaviour, we may end up having to conclude that there is not one but two forms of liquid water.

This is just one of the interpretations currently on offer. More prosaically, it may be that the catastrophe is illusory – that the response functions aren't really blowing up out of control at all. If we could only supercool water those few degrees further, perhaps we'd find that it doesn't really do anything outrageous but just settles down again. In other words, maybe the no man's land hides merely a wobble, not a singularity.

Well, that's possible, but it would be a bit of an anticlimax. The alternative, which is now gaining support from experiments, is altogether more enticing. In 1992, physicist Gene Stanley, his graduate student Peter Poole and their coworkers at Boston University set out to investigate supercooled water through computer simulation. They

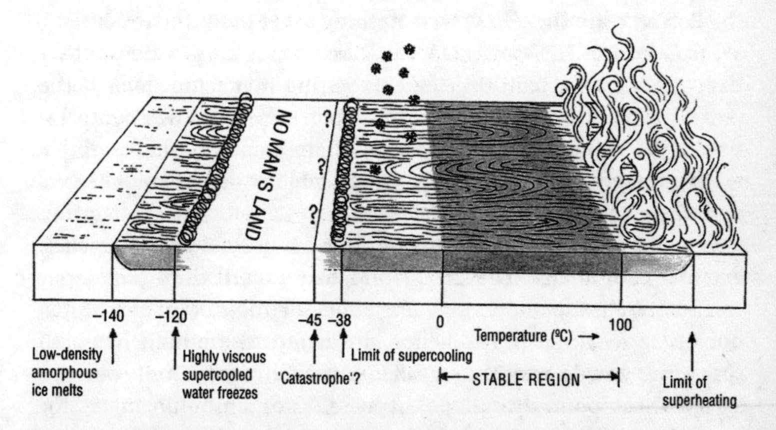

Fig. 7.5 Liquid water exists over a far wider temperature range than we normally perceive. It is stable between its boiling point (100°C at atmospheric pressure) and its freezing point (0°C) – shown here in grey. Outside of this range, liquid water can remain provisionally stable: if nucleation of the gas and solid states, respectively, can be avoided, it can be superheated above boiling point, and supercooled below freezing. But it becomes impossible to prevent nucleation of ice in bulk water below −38°C. If plunged to below −140°C, liquid water is 'arrested' into a glassy form of ice. When this ice is warmed, it first melts to a highly viscous form of water before freezing to crystalline ice above −120°C. So there is a no man's land between −120 and −38°C within which it is all but impossible to make water in a liquid form.

mapped out the boundaries of their virtual Water World by simulating water over a range of low temperatures and different pressures.

At pressures of several thousand atmospheres and temperatures of less than –75°C, they stumbled across an unexpected border. There were *two* kinds of liquid water out there, which could be interconverted in a phase transition. And the boundary between them came to an abrupt end at a critical point, like that between the normal liquid and the gas. In other words, water seemed to have two critical points, and two liquid states. There may be a hidden border between two ancient kingdoms of Liquidland, now lying far within Solidland – a border that ends in the ancient, ruined city of Kritikalov, whose seismic instability, like that of thriving Kritikala, can still be felt in the surrounding region (Fig. 7.6).

The two kinds of liquid were so cold that their molecules were pretty much immobile – they had become amorphous ices. These, the Boston team suggested, were nothing other than the well-known amorphous ices HDA and LDA. But glassy solids are just liquids that have ground to a halt. Because this seizing is not inevitable – the transition from liquid to glass can be postponed to lower temperatures if the liquid is cooled very slowly – one could imagine finding a way to preserve the two liquids, which might then separate like oil and water.

Yet who cares about a putative critical point located at such extreme coordinates in Water World that experiments can't even reach it? Well, the important thing about a critical point is that it is not just a localized phenomenon; it disturbs the terrain of phase space over a wide area, the so-called 'critical region'. This is because at a critical point things go haywire – response functions, for example, become infinite. And just as Mount Everest does not thrust abruptly skywards from a flat plain but is surrounded by the rugged Himalayas for hundreds of miles, so too would the upheaval at a critical point between two forms of supercooled water at high pressure and very low temperature have consequences for the behaviour of water under less extreme conditions.

This led the Boston researchers to propose that Angell's catastrophe temperature of –45°C might not be a point of genuine breakdown at all, but just a ghost of the real catastrophe at the second critical point – a topographic premonition of nearby mountains.

The idea that a single substance can possess two different liquid

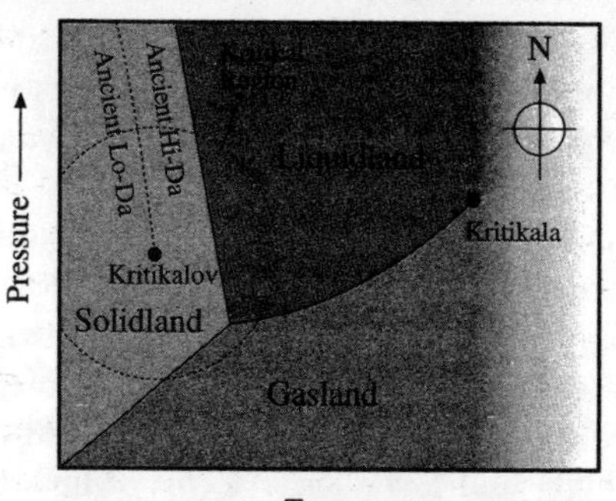

Fig. 7.6 Are there two kinds of liquid water? Computer simulations imply that at low temperatures and high pressures a phase boundary separates two kinds of liquid water with different densities. This boundary ends in a second critical point. In terms of Water World, it is as if an ancient and now barely perceptible border separates two old kingdoms Hi-Da and Lo-Da, now subsumed into Solidland – like Thrace and Macedonia in modern Greece. Repercussions of the seismic instability at the ancient city of Kritikalov (where Hi-Da and Lo-Da once merged) extend far and wide through the Kritikal Region.

states (even if they're exotic, supercooled ones) is a challenging one, and would be hard to credit for any substance that did not share water's unusual structure.[13] If a liquid is crudely like a tank of ball bearings, gently shaken to avoid any orderly packing, it will have the same density everywhere. Disorder is disorder, and that's that. But water molecules are not like ball bearings; they are our elegant dancers, linking up outstretched arm to outstretched ankle in the random hydrogen-bonded network. Computer simulations imply that the high-density state of supercooled, pressurized water contains a highly defective network with lots of strained or broken bonds. This allows some molecules to stray into the empty space between hydrogen bonds. In the low-density state, the network retains much of its integrity, full of the empty space characteristic of normal liquid water. What we're seeing in this split personality is the same aspect of water that leads to its density anomaly at 4°C: the struggle between a desire for compactness and an inclination for optimal bonding between molecules.

Since water is notoriously hard to simulate accurately on a

computer, many researchers won't believe in this story unless we can see the two liquids for real. If we want to know what Water World really looks like in the supercooled region, we need to go there. But the interesting region is a no man's land: however you approach it, the water always freezes one way or another.

In 1998 Osamu Mishima from the National Institute for Research in Inorganic Materials in Japan, along with Gene Stanley at Boston, found a way to cheat the barriers and catch a glimpse into no man's land. They watched tiny droplets of pressurized ice melt as the pressure was increased or decreased. In theory, this melting should generate the distinct high- and low-density forms of liquid water if the coordinates of pressure and temperature are in the right region of the phase diagram. In practice, though, the melted ice refreezes instantly into a different form at such low temperatures. But a signature of a switch-over from one kind of liquid to another can still be spotted even if the liquids don't stick around, because it leaves an imprint on the topography of Water World.

Mishima and Stanley were able to map out the full three-dimensional phase diagram of water in this difficult region: not a two-dimensional map like that in Fig. 7.6, but the version with the height contours included. In phase space, this third dimension corresponds to the density of the substance. Then the two-dimensional map becomes a three-dimensional relief model: a surface. If there is a metastable transition between two liquids at high pressure and low temperature, this would show up as a break in the slope of the surface, where a kind of cliff separates the high-density liquid from the low-density liquid. Mishima and Stanley were able to see a hint of this cliff by tracing out the contours of the surface and interpolating the lie of the land between separate transects. It's tantalizing support for the idea, but stops short of providing incontrovertible proof that liquid water does indeed show two faces in the depths of no man's land.

A winter's tale

If you think that all of this seems terribly esoteric – that, distant clouds notwithstanding, supercooled water is hardly the kind of stuff you encounter every day – that is because you are not an Antarctic fish. The water temperature at the surface of the Antarctic Ocean typically lies between –1.5 and –1.9°C, while fish blood freezes at about a degree above this. Yet the blood of fish that dwell in these waters is not frozen, but is in a permanently supercooled state. If fish conducted scientific research, you might expect them to set up whole institutes devoted to studying supercooled liquids, since their very existence depends on this precarious metastable state.

But how is it that the Antarctic Ocean is itself below freezing point? Is the ocean supercooled too? No it isn't, for the ocean waters are full of salt. Dissolved salt (in fact any dissolved substance or 'solute') lowers water's freezing point. This is precisely why salt is scattered on roads in winter to prevent the formation of ice. It is for the same reason that blood freezes at around –0.5°C, not at 0°C.

The freezing point of the salty waters of the Antarctic Ocean is low enough that at –1.5°C its liquid state is still more stable than ice. But fish blood *is* supercooled at this temperature. And fish, like many other organisms, have evolved a clever means of ensuring that it stays that way.

Natural antifreeze

You don't have to be a polar explorer to know that freezing is fatal; but ice kills in several ways. The danger posed by the freezing of blood and body fluids (such as the watery cytoplasm that fills living cells) is not just that the body's hydraulics seize up, like frozen pipes in a central heating system or frozen diesel in a lorry. When cells freeze, life is not simply 'put on ice' – it is generally disrupted irreparably. Cells are killed by freezing as a result of all sorts of irreversible disasters. The protein molecules that orchestrate the cell's chemical processes unwind and lose their ability to function. Cell membranes are slashed apart by the sharp edges of tiny ice crystals; and the membranes themselves become leaky as their constituent molecules clump together and solidify like cooling fat in a pan. And if the extracellular fluids of the body, like blood, begin to freeze, the cells themselves run the risk of dehydration as their water is sucked out. When a solution freezes, the ice crystals reject the solute, and so the solute's concentration in the remaining water increases (for this reason, the saltiness of sea water increases when part of it freezes into sea ice). The increase in solute concentration when blood plasma freezes sets up a so-called osmotic pressure (explained on page 220) which sucks water out of cells and leaves them dehydrated.

For this reason, most living organisms must avoid getting frozen at all costs. There are two general ways of doing this, which we might call the stable and metastable strategies. The first is a safe bet, but doesn't work much below freezing point. The second confers protection down to lower temperatures, but precariously.

The stable strategy involves depressing the freezing point of body fluids by cramming them with solutes. When temperatures drop, many insects manufacture compounds called cryoprotectants in their blood and cell fluids. Natural cryoprotectants are usually sugars, such as fructose, or organic molecules called polyalcohols, which include glycerol and ethylene glycol. Typically the insect starts accumulating these substances in the autumn, when falling average daily temperatures signal the approach of winter. The creature later clears the cryoprotectants from its tissues and bloodstream in the spring. The antifreeze used in car radiators works on just the same principle, and indeed usually contains one of the same chemical compounds: ethylene glycol. You'll typically use a far higher proportion of antifreeze in your car (mixed with water in a ratio of between 1:2 and 1:1) than insects could tolerate in their blood; but nonetheless cryoprotectants can constitute an appreciable fraction of the insect's weight – up to one fifth at the peak of production. Such high levels of glycerol allow the Arctic willow gall insect to withstand temperatures of –66°C, which seems to be a world record for freeze avoidance. The Himalayan midge even stays up and about its business at –16°C, when most other creatures have become immobile with their metabolism all but shut down.

But this sort of extreme supercooling does not rely solely on cryoprotectants. Insects already have a head start on larger creatures, in that their small size helps to suppress ice formation even in the absence of any biochemical defence: the chance of an ice particle appearing in a small volume of water is smaller than it is in a large volume. The chance that ice formation might be seeded by contact with ice crystals in the external environment – always a danger for creatures with permeable skins – is negated by the waxy outer coating of insects. And they minimize the chance of ice congealing on small internal particles such as food, ingested dust or bacteria in their digestive system by fasting as the cold season approaches and thoroughly evacuating their gut. As winter begins to bite, the insect becomes fully prepared to weather its ravages through this systematic process of 'cold hardening': tucked away in a dry hibernation site, gut empty and body full of antifreeze.

Cold-water fish do not have the option of most of these strategies. Yet since winter at high latitudes can drive sea water colder than the freezing point of fish blood, they run the risk of freezing solid even while the surrounding water remains liquid. One survival strategy is simply to take a winter break: to migrate to warmer climes. Some fish move instead to deeper waters, where there is no risk of sea ice to trigger freezing in their supercooled bodies. But many cold-water fish brave the chilly season by adopting the second freeze-avoidance strategy: going metastable. They manufacture protein molecules (or glycoproteins, which have sugar molecules attached) that suppress ice nucleation and so maintain the blood in a supercooled state. These proteins are called antifreeze proteins. Cold-water fish are not alone in making antifreeze proteins: insects and some other land animals do it too.

Although in principle effective down to –10°C, antifreeze proteins stave off freezing of the blood of Antarctic fish until just below –2°C. That narrow margin is just adequate, for the sea water never gets colder than –1.9°C, the temperature at which it freezes. It is not yet known how these proteins work, but they probably stop nascent ice crystals from growing larger by sticking to their surface. Some fish antifreeze proteins have surfaces studded with regularly spaced chemical groups that can form hydrogen bonds. The position and spacing of these groups match up very closely with the water molecules at the surface of a normal ice crystal. So the proteins seem designed to lock into place on the ice surface through hydrogen bonding, like a multi-pin plug fitting its socket. If a tiny crystal of ice forms in the fish's blood, the antifreeze proteins dock onto the crystal's surface and block the access of other water molecules.

Let it freeze

While freezing-point depression and supercooling are undoubtedly effective mechanisms for winter survival, they are not the only options. Some animals don't resist freezing but instead manipulate it to their advantage. This is known as freeze-tolerance.

To acquire a tolerance to freezing, an animal has to ward off each of the hazards that ice poses. Cells can never survive if their internal fluid freezes, and so any ice must be kept on the outside, in the fluid between cells. The cell membranes must be protected against the leakage that occurs if they become stiff with cold, and also against the ragged teeth of ice crystals. And dehydration must be avoided by ensuring that, as freezing proceeds outside the cells, the concentration of solutes in the remaining fluid does not become much greater than that within the cells, which would trigger extraction of water by osmosis.

To ensure that their cells and tissues are not cut to shreds by ice crystals, freeze-tolerant creatures do something unexpected: they start to manufacture proteins that encourage the formation of ice. With lots of little ice crystals rather than a few big ones, the result is a mush, more like driven snow than prickly icicles. The trick here is to trigger ice formation as soon as possible after the temperature falls below the freezing point: to *avoid* supercooling, since once ice crystals appear in supercooled water they grow rapidly. Ice-inducing proteins are no better understood than antifreeze proteins; but it may be that they act as seeds for crystal growth by binding water molecules to their surfaces in an ice-like arrangement.

Ice formation must be promoted only outside cells; inside, it's bad news. And even then, if *all* the fluid outside a cell freezes, the cell risks dehydration. So both the location and extent of ice growth must be carefully controlled for effective freeze-tolerance – something that insects and larger animals achieve by generating cryoprotectants. Outside the cell these solutes bind to some water molecules and so decrease the amount that can be frozen; inside the cell they lower the freezing point.

The delicate business of freeze-tolerance is demonstrated in a most dramatic fashion by land-hibernating frogs such as the wood frog, the grey tree frog, the chorus frog and the spring peeper. Being so large (relative to insects) and possessing water-permeable skin that is inevitably going to come into contact with ice in winter, the frogs have no chance of escaping freezing. So they have learnt to live with it. In the depths of winter they appear to become frozen solid – although in fact it is only the extracellular fluid, sixty-five per cent of the total, that turns to ice. Whereas insects spend much of the autumn preparing for the big freeze, freeze-tolerant frogs do nothing until ice actually begins to form inside them. Then their bodies go into overdrive, generating the cryoprotectant glucose at a furious rate and doubling the heartbeat rate to pump it quickly around the body. Within hours, the blood glucose soars to a level seen in a severe diabetic attack, 200 times greater than normal. The cell fluid becomes syrupy and its freezing point drops. This increase in cell sugar also helps to counteract the osmotic pressure building up in the extracellular fluid as freezing proceeds, and the concentration of solute inside and out reaches a stable balance before dehydration sets in. As the frog sets solid, its heart stops beating and it crouches in a state of suspended animation.

The recovery from thawing is equally abrupt. Within an hour the heart starts beating again, even before the creature's extremities have thawed. That way, no part of the frog's body is released from the deep freeze before the heart is ready to sustain its tissues with oxygen.

Plants have mechanisms for both freeze-avoidance and freeze-tolerance, but rather little is known about these. Like some animals, plants generally commence a process of cold hardening as winter approaches. Their survival prospects may depend on whether or not they have time to acclimatize properly. Gardeners are all too aware that even a light frost can be fatal when it appears of a sudden in mild weather; but if the temperature falls slowly enough, plants can be remarkably hardy. Many temperate woody plants will die if temperatures fall below 0°C in summer, but the 'killing temperature' falls to –30°C by late autumn. In Siberia, it is not unusual for conifers to be exposed to winter temperatures of –65°C; and many northern conifers, including the white cedar, red pine, black and white spruce and the larch, can brave temperatures below –80°C. Such extremes require freeze tolerance, since there's no avoiding it when the environment gets that cold.

In many plants cold hardening involves the production of a hormone called abcisic acid, which appears to slow down growth and to make the cell membranes more permeable to water – an important property for preventing the freezing of cell water. The extracellular water in plants freezes first, and the resulting osmotic dehydration of cells means that the cell fluids become more concentrated in solutes and their freezing point decreases. Plants do not produce antifreeze as such; but typically sap can be supercooled to around –2°C because of the sugar it already contains. The cell fluid of some plants, however, can be supercooled to an astonishing –47°C.

Stored on ice?

For humans, the superficial allure of an ability to survive freezing is the prospect of cryopreservation: keeping whole bodies in a state of arrested animation for indefinite periods, in the hope that it might prove feasible in the future to revitalize them. Already dead bodies lie frozen in liquid nitrogen, invested with the anticipation that one day medical science will find a way to grant them new life, even immortality. Several US companies provide this service at extravagant cost – with a discount, I understand, if you have just your head frozen. Yet at liquid-nitrogen temperatures almost all of the body's cells are irreparably damaged, and body organs are susceptible to cracking. So the only slim hope of 'survival' is for one of the few surviving cells (if any) to be cloned – which means that saving your head rather than, say, your large intestine is a purely aesthetic choice.

More practical, and certainly more socially useful, is the possibility of using the tricks of freeze tolerance to preserve organs for medical use. Some researchers are investigating the use of the sugar trehalose, employed by some freeze-tolerant insects, to preserve frozen human pancreatic cells for treating diabetes. When they are cooled to 5°C, the membranes of these cells become leaky enough for trehalose to seep into them. The trehalose-loaded cells will still function after being frozen in liquid nitrogen and then thawed. There are also attempts to use a combination of fish antifreeze proteins and the cryoprotectant glycerol to render livers freeze-tolerant. Whether whole people could ever survive the ice-bound rigours of the wood frog, however, remains highly doubtful.

Freeze avoidance has pay-offs too. Supercooled solutions might be used to store enzyme proteins and blood for pharmaceutical and medical applications. Solutions of proteins don't last long on the shelf – they tend to lose their biological function over a period of a week or so. Cooling slows down the deterioration, but not very significantly. Freezing the solution is counter-productive – when it is thawed, its biological activity falls to almost nothing in a few days. Yet storing enzymes in supercooled emulsions at –12°C allows them to retain their activity completely for over a year. The emulsions contain tiny droplets of water suspended in oil which, by virtue of their small size, can remain supercooled virtually indefinitely. A similar trick works for whole cells, which can typically survive supercooling to at least –23°C. And red blood cells continue to function after being supercooled by as much as –38°C.

So while there's no avoiding the fact that ice is deathly stuff, life goes on below zero. And water below freezing, whether solid or fluid, surely retains surprises aplenty for the intrepid traveller in Water

World. Refracted through the prism of metaphor, the words of Hermann von Helmholtz are delightfully prophetic:

> While some content themselves with admiring from afar the dazzling adornment which the pure, luminous masses of snowy peaks . . . lend to the landscape, others more boldly penetrate into the strange world, willingly subjecting themselves to the most extreme degrees of exertion and danger, if only they may fill themselves with the aspect of its sublimity.[14]

PART III

Life's Matrix

8 The Real Elixir

The biology of water

A seed of abundant waters, he comes out of the Ocean.

The *Rig Veda*

He told me he had the sea in his blood, and believe me you can see where it gets in.

Spike Milligan, *A Dustbin of Milligan*

Yesterday it rained, and now the garden will go crazy. It is early summer, and for weeks the ground has been parched. I have feared for the roses, tended them with an irresponsible hosepipe. But yesterday the skies growled, the lightning flashed, and the heavens opened. In my garden, precious lives are no longer at stake; water has returned.

Like any gardener, I overdramatize the situation. Life here in Northern Europe is amply watered; only my arbitrary floral preferences are at risk. Life is tenacious enough that it will make do with absurdly limited, sporadic supplies of water if that is all there is to be had. In parlous circumstances – in the deserts and the dry, rocky lands of the planet – living organisms become mechanisms for sequestering water, little enclaves of the fluid stuff in a parched world. For whether you are a scorpion or a cucumber, a Salmonella bacterium or a bull elephant, water is literally your lifeblood, give or take a few additives.

You don't have to be a biochemist or a science-fiction enthusiast to be familiar with the notion that all life on Earth is based on carbon – that the 'biomolecules' that make up living organisms are constructed largely around a backbone of carbon atoms. This idea is actually a bit of a myth, as atoms of other elements – particularly nitrogen, oxygen and phosphorus – are essential parts of those molecular frameworks too. But the point one tries to convey with talk of 'carbon-based life forms' is that carbon has some very special chemical properties, such as its ability to form strong, stable bonds with itself and to link up into the long chains that feature in the molecules of life.

Carbon is indeed a very remarkable element; but I don't think I shall detract too much from its glamour if I suggest that it is not exactly the ur-substance of life that it is sometimes jacked up to be. If life (one has to say 'as we know it', and live with the cliché) is ever going to get started on any planet, carbon is surely needed[1] – but even before that, life needs flowing water.

Water is life's true and unique medium. Without water, life simply cannot be sustained. It is the fluid that lubricates the workings of the cell, transporting the materials and molecular machinery from one place to another and facilitating the chemical reactions that keep us going. Water is sustenance and cleansing fluid, bearing nutrients to where they are needed and taking away wastes. It is even a structural agent in plants, as we will see – it enables flowers to hold up their heads to the Sun. No wonder that we will quickly die without it, for we need to consume at least a litre a day for long-term health. Some cells can avoid death if their water is withdrawn; but then they shut down utterly until rehydrated.

And we tend to forget, landlubbers that we are, that until comparatively recently an aquatic environment was the sole milieu for life on Earth. The planet has apparently hosted living organisms for an astonishing 3.8 billion years of its 4.6-billion-year history, and yet colonization of the land began only around 450 million years ago. Our watery past is beautifully and strikingly illustrated in a mural of life's pageant in the Los Angeles County Museum of Natural History, designed by geologist and early-life expert William Schopf and rendered by artists Jerome and Elma Connolly. It is a 'Time Ribbon' that begins in fire and ends as an astronaut steps off the sandy-hued strip. But in between fire at one tip and earth at the other is the sea-green stuff that has coddled – in both senses – all of Earth's life almost since the Sun first shone.

Life at sea

That we live on land is, in the grander scheme of things, best regarded as an anomaly, even an eccentricity – albeit with sound evolutionary justification. The story of life on Earth is, if we retain a true sense of proportion, a story of life at sea, and it is there that we must begin in order to really appreciate what water means to life.

Just about anyone who has studied the matter now believes that life began in water. Anaximander of the Ionian school of classical Greece suggested that the first organisms were sea-bound, even going so far as to venture that humans developed from fish. Beyond the centrality of water, however, you'll be hard pushed to find a consensus today about life's origins. Some believe that the setting for the appearance of the first tentative players in life's cast was a warm pool or lagoon, replenished by the tides, where evaporation allowed complex organic molecules present at extremely low concentrations in the seas to be slowly concentrated until they could start to interact. Charles Darwin is commonly credited as the originator of this idea:

> It is often said that all the conditions for the first production of a living organism are now present, which could ever have been present. But if (and oh! what a big if!) we could conceive in some warm little pond, with all sorts of ammonia and phosphoric salts, light, heat, electricity, etc., present, that a protein compound was chemically formed ready to undergo still more complex changes . . .[2]

Other scientists think that it all took place in the heady brew of deep-sea hydrothermal vents, which are hot and murky fountains on the sea bed. Here sea water percolates through fissures in the ocean floor, is heated by an underlying body of magma, and bursts out through mineralized chimneys, bearing a rich load of inorganic minerals and gases, amongst them methane for building carbon compounds. Still others have been content to regard the wide oceans as a suitable medium for life's origins, enriched with the necessary simple organic molecules through the agency of chemical reactions between atmospheric gases stimulated by lightning discharges. 'A hot dilute soup' was how British biologist J.B.S. Haldane described this scenario.

First words

Whatever its origin, we do know where any theory of life's beginnings must lead, because the basis of all known life on Earth is the same. If we get conventional and neglect water, the two most important kinds of biomolecule in the cell are proteins and nucleic acids. In a nutshell – more of a caricature, actually – nucleic acids carry the instruction booklet for the cell, and the proteins read out the instructions. These two components – proteins and DNA – are interdependent, in the sense that proteins are put together largely according to the instructions on DNA, while the copying of DNA which is essential for cells to reproduce and proliferate relies on the mediation of proteins.

Crudely speaking, the information for one protein is encoded in a well-defined stretch of DNA called a gene. But proteins are not constructed directly from the instructions on DNA. Instead, the relevant gene is first copied onto a short-lived nucleic acid called RNA, and this molecule provides a kind of template on which a protein is built. All of these biomolecules – proteins, RNA and DNA – are chain-like assemblies, called polymers, made by linking together small molecules called monomers.

We can regard DNA as the book of the cell – a string of words (genes) which are themselves made up of individual letters (monomers). Proteins are the same words translated into another language with its own alphabet. Each translated word is invested with meaning, but none tells a story on its own. The alphabet of proteins has just twenty characters, and they are all molecules called amino acids. When each amino acid is appended to the character string that will become a protein, a molecule of water is constructed from unwanted atoms in the linking unit.

The alphabets of RNA and DNA are even more rudimentary: they contain four characters each, and three of them are common to both – the languages are almost identical.[3] But the characters themselves are elaborate ones, with a substructure every bit as complex as Chinese or Japanese characters.

To account for life on Earth, the first task is therefore to explain where the alphabets came from and how they might have been joined up into words. Progressing from words to a book is a puzzle of another order, and we can presently only marvel at how the words assembled themselves into stories of such stunning richness and diversity.

The environment in which these alphabets arose at least 3.8 billion years ago was scarcely like that on Earth today. The atmosphere probably contained nitrogen gas and water vapour, but virtually no oxygen gas: the planet's oxygen atoms were bound up instead in compounds. Volcanic activity, rampant on the hot young planet, would have expelled volatile carbon-containing gases into the atmosphere from the deep Earth, but we don't know whether these would have been predominantly methane, carbon monoxide or carbon dioxide. Sulphur-containing gases might have been present in significant amounts too – hydrogen sulphide, perhaps, or sulphur dioxide, both generated in geological processes today.

In 1953 Harold Urey and Stanley Miller at the University of Chicago showed that amino acids can be made by energetic processes taking place within a mixture of simple gases. They passed electrical discharges through a mixture of methane, ammonia, hydrogen and water vapour, and found small but significant quantities of relatively complex organic molecules in the solution formed when the water vapour cooled and condensed. These included the simplest amino acids, organic acids and urea. The experiment indicated that delicate chemistry is not needed to make such compounds – we might expect them to appear in a prebiotic sea struck by lightning under an atmosphere of methane, hydrogen and ammonia.

But the Earth's early atmosphere wasn't like this; in particular, most of the nitrogen was present as elemental nitrogen gas, not ammonia. Similar experiments with a mixture of methane, water, nitrogen and only small amounts of ammonia do, however, also generate ten of the twenty amino acids found in natural proteins. With the addition of hydrogen sulphide, the two natural sulphur-containing amino acids can be formed too. Which is all very well, except that the primitive atmosphere probably wasn't like *this* either: it is more likely to have been composed mostly of nitrogen and carbon monoxide and/or carbon dioxide. Spark discharge experiments that use carbon monoxide or dioxide as the source of carbon don't do half as well – the mixture that results contains little more than a single character of the protein alphabet, and the simplest one at that.

If the origin of the protein alphabet is a headache, making the elaborate characters of the nucleic-acid alphabet is something of a nightmare. Nonetheless, all of them have been made one way or another by the reaction of chemical substances that might, if you're an optimist, have existed on the early Earth. One component of all the nucleic-acid characters is a sugar molecule which can be created by dishing out rough treatment to formaldehyde. Another component is a so-called nucleic-acid base. There are four types of these in DNA, and four in RNA – they are like the crucial brushmarks that distinguish otherwise identical characters. The nucleic-acid bases can be fashioned from reactions involving hydrogen cyanide or a small molecule called cyanoacetylene. But all of these syntheses require indelicate loading of the dice, for example by using concentrations of the reactants greater than could have been mustered on the early Earth.

However, the Earth may not have had to build its life forms entirely from scratch. Simple organic molecules, including amino acids, are found in some meteorites, and an appreciable proportion of the inventory of organic molecules in the atmosphere and oceans of the young planet may have been delivered by planetesimals of the same composition as they crashed about in the demolition derby of the early solar system (see page 16).

Reconstructing the chemical origin of life is just one hurdle after another. Even if you can dream up barely plausible schemes for making the building blocks of life's principal biomolecules, joining them up is an awful business. It now seems likely that minerals may have acted as catalysts that assisted the assembly of monomers of proteins and nucleic acids into long chains.

Even making the links isn't the end of the matter, because water can and will sever these links in a reaction called hydrolysis ('splitting with water'). That's a problem, then, if you're supposing that the chemistry of life took place in ancient seas or lagoons. And it gets worse. Linking up two amino acids has the complication that it spits out a molecule of water, and this means that the more water there is around, the more hydrolysis will dominate over link forging. So even if chains of amino acids – the ancestors of proteins, like words that have not yet acquired meaning – could be formed, would they survive for long enough to take on meaning, to start forming sentences? Says Chris Chyba, a planetary scientist at the University of Arizona,

> Proteins are necessary for all life on Earth, but how could these molecules have formed in the seas of prebiotic Earth, since water acts not to link amino acids together but rather to split them apart? . . . Water-based life must therefore fight a constant battle against destruction.[4]

For nucleic acids, the problem of hydrolysis may be even more severe. In particular, the sugar rings that form a crucial part of the nucleic-acid alphabet are hydrolysed relatively quickly in water. In other words, although water is surely the environment in which life began, that contingency comes with plenty of problems of its own.

With all this to cope with, it is small wonder that some researchers have ditched the idea of a 'dilute soup' of organic molecules and resorted to entirely different scenarios for life's origin. Most prominent amongst them are hypotheses that invoke the special conditions that exist around submarine hydrothermal vents. Proponents of vent theories for the origin of life say that here is everything one could need: water rich in minerals and simple carbon-containing compounds such as methane and carbon monoxide, as well as ammonia (which is not readily formed in other geological environments). There is also a source of energy to drive the chemical reactions: the hot waters of the vent, which can reach temperatures of around 380°C.

There are other attractions to the 'vent hypothesis'. Some of the most primitive organisms on the planet today use hydrogen sulphide in their metabolism; and this compound is abundant in vent fluids. And the sulphur-rich water deposits sulphide minerals such as iron pyrite (fool's gold) in and around the vent's chimney structures. Iron pyrite can be converted to a different form of iron sulphide by chemical reactions that soak up electrons. These electrons can be used to forge links between relatively simple organic compounds, including amino acids. It has been suggested that iron pyrite at hydrothermal vents may have acted as a kind of battery to drive the chemistry involved in linking together monomers into compounds resembling proteins. But this idea remains very speculative, and it demands a chain of serendipity that is scarcely less optimistic than that required for polymerization in a conventional 'prebiotic' soup.

A short history of life

If we cannot know exactly *how* life started, we can at least gain some idea of *when*. Until quite recently, the earliest known evidence for life came from 3500-million-year-old rocks in eastern South Africa and Western Australia. These are amongst the oldest rocks known on Earth, and display abundant signs of early life. In particular, the rock strata contain wavy, irregular layered structures closely resembling those found in younger rocks that are known to be the fossilized remains of complex marine microbial ecosystems called stromatolites. Living stromatolite structures, which can be found today in some parts of Australia, consist of communities of single-celled algae and bacteria, some of which use photosynthesis to harness the energy of sunlight for their biochemical processes and some of which garner their metabolic energy from other sources. The fossilized structures of these communities preserve their characteristic layered forms. Even more striking are the fossilized remnants of individual algae and bacteria within these rocks – microscopic blobs and streaks that could be almost anything to the untrained eye but are, to the knowing gaze, the imprints of ancient organisms.

The primeval communities that we glimpse in these rocks seem to be already rather diverse and advanced, containing so-called blue-green algae that use the sophisticated process of photosynthesis to stitch carbon from atmospheric carbon dioxide into their cellular constituents. So it appears that life must have been up and running for a long time before these organisms left their stony impressions to posterity. And indeed, in 1996 a group of researchers from the Scripps Institute of Oceanography in California uncovered evidence for life in rocks from southwest Greenland that are a good 300 million years older than the African and Australian formations. Life, it seems, is almost as old as the oceans themselves.

What were the earliest of organisms like? In many ways they were probably not so different from some of the simplest species that we see today. They were surely single-celled – bacteria or algae. To sustain themselves, they had the option of joining one of two camps. Autotrophs – literally 'self-feeders' – do the hard work of making organic molecules from inorganic raw materials like atmospheric carbon dioxide. For this reason, they are also known as primary producers. Green plants belong to this camp, along with the humble blue-green algae and photosynthetic bacteria. Heterotrophs – 'other feeders' – live off the garbage of the autotrophs, or off the autotrophs themselves: they consume ready-made, second-hand organic compounds to fuel their metabolic processes. We are heterotrophs, since we don't photosynthesize; but it's a trait we share with plenty of bacteria and algae. Both autotrophs and heterotrophs are found in the stromatolite fossils of 3.5 billion years ago.

One might imagine that autotrophs had to come first in life's roll call, just as vegetables have to be grown before they can find their way into pre-packed TV dinners. Yet autotrophy is a challenging task – to make carbon molecules from thin air – and it requires some elaborate biomolecular machinery for conducting difficult chemistry. Most theories of life's beginnings assume that the first organisms were heterotrophic – that they used relatively complex organic molecules such as amino acids and sugars that were ready-made in the environment. This idea was first clearly expressed in the 1930s by the Russian scientist Alexander Oparin. As long-serving director of the Bakh Institute of Biochemistry in Moscow which specialized in the chemistry of beer and wine brewing, Oparin was well versed in the complexities of cooking up potent substances from insipid ingredients. His book *The Origin of Life* (1936) defined much of the thinking on the subject for the next few decades, and provides one of the first suggestions that modern science can meaningfully address the issue through both theory and experiment. Oparin's ideas about heterotrophy provided the stimulus for Urey and Miller to investigate whether the requisite organic compounds could arise spontaneously from simple gas mixtures.

Yet it is autotrophy that holds the key to the early evolution of living organisms. Carbon drains away from ecosystems all the time, as dead organisms and fecal waste rain down to the sea bed and are recycled into the geological world. That carbon needs to be replaced, and autotrophs pull it out of the air. Carbon dioxide is pretty soluble in water, so it would have been accessible to organisms growing in shallow ocean margins, lakes, lagoons and ponds. For Earth's earliest autotrophs, the challenge was to turn this carbon dioxide into the organic molecules needed for metabolism, using sunlight as the main driving force – a task no different to that faced by photosynthesizing green plants today. In essence, photosynthesis is a way of 'fixing' carbon into the sugar glucose. This, as we know from experience, is an energy-rich molecule, and for a photosynthesizing organism it is a store of chemical energy that can be tapped to power the cell. In plants, glucose is often stored in the form of starch, a polymer of which glucose is the monomer. It is also the basic building block of another polymer, cellulose, which provides the tough structural fabric of plant cell walls.

Glucose is composed of carbon, oxygen and hydrogen atoms. So to make it from carbon dioxide, a source of hydrogen atoms is required. Many of the earliest autotrophs seem to have extracted hydrogen from hydrogen sulphide; some very primitive bacteria, called purple sulphur bacteria, still do. Take away the hydrogens and what's left is sulphur. Many of the bright yellow sulphur deposits in the world today are the waste heaps of primitive autotrophs.

The heterotrophs of the early world fed on the energy-rich glucose produced by the autotrophs. They harvested the energy by fermentation – the same chemical transformation that we use, assisted by heterotrophic yeast, in brewing. During fermentation glucose is split in a process called glycolysis, which releases chemical energy that can be stored in molecules denoted as ATP, the batteries of the cell. The glucose is eventually converted to ethanol and carbon dioxide. Carbon dioxide, the 'food' of autotrophs, is of no use to heterotrophs, who release it into their environment. So the earliest heterotrophs brewed the sugars of the autotrophs into a bubbling, watery wine.

Fermentation is a way of breaking down sugar without air – it is an 'anaerobic' process. The famous French biochemist Louis Pasteur called it 'life without air'. Oxygen-breathing heterotrophs, a select band that includes you and me, conduct glycolysis too; but we generate water and carbon dioxide as end products. In evolutionary terms this 'aerobic' (oxygen-requiring) metabolism is a relatively advanced faculty. On the early Earth there was little if any oxygen gas in the atmosphere, and heterotrophs had to do without it.

The reliance of early autotrophs on hydrogen sulphide posed limitations on the spread of life: only regions of the watery planet that were well supplied with this appreciably soluble gas, or with other rarer sources of hydrogen such as organic acids, could support autotrophy. Then came a discovery that changed the world. It has never looked back since.

Poison in the water

The point is obvious, once you think about it (but no one can expect a single-celled bacterium or alga to have done much of that). Here you are, surrounded by water, and you need hydrogen for photosynthesis. Why rely on mining it from rare sources like hydrogen sulphide when the very medium of your existence is two parts hydrogen? At some point – we cannot be too sure when, but it was probably at least as far back as three and a half billion years ago, perhaps earlier still – some cyanobacteria discovered how to extract hydrogen from water, in a photosynthetic water-splitting reaction. This same basic process is utilized by plants and nearly all other photosynthetic primary producers today.

So water was not just the medium for early life: it was a raw material. An organism that gets its hydrogen from the abundant water all around stands at a distinct evolutionary advantage to one that has to rely on scarcer hydrogen sources such as hydrogen sulphide. So you might imagine that water-splitting photosynthesizers would have rapidly overwhelmed the living world. But there is a drawback. Just as sulphur is left over as waste when hydrogen sulphide is stripped of its hydrogen, so oxygen is produced when water is so deprived. And oxygen is a poison.

This may sound bizarre to us, who would expire within minutes if it was withheld; but oxygen is a potent, reactive and corrosive gas. In particular it is liable to generate free radicals, highly reactive bleach-like compounds that burn up organic molecules. We are able to tolerate an oxygen-rich environment only because our cells possess complex biochemical mechanisms for suppressing its many harmful influences. To early organisms faced with oxygen from water-splitting photosynthesis, it was as though their neighbours had started dumping toxic waste in their back yard. Like the splitting of the atom, the splitting of water was a discovery that threatened all life on Earth. Says geoscientist Lynn Margulis, 'The oxygen release from millions of cyanobacteria resulted in a holocaust far more profound than any human environmental activity.'[5]

Yet the inevitable 'oxygen crisis' seems to have been delayed by over a billion years. Oxygen producers were abundant by at least three billion years ago, and probably much earlier. But the oxygen content of the Earth's atmosphere did not rise above a low level – tolerable for anaerobic heterotrophs, and certainly not breathable by our standards – until two billion years ago. In the interim, something must have soaked the oxygen up and postponed the global pollution crisis.

Where did the oxygen go? Well, for one thing, oxygen-breathing organisms evolved and consumed the gas. The awesomely powerful Darwinian combination of random mutation and natural selection allows life to occupy all sorts of niches, even ones that look as unpromising as oxygen usage. Within the great stromatolite communities of photosynthetic bacteria and algae, heedlessly pumping out oxygen, some primitive heterotrophs began using this toxic gas in much the same way as we do: to burn up sugars to carbon dioxide and water. This process of *aerobic respiration*, you see, is the very reverse of photosynthesis, and instead of consuming energy it releases it as heat. The first aerobic respirers were probably able to switch back and forth between metabolizing by aerobic respiration and by anaerobic fermentation, choosing as circumstances dictated.

Such organisms, which still exist today, are called facultative aerobes.

Volcanoes can also remove oxygen from the atmosphere, by emitting gases such as hydrogen, methane and carbon monoxide that react with oxygen. And the imprint of another geological 'oxygen sponge' is preserved in a striking manner in marine sedimentary rocks throughout the world, in the form of bands of red rock composed primarily of haematite, the form of iron oxide in rust. These rock sequences are called banded iron formations, and they are rich sources of iron ore. They are thought to be the result of the reaction between dissolved iron and oxygen in the oceans, which forms the insoluble oxide. Banded iron formations are found in 3800-million-year-old rocks in Greenland, suggesting that oxygen-producing autotrophs may have existed on Earth even this long ago.

But by two billion years ago these defences against oxygen could no longer hold out, and they were overwhelmed. From this point on, the composition of the atmosphere began to approach the oxygen-rich state we see today. The change provoked major evolutionary upheavals. In an oxygen-laden atmosphere, heterotrophs that cannot respire are doomed unless they are insulated by, for example, dwelling in the deep sea or the deep earth where the toxic gas cannot penetrate. The two-party system – oxygen-producing photosynthesizers and aerobic respirers – came to dominate.

At about this same time, organisms became markedly more complex. They diverged into two classes: prokaryotes, the most 'primitive' form of single-celled organisms; and eukaryotes, whose cells are compartmentalized like the rooms of a house. One of these self-contained, membrane-encapsulated compartments, called the nucleus,[6] contains most of the cell's DNA. Others, called organelles, are the locations for various specialized biochemical processes. Plant cells, for example, contain organelles called chloroplasts where photosynthesis is conducted. Organelles are thought to be the remnants of once fully-fledged prokaryotic individuals, with which the eukaryotes fused in a symbiotic relationship. Lynn Margulis suggests that this symbiosis was driven in part by the oxygen crisis, in the face of which prokaryotes that joined forces found better ways to cope.

Unions of eukaryotes conveyed similar benefits. Some began to cluster into multicellular groups, which would at first have operated as cooperative communities of autonomous cells but which eventually became multicelled organisms. The earliest era of Earth's history – the Archaean – was at an end, and the Proterozoic era, literally the time of the first animals, was underway. And all because of the splitting of water.

High and dry

Well, eventually we left the seas – that much is clear. But it was a slow and difficult transition, and we did not dare chance it until relatively recently in the geological past. By 'we' I mean of course our amphibious ancestors; but they were not the first. No fish stood to gain by dragging itself on ill-adapted fins over the mud unless there was food to be had there. Before animals could colonize the land, plants had to get there first.

The first land settlers were probably photosynthetic cyanobacteria. The earliest so-called higher plants, the multicelled ancestors of liverworts and mosses (collectively called bryophytes) and ferns, took to the land at least 460 million years ago. These plants began to spread into the waterlogged swamps that covered much of the tropical continents, and by 325 million years ago, in the Upper Carboniferous period, these regions had become verdant with large plants.

Life on land brings problems, the most obvious of which is that a water supply can no longer be taken for granted. The earliest higher plants would have kept their roots saturated with water in swampy ground – but even with access to water, there is the difficulty of keeping it. Cell membranes are permeable to water, and so cells exposed to air are in danger of drying out by evaporation. Higher plants have a number of defences to combat water loss. Their leaves and stems are usually coated with a cuticle, a waxy layer that seals the water in; and plant spores have watertight coats of an organic substance called sporopollenin. More primitive forms of higher plants, the mosses and liverworts, exhibit specialized adaptations to the threat of drought: they have short life cycles, blooming and dispersing their spores rapidly in the most favourable parts of the year. The spores can survive total dehydration without being irreparably damaged.

Another of the hardships of land life is that temperature variations are more extreme. Because of water's large heat capacity, it warms and cools more slowly than air, and so seasonal conditions in the sea do not vary as greatly as those on land. Short life cycles help to minimize the hazards of these variations: you live less long, but at least you breed before you die. In addition, greater exposure to sunlight in the open air is not all good, as we know too well: in addition to heat and visible light, the Sun gives out ultraviolet rays, which damage delicate organic tissue. Sea water itself affords marine organisms some protection from UV light, and algae in the upper layers shelter the organisms deeper below. But on land, plants stand in the full glare of the Sun. Many bryophytes and ferns contain pigments called flavinoids which appear to serve a protective function by channelling the potentially damaging energy of UV rays into less harmful forms.

Given these hardships, why bother to colonize the land at all? The simple evolutionary answer is 'because it's there'. A plant blessed with a successful adaptation to the difficulties gained a whole ecological niche unchallenged by competitors, just as the early European settlers in North America felt that they'd have a continent to themselves if they could brave out the harsh process of settling it.[7] (Unlike the Pilgrim Fathers, the earliest land plants were right about this.) But it is probable that organisms dwelling in coastal regions of the seas were given added incentive for adapting to land life by changes in sea level owing to climate shifts such as ice ages (see page 65), which left them stranded high and dry.

The first animals to leave the sea faced comparable problems. But unlike plants, they had their own means of locomotion, and so they would have enjoyed a greater range of survival strategies. In particular, land and sea were not all-or-nothing options. The earliest land dwellers were amphibians, evolved from fishes by the acquisition of limbs that, like the fins of a mudskipper, could propel them along the ground. Thus mobile, amphibians could return to water when the need arose. And it surely did, for amphibians then and now have skins that are permeable to water vapour. This permeability is essential, since it provides an escape route for the carbon dioxide gas produced through respiration. Frogs expel more carbon dioxide this way than through their lungs. But in the open air, water vapour escapes too, and amphibians risk desiccation if they stay out of water for too long – as testified by the occasional sad, dry carcass of a frog on my doorstep. Moreover, the young of amphibians, as well as the larvae of some insects such as dragonflies and mosquitoes, must spend their early days immersed in water, a testament to their aquatic origins. So these creatures must lay their eggs in water, and the eggs perish if the water dries up.

The descendants of amphibians – reptiles – developed strategies to reduce their dependence on immersion in water. To evade evaporative dehydration, they have tough, scaly skins with a waxy coating. And they became able to lay their eggs out of water by providing for the young a ready-made little pool within which to complete their early development: the amnion, a sac filled with amniotic fluid, encased in a protective shell of mineralized protein that prevents evaporation. Shells are, you might say, evolution's way of bringing the sea with you.

Hydraulic plants

Plants need good plumbing. Excepting the bryophytes, higher land plants possess a vascular system for ensuring that all of their cells are well supplied with water and other essential, soluble substances. In trees in particular this vascular system determines the plant's shape. It is a 'fractal' structure, which means that its form repeats over many length scales: the shape of a tree is echoed in the delicate tracery of the leaf's veins. This kind of branching system seems to provide the most efficient way of delivering fluids to many points over a large volume, and can be seen in other biological fluid-distribution networks: lung passages, blood vessels.

A plant's vascular system can be regarded as a bundle of tubes divided at each junction into smaller bundles. These tubes are long cylindrical cells joined end to end, either open or with sieve-like caps. In contrast to the human blood-circulation system, the bundles branch at both ends: at the roots and at the leaves. And despite appearances, the vascular system is not in fact a circulatory system: whereas our blood travels from the heart to the extremities and back again, the flow in the vascular system is one-way. Yet plants need to transport fluids in both directions – down and up – and so there are two independent networks of channels in the vascular system. Tubes called xylem carry water and dissolved nutrients from the roots, which absorb water from the soil, up to the leaves. The cells of the xylem are pre-programmed to die before they begin to carry out their water-bearing function: they are dead channels. The phloem, meanwhile, are the vessels that bear the sugary products of photosynthesis from the leaves, where the sugar is manufactured, to the rest of the plant. The tubes of the phloem are primarily living cells.

The cells of all higher plants are lined with cellulose fibres, which make them very robust. Whereas woody plants grow very thick, rigid cell walls that are able to bear tremendous stresses, even non-woody plants are able to stand upright and to withstand the stress of wind and rain without the benefit of a hard skeleton. Their tough cells become swollen with water to pressures that would rupture our own more delicate membranes.

This uptake of water occurs by osmosis. The cell's interior is typically rich in the glucose sugar produced by photosynthesis. Because the cell membrane is permeable to water, the fluids inside and outside the cell are in contact, and the differing glucose concentration inside and out leaves them out of balance. But glucose molecules are too big to pass through the membrane's fabric. So if the glucose can't disperse outwards, water will have to come in. Just as air will move from a region of high pressure to a region of low pressure, so is water drawn into the cell to try to equalize the glucose concentration. In effect, the difference in sugariness on the two sides of the membrane creates a kind of pressure – the osmotic pressure – that drives water into the cell.

The swollen cells are said to be turgid, and they become taut and stiff like over-inflated balloons. So water, in spite of its fluidity, acts as a mechanical stiffener for plants. The consequence of depriving non-woody plants of water, all too familiar in my home, is that the cells lose their turgor and the dehydrated plants become limp.

Glucose is made in the leaf by photosynthesis. In 1771 Joseph Priestley observed that plants absorb carbon dioxide from the air and give back oxygen: in his anthropomorphic terminology, a plant can 'restore air which has been injured by the burning of candles'. Of course, this is an act of 'restoration' only as far as we and other aerobes are concerned; for the plant carbon dioxide is precious, and oxygen useless.

The sequence of chemical steps by which carbon dioxide is fixed into glucose, water is split and oxygen ejected is a complicated one, and many of the finer details are still being unravelled. It might be best compared with a factory production line: raw, unfashioned materials come in and are manipulated by workers in a coordinated sequence until they are altered beyond recognition. Thus transformed, the products are dispatched and the workers resume their positions ready for the next intake. The workers are proteins, pigments and other molecules in the chloroplast. The whole affair needs energy, and in the chloroplast this is supplied by sunlight.

The glucose diffuses from the chloroplast to the tips of the phloem distribution system in the leaves. Meanwhile, the xylem carry water up to the leaves, where some of it will be split apart. Here the plant faces some difficult engineering problems. First, in order to be able to photosynthesize it must admit air, with its flimsy dose of carbon dioxide, into the leaf interior. And it must be able to get rid of the oxygen produced, which is not only unwanted but positively harmful if it accumulates. This requires passageways for two-way gas transport. But if gases can get in and out, then water can escape by evaporation, and the plant is faced with the omnipresent danger of land organisms: drying out.

To combat this hazard, leaves have developed means of carefully regulating their intake and outflow of gases. The surface of the leaf contains thousands of tiny pores called stomata. One square inch of leaf surface can contain between several thousand and several hundred thousand stomatal openings, depending on the species. The stomata open and close in closely controlled sequences to admit carbon dioxide and release water vapour. They are the only means of egress through the leaf's waxy cuticle, which otherwise prevents evaporation. The extent to which the stomata open or close varies with the plant's environment. Succulent plants such as cacti, which grow in hot, dry climates, keep their stomata closed for much of the time, minimizing their water loss. As a consequence, they can't admit much carbon dioxide either, and so they grow only slowly. The evaporation of water vapour through the stomata has the generally beneficial side-effect of cooling the plant, because of the latent heat taken up when water is vaporized. This provides the plant with a mechanism for regulating its temperature, which is entirely analogous to the function of our own sweat pores. But because succulents have to make very sparing use of this cooling mechanism, they have evolved to withstand higher cell temperatures.

When water becomes scarce, plants have more extreme ways of coping with water loss. In temperate and high latitudes the cold winters mean that water in the environment may remain frozen, unavailable to vegetation, for long spells. Many plants in these regions shed their leaves in winter, so that the evaporative surface area is reduced and the plant requires less nutrients. An alternative adaptation to dryness is to alter the leaf's shape so that its surface area is minimized. Thus evergreen plants with needles instead of leaves are common both in the hot dry Mediterranean and the cold dry north. In the Arctic regions of North America, Europe and Asia, however, where the water may stay frozen all year round, plant life becomes all but untenable, and we find only cold, rocky deserts.

The second engineering challenge for a plant is to pump water up from its roots to its leaves. The hydraulic requirements can be formidable. In a mature redwood tree the water must be lifted over a hundred metres above the ground. A vacuum-pump method can't achieve this. Sucking on a drinking straw creates a vacuum at the top, towards which the water is then pushed up by the air pressure at the bottom. But even with a total vacuum – zero pressure – at the top of the tube, the water will rise at most by just a few metres, which is nowhere near high enough to get it up to the top of a typical tree. If the tube is very narrow, the rise is accentuated by capillary forces (page 41); but even this is not enough.

Trees solve the pumping problem by making active use of water loss from the leaves, turning the potential problem of dehydration to their advantage. Expulsion of water through the leaf's stomata creates an imbalance between the water content at the bottom and top of the tree, which draws water up the xylem in much the same way as the imbalance in sugar concentration inside and outside a cell draws the water inside. As water rises up the xylem, it experiences something that at first sounds bizarre: its pressure falls to less than zero – less than that of a vacuum! Yet there is nothing particularly outlandish about the concept of a negative pressure: it is simply a tension, an outwards pull. The water resists this pull like a stretched piece of elastic. For this reason, water under tension is often called stretched water.

The stretched water in the xylem of tall plants is notable for being in a metastable state (see page 186). Strictly speaking, the water in these vessels is *superheated*. It might seem odd to consider water at the temperature of a comfortable summer's day to be superheated; but because the pressure in stretched water is so low, it ought really to fly apart into a gas. The only thing that prevents this is that the liquid has nowhere to expand into. A liquid under tension is in constant danger of cavitation – of acquiring gas bubbles where the liquid is literally pulled apart by the tension. Gas bubbles in the xylem could be as fatal to the plant as gas bubbles in our bloodstream – though for a different reason. The water is tugged up the tubes only so long as a continuous column of liquid is maintained from top to bottom. A break in this column, due to a bubble filling the passageway, isolates the column of liquid above from that below, which no longer feels the upward pull.

So plants have evolved sophisticated defences against the formation of bubbles. Keeping all the xylem tubes narrow (no more than half a millimetre across), rather than having wide artery-like channels, is one such precaution, since confinement reduces the chances of bubble nucleation in the way that ice nucleation is suppressed in tiny cloud droplets (page 189). The highly interconnected structure of the xylem network means that sap can continue to ascend even if some parts of the network are disrupted – just as there is usually more than one way of getting from A to B on the London Underground. And some plants have special pits in the walls of the channels that trap small bubbles before they can coalesce into larger ones.

The regulated loss of water vapour from the leaves is therefore an important process in the life of a plant. This process is called transpiration, and it can effect a prodigious redistribution of water from the soil to the atmosphere. A typical birch tree transpires around eighty gallons of water each day: much more than would be evaporated from the bare ground or even from a standing body of water of the equivalent area. This means that vegetation is not a mere passive beneficiary of a particular local climate, but may help to engineer it. The humidity of tropical rain forests is largely due to transpiration from the leafy forest canopy. It would be wrong to assume, therefore, that by cutting down rain forest we would be simply clearing land that can be farmed in the same moist, nurturing climate. Rainfall on such deforested land would end up as run-off, draining into rivers, rather than being recycled into the air by the vegetation, and so not only will soils be eroded but the climate itself may become increasingly arid. By the same token, waterlogged soils can be reclaimed for cultivation by planting trees such as alder and willow, which have a high transpiration rate and whose roots can tolerate submersion in water.

Red- and blue-blooded

We have our own 'sap', and it is rich stuff, reddened with iron and invested with mythical association. Blood and water are conflated and transmuted in legend and ancient belief, and once again we are drawn back to the Nile, reddened by suspended clay in its sluggish passage through Lower Egypt. Here is the traveller George Sandys extolling the virtues and mysteries of the river in the early seventeenth century:

> ...the waters thereof there is none more sweet being not unpleasantly cold and of all others offers the most wholesome [draught]. So much it nourisheth, that the inhabitants think that it forthwith converteth into blood retaining that property ever since thereunto metamorphosed by Moses.[8]

Given the evocative pattern of the Earth's branching waterways, and the constant circulation of the liquid in their channels, natural philosophers were not slow in making the sanguine association – Leonardo was not the first. It infuses the writing of Edgar Allan Poe, and poet Paul Claudel gives it explicit voice:

> We feel all water to be desirable; and, certainly, more than the virgin blue sea, this one appeals to that part of us between the flesh and the soul, our human water loaded with virtue and spirit, dark burning blood.[9]

That the veins and arteries distribute blood around the body was clear to sixteenth-century physicians such as Brussels-born Andreas Vesalius and the Spaniard Michael Servetus, although it was not until 1616 that the Englishman William Harvey deduced that the true manner of this distribution is circulatory. Harvey recognized that the heart acts as a pump to send the blood along the arteries; and his investigations led him to 'think whether there might not be a motion, as it were, in a circle' – so that the blood ultimately re-enters the heart. 'Now this I afterwards found to be true,' he announced,[10] a conclusion impressively supported by experiment rather than by the religious or metaphysical arguments of his predecessors. It was not until a good half-century later, however, that this circulatory system was seen by microscopic means to be a closed one – that the blood does not soak into tissues, there to be withdrawn by other vessels, but remains contained within the network of arteries, veins and capillaries.

Equally vital but less celebrated is the body's other sap, the lymph which bathes all of the body's cells. Lymph is a watery fluid related to and derived from blood, but containing no red blood cells. It has its own circulation system, which collects the fluid from the bodily tissues, filters it in the lymph nodes, and then carries the lymph towards the heart where it is blended with blood via a duct adjoining a major vein. The lymph system is an essential component of our defence system: in the lymph nodes, white blood cells purge the fluid of foreign particles and substances.

It is tempting to regard our blood and cellular fluid as a surrogate for the salty sea that once bathed our single-celled ancestors. The composition of blood plasma and the watery cell liquid called cytoplasm does have similarities with sea water: all contain sodium, potassium and chloride ions (which is why blood tastes salty) as well as bicarbonate. But the resemblance is largely incidental: cells simply need these ions to function. Blood plasma is a complex mélange of other dissolved substances too: proteins, sugars, hormones, vitamins, excretion products, fat globules and much else. And for all that our body water is seasoned with salt, sea water's greater salinity renders it poisonous. Paradoxically, drinking sea water can kill by dehydration: if the body fluids outside our cells become highly enriched in salts, osmosis depletes the cells of water. This is why freshwater fish perish in the sea. Conversely, seawater fish fare badly in fresh water: their cytoplasm is more salty, and so the cells take up fresh water by osmosis until they rupture, for they have none of the strengthening fibrous fabric of plant cells. Only a few aquatic creatures, such as eels and salmon, are able to move freely between fresh and salt water, by regulating the saltiness of their body fluids.

Blood is the transport medium of the body: it bears oxygen to cells for respiration, and carries away carbon dioxide and other wastes. Oxygen is not soluble enough in water to satisfy the needs of our cells by dissolution alone. Rather, blood employs an active oxygen-transporting vehicle to sustain respiration: red blood cells, or erythrocytes, which enable blood to carry up to sixty times more oxygen than it can dissolve. Even then, the heart must have a throughflow of something like seven thousand litres of blood each day to supply our cells with the oxygen they need.

Erythrocytes contain large amounts of the protein haemoglobin (*haeme* means 'blood' in ancient Greek), within which oxygen becomes bound to an embedded atom of iron. It is iron that imparts the red colour, akin to that of rust. Haemoglobin has a seesaw-like affinity for oxygen, binding it where oxygen is abundant (in the lungs) and releasing it where it is depleted (in the tissues). Haemoglobin gives up its oxygen to another protein, myoglobin, which stores it for use in respiration. The affinity of iron for carbon monoxide is even higher than that for oxygen, which is why this gas is so poisonous: it blocks oxygen binding and transport by haemoglobin in the bloodstream.

But not all creatures need such a powerful oxygen-concentrating mechanism. Because oxygen is more soluble in cold water than warm, there is more of it dissolved in Antarctic waters, and so Antarctic fish need less haemoglobin in their blood. The icefish *Chaenocephalus aceratus* has refined its oxygen consumption to such a degree that it can make do with no haemoglobin at all: it moves sluggishly enough that the oxygen dissolved in its blood plasma is sufficient to meet its needs. The blood of the icefish is therefore a translucent white. Haemoglobin is the oxygen carrier in vertebrate animals, but others use different carriers: haemocyanin in arthropods such as insects and molluscs, haemerythrin in some marine invertebrates. Haemocyanin utilizes copper instead of iron as the oxygen-binding metal, and so it has a different colour: blue, not red. Arthropods, moveover, have no system of veins to distribute their blood – it is simply pumped into the body cavity by the heart, there to be contained under pressure by the body's hard shell.

Carbon dioxide is relatively soluble in water, and so dissolution can be relied upon to carry this waste product from cells to the lungs. But there is a snag: dissolved carbon dioxide generates an acid, carbonic acid. While the acidity of several bodily fluids, such as saliva, urine and stomach fluids, can vary significantly without adverse effects, blood is not so tolerant. Pure water has a pH of 7; a lower pH denotes acidity, and a higher pH alkalinity. The blood of a healthy person is slightly alkaline (pH 7.4), and its pH does not vary by more than 0.1 in either direction. If the pH falls below 7.3, the blood can no longer take up and carry away carbon dioxide produced in cells. If it rises above 7.7, the carbon dioxide in the bloodstream cannot be effectively released to the lungs. A fall in pH to just 6.8 – which can happen through disorders such as diabetes – can lead to coma and death.

To ensure such fine-tuned stability, blood uses a clever and meticulous means of pH control called buffering. This relies on a kind of balancing act that all chemical processes are known to perform, which is enshrined in a maxim called Le Chatelier's principle. It can be summarized as follows: if you disturb a chemical equilibrium, the system will adapt to oppose and absorb the change. If you heat up a chemical reaction that itself generates heat, it will tend to go backwards so as to *consume* heat. If a process generates an acid, then adding acid similarly drives it in the other direction. This resistance to change might seem almost teleological, but is in fact explained by the basic laws of thermodynamics, the science of transformations.

The pH buffering of blood makes use of Le Chatelier's principle to counteract inputs of acids or alkaline substances into the bloodstream. In general a pH buffer is obtained from a mixture of a 'weak' acid and one of its salts. In Chapter 6 I mentioned that the agent of acidity, in water at least, is the hydrogen ion: acids generate hydrogen ions in water. When carbon dioxide dissolves, some of it combines with water to form carbonic acid, which owes its acidity to a tendency to fall apart into hydrogen and bicarbonate ions. Carbonic acid is classified as a 'weak' acid because it does not fall apart completely into ions: some of the carbonic acid molecules break up, some stay together. 'Strong' acids like sulphuric and nitric acid, on the other hand, are completely dissociated into ions in solution.

Blood contains sodium bicarbonate, a salt of carbonic acid.[11] If the acidity of the blood starts to increase, so that the pH drops below 7.4, the bicarbonate ions soak up the hydrogen ions to form carbonic acid. Some of this in turn 'reacts backwards', disassociating itself from water to form dissolved carbon dioxide. The ultimate result is that carbon dioxide gas diffuses out from the blood into the lungs. In effect, everything shifts so that the input of acid is compensated by a release of acid-forming carbon dioxide. If, on the other hand, the blood becomes more alkaline, a corresponding series of shifts is manifested as an uptake in carbon dioxide gas from the lungs, resulting in an increase in acidity to offset the alkalinity. Thus the reservoir of carbon dioxide gas in the lungs provides the means for keeping the blood's pH in balance. This is just one of the many mechanisms that the body has for self-regulation, or homeostasis.

Just juice

Many people perceive something unnerving in human body fluids. Alien abductees seem to regard them as highly prized amongst extraterrestrials, and General Ripper considered his to be the victim of a Communist plot in Stanley Kubrick's *Dr Strangelove*. Perhaps here is an echo of the ancient belief that good health was a question of finding the right balance amongst the body's vital juices.

The Greek physician Galen was to medicine what Aristotle was to natural philosophy, and he paralleled the Aristotelian scheme of the four elements with the theory that there are four body fluids: blood, phlegm, and yellow and black bile. Galen's idea that personality traits could be ascribed to a preponderance of one or other of these four 'humours' persists in linguistic terms today: his temperaments were respectively sanguine, phlegmatic, choleric and melancholic. In these excesses, said Galen, lie the origins of disease.[12] Medieval scientists such as Jean Fernel and Bernadino Telesio in the sixteenth century struggled to reconcile the concept of the four humours with the emerging chemical and physiological discoveries of the time, such as the role of different bodily organs and the observation that the body fluids tend to be acidic or alkaline. Iconoclasts like Paracelsus and Johann van Helmont, however, helped to steer the thinking of the time away from Galen's humours and towards their new pre-chemical notions; even the sceptical Robert Boyle advised physicians to pay more heed to the salty and 'sulphurous' characteristics of body fluids, features that tally well with Paracelsus's *tria prima* (see page 119).

Be that as it may, bodily fluids retain associations rich in sociological and metaphorical content. Sperm is potency, and jealously guarded by many an eccentric misogynist. It was long regarded as the sole source of the embryonic child, and there is no surprise in discovering that this belief originated with Aristotle, the doyen of Western patriarchy. Milk, in contrast, is the maternal water. For Bachelard,

> Water is a milk as soon as it is extolled fervently, as soon as the feeling of adoration for the maternity of waters is passionate and sincere.[13]

It is sweet, this nourishing fluid, and that is the consequence of the lactose sugar it bears. Some of this lactose gets converted to lactic acid, accounting for the slight acidity of milk: cow's milk has a pH of about 6.5. As milk ages, bacteria relentlessly transform lactose to lactic acid by fermentation, which is why old milk tastes sour. (*Acidus* is indeed the Latin word for 'sour'.) Lactic acid is also an intermediate product in the breakdown of glucose during metabolism. If our metabolic rate is increased by vigorous exercise, lactic acid can be generated faster than it is broken down, and it then accumulates in muscle tissue. This leads to muscle fatigue and the painful symptom known as 'stitch'.

Not all body fluids have such benign associations. Saliva, which breaks down food before it is swallowed, has rather unjustly become a defiling fluid, a medium of insult. 'Always... spit on a well-scrubbed floor', playwright Christopher Fry advises.[14] Drooling is frowned upon, however instinctive a response it might be. It is the prelude to digestion: saliva contains the enzyme amylase, which splits the bonds between linked sugar molecules in carbohydrates such as starch. By this means, bread and potatoes get converted to sugar even while still in the mouth, and acquire a sweetness if they are chewed for a long time.

But foods need more extreme treatment to release fully their bounty of sugars and nutrients. That's a task reserved for the stomach, awash in the corrosive acid of the gastric juices. Their acidity – a pH of less than 1 – considerably exceeds that of vinegar and lemon juice, and approaches that of battery acid. To achieve this potency, the gastric juices don't rely on weak organic acids but use the hard stuff: hydrochloric acid, secreted by cells in the stomach lining. These acerbic fluids kill any micro-organisms in the food, while the stomach's enzymes set about breaking the material down.

How does this potent concoction of acid and enzymes avoid digesting the stomach itself along with all the organic matter we fill it with? The enzymes can't discriminate at all between a steak and stomach tissue; they are simply kept at bay by an alkaline lining of mucus on the stomach wall. If this lining is breached, the acid and enzymes go to work on the stomach, and the result is a gastric ulcer. Much less serious are the discomforts occasioned by heartburn, due to the leaking of stomach acid into the oesophagus, or by acid indigestion, caused by an excess acidity in the stomach fluid. Remedies for indigestion consist of mild inorganic alkalis such as sodium bicarbonate or magnesium hydroxide (milk of magnesia), which neutralize the excess hydrochloric acid in the stomach.

Most reviled of all, of course, is urine, our fluid purgative. Urine is the body's way of dealing with its unwanted nitrogen. Carbohydrates contain only carbon, oxygen and hydrogen, and they are converted by metabolic processes into carbon dioxide and water. But proteins and nucleic acids contain nitrogen too, and when they are broken down one of the waste products is ammonia, a combination of nitrogen and hydrogen. Ammonia is toxic, and lethal at even small concentrations. So it must be expelled efficiently, and what mammals and amphibians do is to convert it first to urea and uric acid. These substances dissolve in water, and so they can be flushed away in solution. But not indiscriminately – for the body cannot afford to be profligate with its precious reserves of water. The mammalian kidney contains a remarkable plumbing system that extracts urea and uric acid from the bloodstream and concentrates it to a high degree before transporting it to the bladder, ready for expulsion.

The amount of concentration is adjusted to reflect the animal's water needs – we are familiar with the idea that more concentrated, more strongly coloured urine is a sign of dehydration. Reptiles and birds do not produce urine at all, but conserve their water by excreting uric acid in solid form. Even marine bony fish have to be heedful of their body water, which is less salty than sea water; and so they generate rather little urine. Sharks and rays are able simply to retain levels of urea in their bodies that would be fatal to any other vertebrates. Frogs, on the other hand, may produce huge amounts of dilute urine which they store in outsized bladders – not in preparation for an overenthusiastic micturition but to act as a water store during dry periods. The Australian aboriginals have long known that these pent-up frogs, buried under the ground in wait for the rains, can provide a source of drinking water in time of need. In the desert, even frog urine is a welcome refreshment.

9 Inner Space

Water in the cell

Everything the heart desires can always be reduced to a water figure.

Paul Claudel, *Positions et Propositions*

Anyone attempting to describe the nature of water in the cell must realize that, because our understanding is so incomplete, the description will probably be outmoded in a decade or so.

Philippa Wiggins (1990), *Microbiological Reviews*

If we're two thirds water, where is it all? Why do we not slosh and wobble like a bulging wineskin? Some of this fluid gurgles in our guts, surges in our veins, lubricates our palate and our eyelids and joints. But much is bound up in individual parcels smaller than the droplets of a fine mist. Have you seen what a sad, shrivelled brown thing a banana becomes when it is dried? That's what our flesh would look like if drained of the fluid that fills its cells.

Apart from water's extraordinary capacity to dissolve a wide range of substances, you might imagine that any old liquid could fulfill the role of solvent and transportation agent for the body's cellular machinery. Maybe this just happens to be water because there is a lot of it about. But when the biologist and Nobel laureate Albert von Szent-Györgyi described water as the 'matrix of life', he had in mind something more profound than a backdrop on which life's tapestry is embroidered.

The fluid medium inside all living cells, called the cytoplasm, is mostly water. But what a cocktail! – spiced with proteins and DNA, sugars, salts, fatty acids, seething with hormones. It is all too tempting to regard the relationship of this fluid to the biomolecules it contains like that of the paper of this page to the words printed on it: as a carrier, a bland background on which the important business is displayed. But this won't do. Water plays an active role in the life of the cell, to the extent that we can consider water itself to be a kind of biomolecule. Without it, other biomolecules would not only be left stranded and immobile, like beached whales – they might no longer truly be biomolecules, unravelling or seizing up and losing their biological function in the process. To scan through the literature of modern molecular biology, you could be forgiven for concluding that the subject is all about proteins and genes, embodied in the nucleic acid DNA. But this is only a form of shorthand; for biology is really all about the interactions of such molecules in and with water. Biologists Mark Gerstein and Michael Levitt put this in vivid terms:

> When scientists publish models of biological molecules in journals, they usually draw their models in bright colors and place them against a plain, black background. We now know that the background in which these molecules exist – water – is just as important as they are.[1]

Which is to say that biological structures and processes can only be understood in terms of the physical and chemical properties of water. Biology starts with water, historically, ontologically, pedagogically.

Community service

In the previous chapter I risked portraying cells as bags of dissolved chemicals – like a solution of salt and sugar, with the added savour of proteins and all the rest. But to understand water's deepest role in life, we'll do far better to see cells as enclosed communities, as microcosms peopled by workers all striving (well, nearly all) towards a common goal: maintaining the community. That means making food and heat, obtaining raw building materials, exchanging information, policing the neighbourhood and discouraging miscreants both local and foreign, cleaning up mess and waste, sending messages to other communities nearby. By far the most populous members of the community are the water molecules, and they have many subtle skills.

It's a crowded place, there in the cell (Fig. 9.1). The most prominent citizens are the large globular and fibrous proteins, which dominate the scene and order the other denizens about. In this densely populated environment, it is essential for nature to make her molecules highly selective in their interactions, so that each protein ignores its jostling neighbours unless specifically 'designed' to interact with them.

No attempt has been made in this image to depict the water molecules and ions that permeate the empty spaces between the

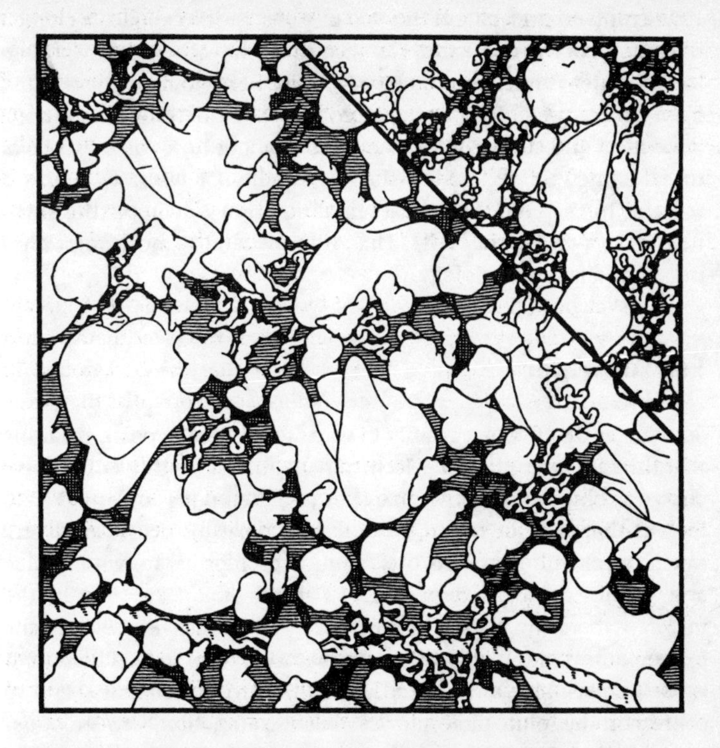

Fig. 9.1 A cartoon snapshot of the inside of a bacterial cell. The large, bulbous objects are protein molecules; the stringy and rope-like ones are nucleic acids. The smaller molecules, such as sugars, ions and amino acids, are shown only in the top right-hand corner, for clarity. Water molecules are not shown at all: they fill up the rest of the available space. (D. Goodsell (1993), *The Machinery of Life*, Springer-Verlag, New York)

bigger biomolecules, since that would render the picture far too busy to be intelligible. But a water molecule is, on this scale, about as big as this full stop. The first thing you'll notice is that there isn't much *room* for the water – the big molecules are packed pretty closely together, so that the liquid of the cytoplasm is confined to the small gaps and channels between them. Nowhere is there a region voluminous enough to accommodate a lump of 'bulk' water – most of it is relatively close to some protein or other. The water, then, is largely confined 'between walls'. It is one of water's eccentricities that confinement in narrow spaces can alter its character profoundly.

Attempts to understand the role of water in the cell face a circular dilemma. On the one hand, we need to ask how the solvent properties of water influence the character and behaviour of the solutes, from simple metal ions to huge proteins. On the other hand, there are critical and still mysterious questions about how the presence of these solutes might modify the very ability of water to act as a solvent. In the end, the fundamental question is: what is the structure of water in the cell? This is one of the most important unresolved issues in biology.

Fear of water

Substances that dissolve readily in water are generally polar, meaning that they are not uniformly electrically neutral. If we think of positive charge as blue and negative charge as red, then polar molecules have regions that are blue or red, whereas non-polar molecules are pretty much purple all over. Ions of sodium and chlorine are wholly blue and wholly red balls respectively. Carbon dioxide is blue in the middle, where the carbon atom is, and red at each end, where the oxygen atoms are. Water dissolves polar molecules so well because it is itself polar—blue at each end, red in the crooked middle – and so confers on the solute molecules a stabilizing attraction between opposite charges.[2] Some biological molecules, such as sugars, are also considerably polar and therefore highly soluble in water.[3]

Many biological molecules have parts that are polar – *hydrophilic* or 'water-loving' – and other regions that are non-polar – *hydrophobic* or 'water-fearing'. Cell membranes are made primarily of tadpole-shaped molecules called phospholipids, which have an electrically charged, hydrophilic head and a non-polar, hydrophobic tail.[4] In cell walls phospholipid molecules are packed side by side in sheets with their heads and tails aligned. The sheets are laminated back to back in double layers, the tails of the molecules in one sheet tickling those in the other. In this arrangement the hydrophobic tails are shielded from water by the hydrophilic heads. The glue that holds cell walls together, therefore, is not chemical bonding in the usual sense, but the aversion of a part of phospholipid molecules to water. Cell walls are peppered with other kinds of biomolecules, such as membrane proteins that regulate functions like cell-to-cell communication and adhesion, and the transport of ions in and out of the cell.

Proteins have an abundance of hydrophilic and hydrophobic regions – they are a motley of our metaphorical red, blue and purple. This is because their amino-acid building blocks have varying degrees of solubility. Surrounded by water, proteins employ much the same strategy as phospholipids: they gather together the hydrophobic parts in a bundle and cover them over with hydrophilic parts. But unlike phospholipids, proteins can conduct this process within a single molecule: the protein chain folds up into a well-defined shape with the hydrophobic regions packed into a buried core and the hydrophilic groups arrayed at the surface, exposed to water.[5] In this way, the water in which a protein is immersed acts as a kind of cement for the protein's shape: aversion to water helps to guide and secure the folding of the protein chain.

The reason why this is crucial for all of biology is that a protein's behaviour is determined largely by its three-dimensional shape – by the way it folds. Enzymes, for example, are proteins that act as catalysts for the chemical reactions of the cell: that is to say, they encourage the reacting molecules to come together and interact. A very crude analogy is that of a molecular vice: an enzyme might hold two molecules in the correct orientation to enable them to stick together. The enzyme offers slots, grooves or hollows, called binding pockets, of just the right size and position to hold the molecules in place while the bond forms.[6]

The moulding of the peptide chain to create the enzyme's binding pocket is the precondition to its biological role. If the enzyme loses its three-dimensional shape, becoming instead strung out into a looser, floppy chain (a process called denaturation, since the protein's folded state is called its native state), it might as well be just any old polymer, no more able to act as a selective biological catalyst than is a lump of rubber.

Not all proteins are enzymes, but they all need to acquire highly specialized shapes to perform their role in life. The antibodies of the immune system, for example, are proteins that have binding pockets sculpted simply for seizing onto foreign particles in the body, rather than for catalysing their transformation. And the coiled shape of so-called structural proteins in the body's tissues, such as collagen in bone and retinae or keratin in skin and horn, determines the strength, flexibility and elasticity of the fibrous tissue.

Yet there is no intelligent force to guide the strand of a protein into its folded native state. Rather, the amino-acid sequence in the protein's chain contains the encoded information needed for its proper folding: the protein comes with its own assembly instructions. The blind forces of physics and chemistry are enough to direct the folding process, when the protein is in water. Most proteins don't work at all in other solvents. Implicit in the information programmed into the protein chain is the instruction: 'Fold carefully in water.' If this is ignored, the information may be misread. And even when the 'instructions' have been followed and the protein has folded to its native state, its usefulness to the cell relies on the continued presence of water. The clustering together of hydrophobic regions of proteins is considered by some biochemists to be among the dominant driving forces for the adoption and maintenance of a protein's native state.

Yet there are undoubtedly other factors, aside from hydrophobicity, that help a protein to fold properly and to maintain that fold. In particular, hydrogen bonds can form between different parts of the chain, clipping them together. But water molecules can compete for the privilege of sticking to a protein's hydrogen-bonding groups. So there is a delicate balance between the affinity of the protein chain for its own hydrogen-bonding (hydrophilic) regions and for the surrounding water. Loops in a peptide chain can also be secured by strong chemical bonds between the amino acid cysteine, containing sulphur atoms that can hook up together.

But regardless of the relative importance of these and other contributions to the maintenance of a protein's folded shape, it's clear that an understanding of the cell's molecular mechanisms cannot afford to neglect the envelope of water around its molecules. The problem is, however, that this envelope does some deeply puzzling things. A good way to start an argument amongst cell-biologists is to ask them what cell water is like. It's time now to peel back another layer, and see what these arguments are about.

Not as we know it

Everything I've said so far about the question of *protein hydration* – the interaction of proteins with water – has implicitly assumed that we can treat these biomolecules like lumps of putty dumped into a beaker of water, which goes on acting like normal liquid water right up to the protein's surface. Let's think about this assumption for a moment. Water derives its unusual character from its high degree of structure: the molecules are joined into a vast, random network of hydrogen bonds. Where the network is disrupted by driving a wall through it, the water molecules at the boundary are robbed of their opportunity to form hydrogen bonds with molecules that the wall has displaced. And recall that the network is pervasively cooperative – one water molecule seizing another by the ankle is sensitive to the wider game of molecular Twister in the vicinity. So the repercussions of the wall's disruptive influence might be felt far afield. In short, there is no justification for treating the water around a lumbering great biomolecule like a protein as if it were no different from pure water in the middle of a beaker.

So what does water really look like around a protein? It might seem like splitting hairs to be wondering whether this water is or isn't like 'normal' water; but the question goes right to the heart of water's uniqueness as a liquid and how this is manifested in the community of the cell.

One of the conventional ways of describing the hydration environment of a protein is to assign to the water envelope two kinds of water molecule: 'bound' and 'free'. 'Bound' water is attached quite securely to the protein, typically via hydrogen bonds. These water molecules are not 'bound' irrevocably: any individual water molecule can and does escape rather rapidly. But at any instant the bound water molecules are relatively immobilized – they are no longer part of the 'true' liquid, their freedom to rotate and tumble is hampered or suppressed entirely. When the protein is crystallized, this bound water crystallizes with it and typically accounts for nearly half of the crystal's mass. 'Free' water, in contrast, remains largely indistinguishable from water in the liquid far from the surface of the protein: the molecules retain their freedom of movement. Bound water molecules are the bees crawling over a flower's petals, rubbing against the stamen; free water molecules are those buzzing around the flower head.

This bipartite division of the hydration envelope has some value for cell biologists, but it's far too simplistic. The boundary between 'bound' and 'free' molecules is a blurry one; and even 'free' water near a protein may not be like water 'as we know it' at all.

One of the earliest suggestions that the water of the cytoplasm differs from ordinary water came from the Russian scientist A.S. Troschin in 1956. Troschin's idea was given more concrete form in the 1960s by the Chinese biologist G.N. Ling, who suggested that water in the cell forms organized, layered structures on the surfaces of proteins. This bold notion was received sceptically in the West, but succeeded in making explicit the idea that proteins do something peculiar to water structure.

We can break the matter down into two separate questions: how does water accommodate a foreign particle in its midst, and what does water do close to surfaces? Both questions remain contentious.

Making space

The fact is that *nothing* – not a protein, not a sugar molecule, not a mere sodium ion from a salt crystal – can be added to water without shaking up and disturbing its exquisite, dynamic structure. Take the deceptively simple case of a lone positively charged ion like sodium. To get the maximum stabilizing effect from the ion's electric field, the water molecules around it will become reoriented so as to present their 'best' side to the ion – the side with opposite charge, the oxygen in the cleft of H_2O's kink. Water molecules point their rears towards positive ions, but with the best of intentions. This creates a so-called 'primary hydration sphere' of disturbed, reoriented water molecules around the ion, which typically consists of a cordon of between four and eight water molecules. Water molecules can come and go from this hydration sphere on timescales ranging from small fractions of a second to several hours.

So solute molecules don't just sit immersed in water: they restructure the water in the vicinity. The primary hydration spheres of some metal ions are virtually crystalline, in the sense that the water molecules are assigned to fixed, geometric positions in space. At first glance, this might lead you to think that all ions *enhance* the degree of structure of the liquid water in their vicinity. But if an ion disrupts the already highly structured hydrogen-bonded network by forcing some of the water molecules to adopt a new configuration, it's by no means obvious whether the net result is an increase or a decrease in structure. The orderliness in a primary hydration sphere might turn out to be like a jigsaw piece from the wrong puzzle – it doesn't fit within the jigsaw of pure water.

This consideration has given rise to a distinction between so-called 'structure-making' and 'structure-breaking' ions. Rearrangement of water's hydrogen-bonded network is greatest for small, highly charged ions such as those of lithium, fluorine and magnesium. The degree of ordering in the hydration sphere around these ions can outweigh the disordering generated in the hydrogen-bonded network further afield, and the ions can be considered structure-makers. On the other hand, big ions with small charge are like the proverbial bulls in a china shop: they mess up the water network but don't offer much local reordering in return. Rubidium, caesium and iodide ions are structure-breakers. But because it's not obvious how one should add up the structure that is made or broken, water expert Felix Franks says that 'much has been written about this concept, much of it misleading'.[7]

All the same, structure-making and structure-breaking has some important consequences. It appears to be bound up, in a way that is still unclear, with the effect that salts have on the solubility of proteins. In 1888 Franz Hofmeister noticed that certain ions make proteins more soluble whereas others reduce their solubility. He arranged these ions in an order, now called the Hofmeister series, that reflects their propensity to help proteins dissolve. Highly charged or small ions like sulphate and fluoride – which are thought to be structure-makers – promote precipitation or 'salting out', while big, low-charge ions like iodide and perchlorate – structure-breakers – promote solubility or 'salting in'. Despite its discovery over a century ago, there is still no satisfactory explanation for the Hofmeister series, although most scientists believe that it is intimately connected with the subtleties of water's hydrogen-bonded structure.

One of the most dramatic illustrations of the effect of water structure on solubility is provided by sugar molecules, which possess hydroxy groups – an oxygen atom capped with a hydrogen, a fragment of water welded onto another molecule. Hydroxy groups can engage in hydrogen bonding with the water network. Very subtle differences in the positioning of the hydroxy groups on different sugar molecules can have profound effects on their solubility. The sugar D-talose, for instance, is fairly hydrophobic, whereas D-galactose, which looks barely distinguishable to the casual observer, is far less so (Fig. 9.2). It seems that the flipping of hydroxy groups in D-talose relative to D-galactose makes the former fit less comfortably into the three-dimensional hydrogen-bonded jigsaw of water.

The forces of disorder

Proteins are blotchy amalgams of the red and the blue – polar, hydrophilic regions – along with swathes of purple, which correspond to non-polar, hydrophobic parts. To understand how water responds to their presence, we therefore need to know also what water does when faced with purple invaders.

Small, wholly hydrophobic molecules such as methane and krypton (a self-contained one-atom molecule) don't form hydrogen bonds, and they don't have electric dipoles that enable them to exploit favourable electrostatic interactions with a polar solvent like water. At face value it would seem that, if you put molecules like this into water, all you are doing is breaking up part of the network of hydrogen bonds to make room for it, without getting any energetic stabilization back in return. No wonder, then, that these gases are not very soluble in water.

But if we look more carefully, we see that this is not really what happens at all. When methane dissolves in water, the water warms up: energy is released. That's a sign that 'bonds' are formed in the broadest sense – that the methane molecules are welcomed by favourable interactions. In other words, inserting a methane molecule into water's hydrogen-bonded network does not just break bonds without recompense – there is a considerable payback. Any breaking of hydrogen bonds as the methane forces out a cavity in the network is more than balanced out by other favourable interactions, which can only be due to van der Waals forces (page 153), between the methane and the surrounding water.

And yet despite all this, methane is still pretty insoluble in water.

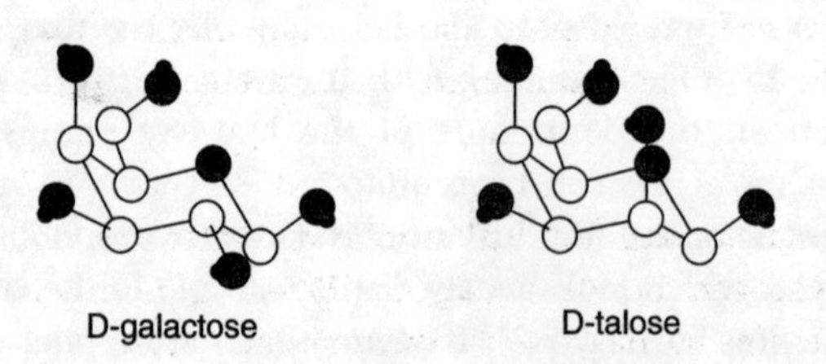

Fig. 9.2 Spot the difference. The subtle distinction between the two sugar molecules D-talose (right) and D-galactose (left) is 'sensed' by the hydrogen-bonded network of water. As a result, D-talose is more hydrophobic – more water-fearing – than D-galactose.

Why so, if it seems to be energetically favourable to surround methane molecules with water? The answer is that the heat change is only half the story. The other half concerns *entropy*, a measure of the disorder in a system. In a crystal all of the atoms are regularly arranged, whereas in a gas they can go anywhere, so long as they don't overlap. So there is more disorder – more entropy – in a gas, and a substance's entropy increases when it is vaporized.

Whether or not a process of change will occur depends on the balance between the change in heat[8] and the change in entropy that it entails. By 'a process of change', I mean *anything*: the reaction of two chemical compounds, the freezing of water, the toppling of a tree, the formation of a star. The scales that govern all change in the Universe weigh up one of these two quantities against the other.

Methane does not dissolve well in water because the favourable heat change – the fact that heat is released – is counteracted by an unfavourable entropy change: the entropy decreases. This tells us that when methane dissolves in water, the two substances are together *more* orderly overall than they were in isolation. The presence of a hydrophobic molecule like methane amidst the random hydrogen-bonded network of liquid water somehow increases its order.

No one truly knows how to interpret this fact in terms of molecular structure. Indeed, it remains one of the most debated issues in physical chemistry. For a long time, most researchers believed in a picture of the ordering process that was sketched out in 1945 by Henry Frank and Marjorie Evans, which came to be known as the iceberg model. Frank and Evans had this to say:

> When a rare gas atom or nonpolar molecule dissolves in water at room temperature it modifies the water structure in the direction of greater crystallinity – the water, so to speak, builds a microscopic iceberg around it.

The rationale for this idea was as follows. Water aims to maintain as many hydrogen bonds as it can. When a non-polar molecule punches a hole in the network, the molecules at the edge of the void have to adopt very specific and rigid orientations to retain all or most of their hydrogen bonds. So water becomes more structured – perhaps even ice-like – around the solute molecule. It's like the intrusion of a notorious bore into a crowded party. No one particularly wants to split off from the crowd in his vicinity, or they might be forced to talk with him. So all those around this unwelcome individual carefully turn their backs and engage even more ardently in conversation with their neighbours, standing their ground and adhering to each of their own little groups for safety.

Decades after the iceberg model was proposed, researchers came to feel that the term 'iceberg' was potentially misleading. It implies that the proposed solvent cage is both virtually crystalline and composed of ice-like patterns of hydrogen bonds. But if the cage exists at all, it is surely a looser, more dynamic structure. Moreover, rather than being ice-like, formed from rings of six water molecules (see page 166), the cage might be more like the cavities that form around hydrophobic 'guest' molecules when their solutions are solidified. These crystalline mixtures of water and gas are called gas hydrates, and they bedevil natural-gas pipelines. If the pressurized gas is 'wet', containing water mixed with the methane in the gas field, hydrates can solidify in the pipe and block the flow even above water's freezing point. In gas hydrates the guests are accommodated in polyhedral cages whose faces are made up largely of five-membered rings of water molecules, which are more akin to those believed to exist in liquid water.

Regardless of the shell's exact structure, one of the implications of the Frank–Evans model is that hydrophobic solutes should attract each other – that they will cluster together in water. Why is this? Earlier I rationalized the tendency of hydrophobic parts of proteins and phospholipids to stick together in terms of their 'water-fearing' nature – their relative insolubility. Clustering, I said, is a way of shielding hydrophobic regions from water. But this explanation is oversimplistic, though treacherously seductive. Not only is it circular – some molecules are hydrophobic because, well, they avoid water – but it's also nonsensical if we think about energetics alone and neglect the effect of entropy. Methane, after all, *gains* bonding-type energetic stabilization when immersed in water.

In 1959 Walter Kauzmann recognized that the true origin of hydrophobic attraction and clustering must lie with the entropy of hydration. According to the Frank–Evans picture, which Kauzmann broadly accepted, hydrophobic particles in water are surrounded by hydration shells in which the water molecules are more highly structured, and so have a lower entropy, than their counterparts in the rest of the liquid. If we bring two such particles together so that they touch, the hydration shells overlap. This means that some of the constrained water molecules in these shells can be released – the total amount of water needed to hydrate the two particles is less when they're side by side, just as you can seat fewer people around two tables when the tables are pushed together. In other words, the amount of enhanced water structuring – of entropy reduction – imposed by the hydrophobic solutes is decreased when they are brought together, and this clustering will happen spontaneously. It is as if some kind of phantom attraction exists between the particles. This apparent force of attraction is called a hydrophobic interaction, and it arises from the preference for maximal disorder.

Hydrophobic interactions clearly do exist – they keep the bricks of cell walls together and mould proteins – but it remains uncertain whether they work in the way that Kauzmann proposed, by release of 'structured water'. For all that his idea defined the accepted paradigm for many years, it now appears to have dubious merit – yet it carries a legacy that is hard to shake off.

Kauzmann's model makes sense only if the hydration shells of hydrophobic particles really do contain more highly structured water. It has proved remarkably difficult to find out if this is so. Small hydrophobic molecules such as atoms of the inert gases neon, argon and krypton, are the simplest cases to study. But they are so insoluble that it is hard to get enough of them into solution to make their hydration shells detectable in experiments. Measurements made in the past few years do seem to show some ordering of water molecules around inert-gas atoms like krypton. But the real question is whether this ordering is any greater than that around a water molecule – whether the hydrophobic atom *enhances* the ordering that is already there. The most recent experiments indicate that the hydration shell is not nearly as orderly as the crystalline cages of a gas hydrate, and is probably no more ordered than water in the pure liquid.

But how else can one account for hydrophobic interactions, if not by increased structuring of water? The decrease in entropy – the loss of disorder – when hydration occurs might be accounted for partly by the fact that the water molecules in the hydration shell become less free to rotate, rather than because they shift into new, more orderly positions. There is some indication that water molecules around a hydrophobic particle are oriented with their O–H bonds tangential to the particle's surface, as if the water molecules are resting an arm on the particle. Maybe the water molecules thereby accommodate the particle not by increasing their degree of structure but by steadfastly maintaining it: a little reorientation of the molecules could be all that is needed to avoid sacrificing hydrogen bonds. But here we are in the realm of speculation and can, at present, venture no further.

On the surface

As I said earlier, an important issue in understanding how water receives a large biomolecule like a protein is that the protein presents a physical barrier – a wall. The molecular-scale structure of *any* liquid can be altered profoundly when it runs up against a wall. The key consideration here is Desmond Bernal's 'impenetrability' (page 152): a wall is a space that the liquid's molecules can't enter. You might imagine that, if there is no force of attraction between the molecules and the wall, the molecules would remain unaware of the wall's presence right up to the point at which they hit it – just as you don't detect the French windows until your nose bounces off the glass. But that's not so. Close to a wall, a liquid generally becomes organized into layers parallel to the wall.

This ordering of the liquid happens for the same reason that each individual molecule in the liquid is surrounded by concentric shells where the fluid is alternately denser and less dense on average (page 151): the need for efficient packing of molecules. A wall compels the molecules to pack together more efficiently in its vicinity, creating orderly layers. Given that any large biomolecule has surfaces, we can expect that it will exert a similar structuring influence on the surrounding water. G.N. Ling's early ideas about layering of water at protein surfaces may thus have some validity. Indeed, given that in the cell many of these surfaces are in very close proximity to one another (see Fig. 9.1), we must wonder if any of a cell's water is *not* affected in this way.

Layering of liquids near surfaces happens for *any* liquid. But you should not expect by now that water will behave just like any old liquid – and indeed it does not. Water close to surfaces does things that seem utterly baffling, and which remain impossible to explain without stretching credibility to breaking point.

Water near hydrophilic surfaces seems normal enough: it appears to be organized into a series of layers, each one molecule thick. Even this, however, does not go wholly uncontested.[10] Biologist Philippa Wiggins argues that surfaces studded with many charged (ionized) groups – as the surfaces of proteins and nucleic acids generally are – sequester a local cloud of ions of opposite charge from the salty cytoplasm, leaving the more remote liquid depleted in ions. This, says Wiggins, sets up an osmotic pressure imbalance between the ion-enriched water near the surface and the ion-depleted water beyond, which in turn makes the former become denser than normal water and the latter less dense and more ice-like. Some studies of water in cells have indicated that indeed it seems able to adopt both higher and lower average densities than ordinary water – one study in 1984 reported an average density midway between that of normal water and that of ice. Wiggins argues that the 'stretched' water is more ordered and less adept at solvating ions – in other words, that the concept of water as the ideal solvent for salts might break down inside the cell, where the solvating power of the liquid varies from place to place. It is a provocative idea that would have profound consequences for cell biology if it proves to be true.

But the biggest controversies rage around water at hydrophobic surfaces. In the 1980s, when sensitive methods for measuring the forces between two very close surfaces were developed, evidence began to emerge that two hydrophobic surfaces with water between them attract one another over distances of up to 300 nanometres. That may sound like a short-range – it is hundreds of times smaller than the width of a human hair – but it is far, far longer than the range of any known interactions between neutral molecules, which don't extend beyond ten nanometres or so at most. An attractive force extending over 300 nanometres between hydrophobic surfaces in water is almost absurdly long-ranged – it must be mediated by something like a thousand intervening water molecules. To understand how surprising this is, imagine how you'd feel if you were to be knocked over in Central Station at rush hour by a porter striding past on the opposite side of the hall.

Explanations for this apparent long-ranged transmission of short-ranged forces have tended to focus on water's uniquely structured character. One suggestion is that the hydrogen-bonded network conspires to propagate the layer-like structuring acquired near the surface over many more layers than usual. But there is no evidence for enhanced structuring of water at hydrophobic surfaces beyond a few molecular layers. Another idea is that the long-ranged hydrophobic force is related to the formation of bubbles of dissolved gas between the surfaces. When a bubble is wedged between two surfaces, they are linked by a meniscus which pulls them together by surface tension. But the force remains even when the water contains very little dissolved gas.

Alternatively, maybe the water itself just vaporizes between the surfaces. The rationale here is that, while a little rearranging of the hydrogen-bonded network may be feasible around a small hydrophobic molecule like methane, cramming water between two large hydrophobic surfaces is just too destabilizing. The edges of water's jigsaw are disrupted over such a wide area that the rest of the puzzle simply won't fit together in between. Water can't do its dance in too narrow a space. Then the water simply vacates the premises, leaving only its vapour behind. Because the pressure in the vapour is lower than that in the liquid, the surfaces are pushed together. This phenomenon is called, reasonably enough, 'drying', and calculations indicate that it is expected between hydrophobic surfaces separated by a hundred nanometres or so.

Whether this mysterious long-ranged attraction between hydrophobic surfaces has any biological consequences isn't known, but that seems entirely possible. Physicists David Chandler, John Weeks and Ka Lum suggest that drying, rather than Kauzmann-style hydration, might be the dominant hydrophobic interaction driving protein folding and the aggregation of individual protein molecules into the gangs in which they often operate. The case is far from closed, however, on just how water in narrow spaces acquires its extraordinarily long grasp.

Switches and wires – cell water at work

Despite all this uncertainty about what water in the cell looks and behaves like, it is emerging as a kind of biological odd-job man, finding all sorts of useful little tasks in life's business. To an evolutionary biologist, this might be no more than you'd expect: if water is the matrix of life, then it is natural that the bigger biomolecules may have evolved to exploit the peculiar capabilities of their solvent. But it isn't just any old molecule that could perform this variety of functions. Indeed, in 1913 biochemist Lawrence Henderson argued that water is particularly 'fit', in Darwin's (more properly Herbert Spencer's) sense, for its biological role.[11]

I've indicated already that some of the water molecules associated with proteins are not just part of a vague hydration shell – they are in some, none too rigorous, sense 'bound' to the protein's peptide chain by hydrogen bonds. Situated in very specific positions that can be pinpointed by experimental methods of molecular structure determination, they can be considered a part of the protein's machinery – bolt-on modules that can be added and removed in solution to enable the protein to perform its task. Water molecules can find their way deep inside a folded protein, sometimes linked into hydrogen-bonded chains – bridges or wires one molecule thick – that thread their way through the protein to connect inside to outside. These water wires provide transportation channels for hydrogen ions, which appear to hop from one molecule to another by the flipping of hydrogen bonds (see page 168). In this way, the wires 'conduct' hydrogen ions deep into the biomolecule, where they may participate as a crucial ingredient in some catalytic chemical reaction.

Water molecules attached to a protein can play a central role in the way that it works. The haemoglobin in the blood of the bivalve mollusc *Scapharca* is different from that in our own blood, in part because it consists of two identical protein molecules, called subunits, bound together. At the interface between the two protein subunits there are seventeen water molecules that occupy well-defined positions, and the hydrogen bonds between these molecules and the protein chains seem to provide much of the glue binding the two subunits. But the waters do more than this.

Each of the subunits possesses an oxygen-binding entity. But they don't act independently; the binding of the two oxygen molecules is cooperative. That's to say, once one oxygen is bound, it is easier for the other subunit to bind a second oxygen. When either one of the subunits binds an oxygen molecule, its structure changes in a subtle way; and it seems that the water molecules at the interface between the two subunits might act to transmit the influence of this change across the interface to the other subunit, allowing it to adopt a structure that is more accommodating to the second oxygen. This cooperative binding is an important aspect of the haemoglobin's function, allowing it to regulate its capacity for oxygen depending on how much there is around. The bound water molecules are acting as a kind of molecular transmission system for this cooperativity, enabling changes in one part of the protein to effect changes in a more distant region.

Bound water molecules have been shown to help a protein decide which molecules to seize in its binding pocket. Certain types of bacteria possess proteins that bind and transport the sugar molecule L-arabinose. This arabinose-binding protein has far less affinity for binding another sugar, D-fucose, whose structure is very similar to L-arabinose (in a manner comparable to that depicted in Fig. 9.2). So the protein's binding site must be very adept at 'feeling' the shape of the molecule it is binding. It appears that bound water molecules are an essential part of this sensing apparatus. There are several water molecules held within precise locations in the binding pocket by hydrogen bonds, and some of these act as bridges between hydrogen-bonding groups on the protein chain and those on the captured sugar molecule. Two of these water molecules are in just the right place and orientation to engage in hydrogen bonding with L-arabinose but not with D-fucose.

It is hard to overstate the importance of this kind of delicate specificity of binding interactions for proteins. You've seen how crowded the cell is; how easily, then, might a protein pick up the wrong target molecule, unless its binding site is exquisitely tuned. Water has been co-opted during the course of evolution to help in that tuning: its propensity for forming hydrogen bonds that point in very specific directions makes this possible. A solvent that lacked this ability wouldn't be half so useful for the cell's delicate requirements. That the only solvent with the refinement needed for nature's most intimate machinations happens to be the one that covers two thirds of our planet is surely something to take away and marvel at.

Part IV

Strange Brew

10 Pride, Prejudice and Pathology

The allure of weird water

There is no greater folly than to be very inquisitive and laborious to find out the causes of such a phenomenon as never had any existence, and therefore men ought to be cautious and to be fully assured of the truth of the effect before they adventure to explicate the cause.

John Webster (1677), *The Displaying of Supposed Witchcraft*

In the end, nothing was salvaged. Why should there be? There isn't anything there. There never was.

Irving Langmuir (1953), Colloquium presented at General Electric

A question I, as an editor at *Nature*, am often asked by friends who are not scientists is this: how can you know, when you refuse strange, hand-written papers that make dramatic and far-reaching claims (typically involving supposed errors in the theory of relativity or quantum mechanics), that you are not dismissing an unacknowledged genius? Galileo was mocked and persecuted, they might say; Einstein was little known when in 1905 he made his major contributions to science. Sure, there are plenty of cranky ideas out there – but how can you be certain that this is just one more of them?

How does one express intuition, in this situation, without projecting arrogance, dogmatic orthodoxy, even the possibility of professional laziness? The shorthand of 'You just know' is not easy to interpret for non-scientists; something more is warranted.

So I might talk about Occam's razor, the invaluable scalpel that scientists are expected to wield (albeit in practice with all the encumbrances and prejudices of human nature). When, for example, we are asked to believe that the tiny differences between two huge numbers, measured by experimental procedures documented in impenetrable recipes and quoted without estimates of errors, reveal a fundamental flaw in the laws of physics that predict the behaviour of the rest of the world with utter reliability – is it then more economical to assume that the laws are wrong or that the experimental techniques aren't up to the task, that the discrepancies lie within the margins of random or systematic error? Even this, however, is not really good enough justification for being dismissive. Quantum mechanics grew out of niggling inconsistencies, general relativity was originally validated by tiny, barely detectable effects of – in retrospect – questionable conclusiveness. So there is more to the unconscious assessment process than Occam's razor, more that is hard to express concisely: a need for congruence with accepted modes of enquiry, a connection with previous scientific endeavour (Galileo and Einstein both had that), an obvious criterion of falsifiability.

What I might do instead, therefore, is to offer the following parable. In making these assessments of the more, shall we say, challenging contributions sent to the journal, I will be exercising nothing more than the same kinds of mental processes we display routinely in our lives. Let's say that an acquaintance, someone whom we do not know well enough to trust implicitly, proffers to us the information that one can buy blue apples in Peru. (Or, if you are a Peruvian, you might like to substitute 'Watford'.) We could choose to believe this at face value – but I submit that most of us would not. Why? Because it requires too much reorientation of facts we already feel to be secure. That blue apples might exist is not obviously an impossibility; but it stretches credulity to think that we would not have already known of so remarkable a thing, if it were true. Of course, we all have vast vistas of ignorance even of the most mundane of topics, but I suspect we feel rather secure in the notion that we have not been labouring all our lives under a fundamental misconception about the colours of apples. It is far more reasonable – though not necessarily correct – to assume that we are being spun a yarn.

To those of you who have exclaimed 'What? You wouldn't even check it out?', I should tell you another element of the tale: the last twenty times you did decide, for whatever reason, to follow up the claim, it turned out to be moonshine. Are you still interested? Blue apples are sighted more or less weekly at *Nature*, and with a rigour that is generally proportional to their plausibility.

There is, however, a much more difficult problem that arises from time to time, and I might express it like this. What if your acquaintance tells you instead of apples that are bright orange? Now, that sounds unlikely – but you are probably less inclined to dismiss it outright. Fruit can most evidently possess this hue, though I have never seen an apple so tinted. Orange apples are not easy to credit, but our instincts do not so forthrightly rebel against the idea. We can imagine that, just possibly, we might be presented here with a truth that we would never have guessed at, which reveals a surprising gap in our knowledge, but which is none the less true for all that.

Scientific orange apples come along every so often, and they pose a thorny challenge for the scientific community. Does it maintain its habitual (and necessary) conservatism, and dismiss them out of hand? Or does it embrace their possibility, and seek verification? The apple tempts – because every scientist dreams of uncovering some new and startling aspect of the world. The risk of a closed mind is a dull and ineffectual career. The risk of gullibility, meanwhile, is a ruined one.

And sometimes the stakes are raised still higher, for the orange apple might, if it exists, be bright with commercial potential. The market for an orange apple might be immense. There is nothing like the prospect of a financial killing to raise the temperature of a scientific debate.

Every field of science is susceptible to orange apples, and most have seen them come – and either be absorbed into the accepted wisdom of the field, or sink without trace. But there is no enterprise in science that is more vulnerable to the fool's gold of these vivid fruits than one in which the most basic concepts still remain uncertain and disputed, in which mysteries and controversies still lurk. You will have heard enough now to appreciate that the science of water is going to be a rich hunting ground for those seeking oddly coloured fruits. In this and the following chapter I will tell you about some of the most notorious. Water has some strange relatives, to be sure, but stranger still are those that never were.

Pleasing your peers

In exploring some of the strange research that the study of water has produced, it will serve us well to take a few steps back to look at the way science conducts itself. Scientists are ambivalent about such scrutiny, particularly when it is focused on outbreaks of questionable science – although these can be amongst the most revealing! Part of the discomfort stems from the fact that, unlike most science, there does not exist any agreed framework for making the analysis.

I do not hope to provide any definitive answers to the question of What Went Wrong in these cases. Rather, I think it might be more useful to try to convey to you how science's error-correcting mechanisms work, and why scientists have come to select these and not others. Perhaps more importantly still, I would like to show that there are varying degrees of wrongness in science. This is, I believe, still a puzzle to the outsider. Why do scientists still employ Isaac Newton's classical physics even after Einstein's relativistic physics showed it to be in some sense wrong? And why do scientists choose to take one far-fetched theory seriously, at least to the extent of devoting time and money to checking out whether there is anything in it, while dismissing others without a second thought? I hope to show why it might be that research associated with water may be especially susceptible to false leads and controversial ideas.

Publish or be damned

What does the scientist do with his or her results, be they derived from benchtop experimentation, computer modelling, observation of nature or pen-and-paper calculations? The catchphrase in today's scientific community is 'publish or perish': there is a tremendous pressure to maintain a high output of published work, ideally in the most visible journals

A paper submitted for publication is generally sent by one of the journal editors to one or more 'referees' – scientists who are considered to have sound judgement on the topic that the paper addresses. The referees will send the journal reports on the quality of the paper, on the basis of which the editors will decide whether or not to accept it for publication. Referees' reports cover a broad spectrum of styles and opinions. Some are brief and enthusiastic ('Great paper. Publish as is'). Some are equally short but utterly dismissive ('This paper is deeply flawed and should not be published in any journal'). Commonly the report will recommend that publication be conditional on several changes being made, to improve the clarity or to correct minor errors. Needless to say, it is not uncommon for an editor to receive reports that give totally opposite recommendations – and then it is up to the editor to find a way to resolve the discrepancy.

The justification for peer review is that it provides a normalizing, corrective and improving influence. When it works as it should, it looks like a splendid system, offering supportive encouragement and advice to good contributions while screening out those that would not add substantially to the literature. But the peers are only human, and the system is not infallible. Most scientists figure that, on the whole, the good outweighs the bad – and that in any case, no one has come up with a better suggestion for policing the scientific literature.[1] Peer review seems at least a preferable option to the practice of the early twentieth century and before, when papers might often be accepted on the whims of a journal editor with no specialized knowledge, or at the demand of a particularly influential yet opinionated member of the scientific élite.

Pathological science

It is not hard to see how the system of peer review can let through some rotten papers – from time to time, referees are sure to be caught napping. But mostly these get buried and forgotten – everyone but the author realizes that they have little intrinsic merit, and ignores them. Occasionally, though, a single misleading publication, or a flawed body of work from one laboratory, can provoke a stampede in which an entire subset of the scientific community gets gripped with a fever and goes rushing down a blind alley only to discover sheepishly that there is nothing to be found there. Why does this sometimes happen and sometimes not?

If we knew that – if we had an infallible way of distinguishing productive science from unproductive (let's not say 'good' from 'bad', or even 'right' from 'wrong', for neither is quite the same thing) – the work of the scientist would be considerably simplified. But I don't know of any such method; and moreover, I don't believe that one can exist. Science cannot work at all if it cannot make mistakes. In fact, as Brian Appleyard has pointed out,[2] science derives much of its formidable strength from the ability to make – and to live with – mistakes. Many other systems of belief have great problems with mistakes – admission of such in religion and politics is seen to be a mark of weakness. Much of the practice of science, meanwhile, consists in the gradual correction of the mistakes of previous generations. You could say that science progresses by bringing mistakes to light, not by trying to hide them. Michael Faraday voiced the hope that, fifty years after his death, everything he had said would be shown to be untrue – for that would provide a measure of real scientific advance. Any attempt to eradicate mistakes from science will then be at the expense of the 'scientific method'.[3]

Science operates at the focus of the tension between opposing forces. It must be prepared to get things wrong – but it must have mechanisms for eventually bringing such errors to light. The larger the error, the more that is at stake, and so the greater the need to identify it quickly. The concept of phlogiston kept chemistry on the wrong track for the best part of a century, and although personally I don't believe that it 'held chemistry back' to any great degree, scientists do not want to spend their lives working under such misapprehensions. Science must find a balance between conservatism – scepticism for new ideas – and innovation. Both are essential to its well-being. Those who criticize science for its refusal to accept certain leaps of faith (to embrace, for example, extrasensory perception) must realize that, if it did not behave in this manner, it would not be science. You might as well berate a whale for having no wings.

Science has the greatest difficulty in knowing which way to tip the scales when discoveries are made that push at the boundaries of conventional thought without obviously stepping beyond the plausible. In this and the next chapter I consider three such cases in some detail. Where the boundaries of credibility lie in each case is a matter of opinion, and there are always voices who say (in retrospect) 'I knew it was nonsense from the start'. One thing is sure – the boundaries contract pretty smartly once the business is over.

Few scientists have dared to dally long in the nebulous regions at the fringes of science, for that is a thankless occupation. Picking over the bones of discarded and discredited ideas is regarded as a distasteful exercise to many scientists, who would rather leave the remains buried and not risk being tarnished by association. It is bad form to dwell over what goes on out there, and there is no shortage of once-respected figures in the scientific community whose dabbling at the fringe in later life serves to remind others of the kind of polite ostracism that lies in store if you stray too far down that road. These are the uncharted netherworlds of the scientific map, which the supremely rational scientist peoples not with dragons but with charlatans and New Age mystics. Sadly, by doing so he fails to notice that much of society is out there too, unaware that any such boundary exists.

Yet one eminent scientist who did dare to turn his critical faculties on these grey areas of science was the great American chemist Irving Langmuir. He never published his thoughts on the matter, but they are preserved for posterity in a poor-quality sound recording of a lecture that he delivered in 1953 at a laboratory of the General Electric corporation. Here Langmuir tried to identify some of the distinguishing features of 'the science of things that aren't so' – by which he meant the scientific claims of illusory phenomena. He called this *pathological science*. The study of water has proved itself unusually vulnerable to such pathology.

To Langmuir, pathological science has six criteria:

i. The phenomena responsible for the claims made are barely detectable.
ii. The phenomena are provoked by a cause of barely detectable intensity, and the magnitude of the effect is independent of the intensity of the cause.
iii. The observation of the phenomenon is claimed to great accuracy.
iv. The explanation offered is extraordinary and conflicts with previous experience.
v. Criticisms are met with *ad hoc* excuses, often thought up on the spur of the moment.
vi. The ratio of supporters to critics rises to about 50:50, and then declines to virtually zero.

Langmuir arrived at these criteria by looking at previous examples of striking scientific claims that could never be repeated reliably and faded to obscurity – in particular, the N-rays 'discovered' in 1903 by the Frenchman René-Prosper Blondlot, a respected member of the French Academy of Sciences. These were supposedly a new type of radiation possessing peculiar properties: penetrating some metals but not others and being emitted by humans. N-rays were eventually seen to be a product of Blondlot's self-deception – but it is notable that their 'discovery' followed on the heels of Wilhelm Röntgen's genuine discovery of X-rays, a phenomenon in some ways as remarkable. To an outsider, N-rays might have seemed at the time no more intrinsically improbable than X-rays. With characteristic dry humour, Langmuir explained that,

> You see, N-rays ought to be very important because X-rays were known to be important, and alpha rays were, and N-rays were somewhere in between, so N-rays must be very important.[4]

But they were not – they were an illusion, a nonsensical concoction of subjective impressions and improvised explanations. They fitted Langmuir's criteria for pathological science like a dream.

Langmuir's criteria provide some very useful guidelines for distinguishing worthy science from false leads.[5] But to a non-scientist, they might appear mystifying. For one thing, they seem rather arbitrary. Individually, all of them can be applied to elements of the scientific canon that are well-established. Relativity and, to some extent, quantum mechanics, both radically challenged the existing foundations of physics, and at the outset both either arose from or predicted phenomena (that is to say, deviations from classical physics) on the border of detectability. Scientists routinely claim measurements of great accuracy, and *ad hoc* defences from criticism are sadly not uncommon. For all that Langmuir's ideas on pathological science give scientists some rules of thumb, which are in any case applicable in a consensual manner only in retrospect, they may offer little insight to an outsider who wants to understand why scientists so resolutely dismiss some ideas while being prepared to entertain notions as fantastic as quantum teleportation.[6]

In my excursion into wild waters, I hope to shed some light on the incongruence of world views that we see increasingly today between some practitioners of science and those who believe that there is more to the world than hard rationalism.[7] In the end, we must each make up our own mind.

Perilous gum

If there is one spectre that haunts water research today more than any other, it is the one unleashed at the end of the 1960s which bears the name 'polywater'. Three decades on, this word can still provoke a wince or a smirk, depending on which side one sits. Neither before nor since has water so thoroughly taken science for a ride.

Polywater was supposed to be a new form of water. Its discovery was announced in the late 1960s by a group of scientists working at the Institute for Physical Chemistry in Moscow. This was no hypothetical stuff deduced by tortuous reasoning from arcane graphs and calculations; they had it on photograph, and could prepare tiny droplet-sized samples in thin capillary tubes. At first they called it, with commendable caution, 'anomalous water'. That bland and cryptic description sounded no more enticing then than it does today, and few people took any notice until a more glamorous title was devised.

The 1960s was a curious decade for international science. Great advances were taking place on many fronts – in molecular biology, in computer science and space technology, in fundamental physics. Yet the communication of these ideas left much to be desired, for the wish of the scientific community to operate as a global enterprise was compromised by the Cold War, by a cloak of paranoia between East and West that has been unmatched since. Science in the Soviet Union was burgeoning, and US scientists had good reason to be worried, as they are no longer, that any important discoveries or inventions they were making were being mirrored, perhaps even anticipated, in the Soviet Union. American citizens and politicians were left reeling from the shock of taking second place to Sputnik and Yuri Gagarin's sojourn in planetary orbit.

Westerners published in the British and American journals; Soviets published in their own Russian-language journals. Often the transmission of ideas from East to West and vice versa relied on transient, chance encounters at conferences in Russia and Europe. This is how news of anomalous water came West.

Down the tubes

The messenger was one of the Soviet Union's most distinguished experts in surface chemistry, Boris V. Deryagin. When he first spoke about the discovery in the West in 1966, his team had already been studying anomalous water for four years, and had published ten papers on the subject – but all in Russian journals, where they went unread by Western scientists.

The discovery was not Deryagin's, however – he was always quite explicit about that. It had been made by Nikolai Fedyakin, working in relative isolation at the Technological Institute in Kostroma, northeast of Moscow. Fedyakin had observed that when certain liquids, including water, were sealed in very narrow glass capillary tubes, secondary columns of liquid appeared in the upper part of the tubes and seemed to grow at the expense of the primary column. It appeared that the liquid evaporated from the primary column and then condensed further up the tube. This was very odd, because the temperature and pressure in the tubes should have been equal everywhere, and so each liquid should simply have coexisted with its vapour at the appropriate vapour pressure. There was no incentive, if you like, for the liquid to evaporate and then condense somewhere else.

The only way to account for the observations was to suppose that the liquid in the secondary columns was different from that in the primary column, with a lower vapour pressure. For one thing, the secondary liquid seemed to be denser. Could a liquid adopt a second structure, identical in chemical composition but denser than the familiar form?

Fedyakin published his observations in a Russian journal of colloid science, but he appeared to have little idea what to make of them. It took Deryagin to appreciate the stunning implications, and on seeing Fedyakin's report, Deryagin seems to have effected a collaboration that was to all intents and purposes a takeover by the powerful Moscow group. Fedyakin appears on most of the subsequent papers from Deryagin's team, but his role after the initial publication was not a major one, and from 1963 no mention is made of the Technological Institute of Kostroma.

Focusing on the case of water, Deryagin realised that what he seemed to have on his hands was a form of water with an anomalously high density. The word 'anomalous' is perhaps misleading here, because it turns out that the form with the lower vapour pressure – the 'new' form – must be the most stable. In other words, normal, everyday water, the stuff that has flowed across the planet since time immemorial, should be metastable with respect to the anomalous form, water's 'true' manifestation.

But why might water adopt a modified structure in these tiny capillaries? Deryagin was aware that the presence of surfaces modifies the structure of liquids, as I explained on page 244. Exactly *how* the structure of water is modified by surfaces remains controversial even now. But few would question that it is modified in *some* way. Deryagin proposed that this modification is inculcated in *all* the water when it condenses from the vapour inside the capillary tube.

The trouble is that no one would have expected this modification to extend much beyond a few molecular layers from the surface – whereas molecules in the middle of the capillary columns could be about 5000 molecular diameters from the surface. And what is more, it was stretching credulity to expect water to 'remember' this surface-modified structure after becoming vapour. But all the same, the

Soviet scientists chose to accept this as a working hypothesis. They seemed at first, and quite sensibly, more concerned with finding ways to make anomalous water reliably than with explaining its existence.

Deryagin's team developed a scheme for growing columns of the stuff in capillary tubes more quickly than had Fedyakin: his secondary columns took weeks to appear, whereas the Moscow team could eventually do it in a few hours. The big worry was contamination. The amounts of anomalous water grown in each experiment were almost microscopic: the columns grew to just a millimetre or so in length, in tubes no more than a tenth of a millimetre in diameter, and it would take only a minuscule quantity of dissolved or suspended contaminants to change the properties of that much water substantially. In particular, there was the possibility that inorganic material within the matrix of the glass capillary tubes might be leached out of the walls into the water. Ordinary glass is primarily silica (silicon dioxide), which is rather insoluble in water; but it typically contains a range of other inorganic substances too, which might dissolve more readily. So Deryagin and colleagues took the precaution of using capillary tubes made from pure crystalline silica (quartz) rather than ordinary glass. They were also careful to check rigorously the purity of the water they were using.

What manner of substance was the anomalous water? It was available only in tiny amounts, severely restricting the experimental tests that could be conducted. The researchers found nonetheless that it did not boil at 100°C, but more typically at around 200°C. It froze gradually over a wide range of temperature, beginning around –30°C. The density of the secondary columns was generally ten to twenty per cent greater than that of water; but the Soviet team deduced that these columns were usually mixtures of anomalous water and normal water. When the latter was largely removed, the density of the anomalous fraction was about forty per cent greater than ordinary water, and the viscosity was perhaps fifteen or so times greater, giving the substance a consistency like paraffin wax. It was as though water had somehow coagulated in the tubes.

Call and response

A few Western scientists stumbled across anomalous water when Deryagin spoke at a chemistry conference in Moscow in 1965 – but the low-key way in which he chose to present the findings, coupled with the poor quality of the translation facilities at the conference, meant that the revolutionary nature of the claims passed unnoticed. Harder to explain is the fact that the announcement met with a rather similar response when Deryagin spoke about the work for the first time in the West, in English, at a meeting in Nottingham, England, in September 1966. Possibly the mainly British scientists present at the conference felt highly sceptical about these astonishing claims – it is a sad but undeniable fact that a speaker using an unfamiliar language usually has a tougher job to sound persuasive – but it may also be that Deryagin's audience was not at its most alert and attentive that day.

Be that as it may, a few British scientists did sit up and take note. Amongst them were Brian Pethica, director of the Unilever Research Laboratory at Port Sunlight in northwest England, and none other than Desmond Bernal, the famous crystallographer who had speculated influentially on the hydrogen-bonded structure of liquid water in the 1930s (page 161). Bernal invited Deryagin to visit his laboratories at Birkbeck College in London, where he and his PhD student John Finney engaged in eager discussion about how to make this new form of water. Deryagin confided to Bernal during this visit that he thought the anomalous water might be some polymerized form, in which the molecules were linked via a modified, more robust form of the hydrogen bonds that join them into a random network in the normal liquid state. Bernal, for his part, offered the opinion that 'this is the most important physical-chemical discovery of the century'; and he set Finney the unenviable task of reproducing the Soviet experiments.[8]

Pethica, meanwhile, set to work with his own team after the Nottingham meeting. By 1968 he was ready to announce at a meeting in the USA that his group had also made columns of anomalous water. Deryagin had already told the Americans about his work during a visit in 1967, but had been awarded, if anything, still greater scepticism than had greeted him at Nottingham. Pethica's announcement provoked no mad rush to the laboratory either, but at least one of the scientists who heard it, Robert Stromberg of the National Bureau of Standards in Gaithersburg, Maryland, considered the phenomenon intriguing enough to warrant further study.

All the same, throughout 1968 what small degree of work there was in the West on Deryagin's anomalous water (for soon it was his name, not Fedyakin's, that was firmly associated with the stuff) took place largely in Britain. Some rumour of this activity found its way back to the USA via Ralph Burton, a liaison scientist stationed in London for the US Office of Naval Research, whose job it was to sniff out developments in Britain deemed worthy of US interest. In April of 1969, the momentum began to gather: Pethica's team at Unilever published in *Nature* a report of their studies, making clear how these were stimulated by Deryagin's work. Dramatically, they included photographs of the anomalous-water columns – visual evidence that something had condensed in the capillary tubes. But they were cautious about any interpretation, saying 'it is possible that a gel could be formed by the leaching of silicates from the glass'.[9] (They used normal Pyrex glass tubes, but noted that this kind of leaching was unlikely for the quartz tubes of Deryagin's team.) Pethica and colleagues concluded, reasonably and indeed obviously, that a lot of the stuff was needed before one stood much hope of figuring out what on earth it was. 'If only we could make a thimbleful,' John Finney lamented.

The development that turned anomalous water into an elusive international star came two months later, when the fruits of Robert Stromberg's interest were revealed in the American journal *Science*. Stromberg had teamed up with Ellis Lippincott, director of the Center for Materials Research at the University of Maryland. Lippincott had already published a study of anomalous water in May, in collaboration with Lionel Bellamy of the British Ministry of Technology and two British colleagues. Bellamy had enlisted Lippincott's help to study the structure of anomalous water using the technique of spectroscopy. This, perhaps the most valuable of the chemist's battery of analytical tools, involves measuring the absorption of electromagnetic radiation (for example, visible light, infrared or ultraviolet radiation) by the substance under scrutiny. A molecule may absorb the radiation at frequencies that correspond to the natural vibration frequencies of the molecule. Even a simple molecule like water has several different vibration frequencies, corresponding to different types of vibration, and so the absorption spectrum – a graph of the intensity of transmitted light as the frequency is changed – consists of a wiggly line with dips whenever the frequency coincides with a vibration frequency of the molecule.

Lippincott, Bellamy and colleagues reported an absorption spectrum for anomalous water that looked markedly different from that for normal water – the first real indication that anomalous water might have a dramatically different structure. They concluded that the anomalous water 'must consist of polymer units', and even suggested a possible structure for these joined-up water molecules in which four were linked in a square. This structure had been considered by Deryagin too.

But the paper by Lippincott, Stromberg and coworkers in June was altogether a more striking and declarative piece of work, and it was the catalyst that awoke the US community to the existence of this peculiar form of water. 'Vibrational spectra indicate unique stable polymeric structure', the paper's subheading declared, and most significantly, the authors produced a catchy name for this new form of water, which gave the paper its one-word title: *polywater*. But beyond that, the paper added little more to the experimental picture than some fresh spectra. It also offered, however, a new interpretation: Lippincott's square assemblies of the previous month were abandoned without mention, in favour of a new hypothesis: that polywater had a structure based on hexagons of six water molecules, linked into sheets or, in a more disorderly manner, into branched chains. Remarkably, the hydrogen bonds in these proposed structures were depicted as symmetrical: whereas in normal water the hydrogen bonds contain a hydrogen atom attached asymmetrically between the two oxygens it unites, in polywater it was supposed that the hydrogens somehow bridged the two oxygens symmetrically, sitting midway between them. This would account for the stability of the polymerized form. There was no clear theoretical rationale for invoking such a bonding arrangement, although the researchers offered some sketchy justification for it.[10]

Sweating it out

Duly christened, there was no holding polywater back. Now that it had been given the imprimatur of a respected US chemist, it was deemed fit for serious research. And suddenly everyone realized just how shattering were the implications of Deryagin's work. It is a salutary and perhaps disturbing lesson that is still relevant for science today: if you want to get noticed, you must get your work into a high-profile journal and describe it in bold, confident terms, ideally with a neat neologism attached. And it helps if you work in the West.

But probably most significant of all for what followed was the fact that in the summer of 1969 the world's news media woke up to polywater. Once that happened, the temperature of the debate was raised by several degrees, the claims expanded to absurd proportions, and the popular media, rather than the scientific literature and conferences, became the preferred vehicle for some of the more outspoken scientists on both sides of the controversy to engage in dispute. A flurry of possible applications for polywater were proposed in newspapers: a lubricant, a corrosion inhibitor, a moderator material for nuclear reactors. Even 'respectable' newspapers seemed prepared to abandon all restraint or semblance of discretion: 'A few years from now living room furniture may be made out of water,' exclaimed the *Wall Street Journal*. The Moon landings that year had shown dust sticking to the astronauts' boots on the supposedly dry surface, leading to speculation that water might have persisted in the Moon's almost non-existent atmosphere as polywater, then mixed with Moon dust into a kind of lunar polymud.

And most provocative of all were suggestions that polywater might hold the key to some of life's mysteries. Was not water in the cell also highly confined by surfaces, like that in the capillary tubes? *Newsweek* announced, apparently without any real evidence, that researchers were on the trail of polywater in living cells. Elsewhere it was implicated in the transport of water columns up the vascular systems of giant trees (page 222), a phenomenon that at that time was still not fully understood.

As if that was not enough, in October 1969 polywater turned from notorious rogue to villain. F.J. Donahoe of Wilkes College in Pennsylvania published a short note in *Nature* whose grim warnings still make astonishing reading today:

> After being convinced of the existence of polywater, I am not easily persuaded that it is not dangerous. The consequences of being wrong about this matter are so serious that only positive evidence that there is no danger would be acceptable. Only the existence of natural (ambient) mechanisms which depolymerize the material would prove its safety. Until such mechanisms are known to exist, I regard the polymer as the most dangerous material on earth . . . Even as I write there are undoubtedly scores of groups preparing polywater . . . Treat it as the most deadly virus until its safety is established.[11]

What was Donahoe so terrified of? It was the *Cat's Cradle* syndrome, rising from science fantasy like a phantom to stalk the Earth for real.

Polywater appeared to be the more stable form of water. Place a small amount of it in contact with normal water, and who is to say that it will not act as a seed for the nucleation of more polywater, transforming all of the liquid to a waxy substance? Donahoe could, it seems, envisage the oceans gradually gumming up, destroying one of the central organs that maintain the planet's life-friendly environment. Here was a doomsday material more potent than plutonium.

Donahoe was taken sternly to task just two weeks later by weighty members of the British scientific establishment: Bernal and his colleagues at Birkbeck, and Douglas Everett at Bristol, one of the giants of surface science. They denounced his letter as 'unduly alarmist and misleading'. Not only was there no evidence in the laboratory that polywater grew at the expense of normal water when the two were in contact, but the latter had existed for billions of years in the geological environment in intimate contact with quartz surfaces – ideal conditions for growing polywater. Yet the oceans were as fluid as ever. Said Everett with a homeliness that seems curiously British,

> Robert Burns's affections were guaranteed to remain constant 'till all the seas run dry'. While he may not have envisaged the possibility that the oceans might instead become anomalous, we feel that his shade may derive some consolation from the fact that they have not already done so.[12]

But such calm and sound advice is no fun for the media, who made a predictably elaborate meal of Donahoe's misguided warning. From then on, says Felix Franks in a commentary on the polywater affair, 'objective and dispassionate discussion [was] almost impossible.'[13]

We can witness, from mid-1969, a strange process being acted out within the scientific community. It is as if the cautious, conservative mantle of science had suddenly been whipped away, as if critical faculties and rigorous thinking were temporarily shelved, while scores of researchers worldwide dropped tools and rushed to join the bandwagon, no matter how hasty and slapdash that required them to be. It is worth bearing in mind that by the end of 1969 not a single decent chemical analysis of the stuff that appeared in the capillary tubes had been published. That is, no one knew for sure that the mysterious secondary columns were truly water! Yet already scores of theorists were hard at work proposing structures for this new form of water, some supported by calculations that were as detailed as the state of the art would allow. Why expend such effort (and research funds) on a substance that might or might not exist? Odd as it may seem, this is standard practice for theorists in chemistry. It is an exploration of the possible, a stimulus for future research – if a compound with interesting properties appears to be stable in theory, it might be worth trying to make it. To what extent such exploratory theory has successfully guided experimental work remains a matter of some dispute; but the theoretical tools are today sufficiently honed that the predictions may be accorded some degree of respect. In 1969 that was less evidently the case.

So the fact that theorists were chasing a potential will-o'-the-wisp was nothing new, nor reprehensible or particularly unusual – although normal standards of rigour may have suffered amongst the frantic scramble. What is odd is that so few people seemed eager to look really closely at what was actually in the capillary tubes.

One who advocated doing so was Arthur Cherkin of the Veterans Administration Hospital in Sepulveda, California. Cherkin had considerable experience with the leaching of materials from glass into water, a problem of much concern in hospitals where high purity is needed and solutions are often stored in glass bottles. Cherkin charged that the frequent dismissal of possible contamination – by Deryagin, Lippincott and other polywater advocates – was not convincing, and that polywater could be nothing more than a dispersion of silica as microscopic particles or as a gel. Deryagin's standard line on the contamination issue had hardened into a lofty disclaimer of responsibility for any polywater samples not prepared in his own laboratory. Of course, Deryagin could ill afford by then to broach the possibility of his being misled by poor experimentation. Yet it is amazing that so few others seemed ready to tackle that problem either. It was as though they wanted to believe in polywater to a degree that blinded them to glaringly obvious questions.

Yet those questions were not going to go away, and as the new decade dawned, ominous signs of problems began to surface. In March of 1970, Dennis Rousseau of the Bell Telephone Laboratories in Murray Hill, New Jersey, and S.P.S. Porto of the University of Southern California published a chemical analysis of polywater samples 'prepared by the standard methods' and having 'the same physical and spectroscopic properties as those made in other laboratories'. What they found was a bombshell: relatively high concentrations of inorganic substances such as sodium (up to twenty to sixty per cent by weight of the entire sample), potassium, sulphate and chlorine (which collectively accounted for a further thirty-three per cent), and trace amounts of various other substances. 'Nearly all of the material in these samples,' they said, 'consisted of the contaminants with relatively few H_2O molecules or H_2O polymers present.'[14] Surprisingly, this study seemed to rule out contamination by a silica gel, but it implied that all the excitement of the previous months had been over wet salt.

But it takes more than one publication to stop a bandwagon. In 1970 Deryagin published an authoritative retrospective on 'anomalous water' in *Scientific American* magazine, whose articles generally concern topics that have become established scientific fact. Most polywater proponents pursued their work undaunted by Rousseau's results – after all, they could always tell themselves that it wasn't *their* polywater he'd analysed.

Yet bit by bit, the 'impurity' camp, with Rousseau at its head, steadily gained ground. Not that this initiative had a particularly united front – there were several different theories about the nature of the contaminants. As well as the silica and salt ideas, a prominent hypothesis was that organic materials were responsible for polywater. Coordinating this offensive was Robert Davis of Purdue University in Indiana, who had seen and translated into English a report of the chemical analysis of Deryagin's polywater samples by a Russian scientist. These analyses had been conducted in 1968, but had been later published only in an obscure Russian-language journal, and it seems clear that Deryagin had been far from happy with them. The Russian report spoke of organic contaminants 'in quantities comparable with the amount of modified water itself'. Where was this organic matter from? Davis conjectured that it originated in the sweat of the researchers, and he made this view known in an interview with the *New York Times* in September 1970. A letter from Davis in the US trade magazine *Chemical & Engineering News* was more outspoken: it called polywater 'polycrap', and denounced the whole affair as a waste of time.[15] Davis's entry into the debate demonstrates the extent to which both the vehicles and the tone of new findings had by then deviated from scientific norms.

Whatever the real contaminants were – and they probably differed from one laboratory to the next, depending on the preparation procedures – by early 1971 most polywater researchers could see the writing on the wall. Pethica's team backed off, admitting the likelihood of impurities, as did Leland Allen of Princeton University, a theorist and once prominent polywater advocate who had proposed a graphite-like hexagonal structure for polywater on the basis of very detailed calculations. 'Polywater drains away' proclaimed *Nature* in March 1971. The tide had turned, and some researchers began salvaging what dignity they could from the whole affair. Research on polywater continued at least into 1972, and publications, lagging by the delays of the peer-review process, continued to appear into 1974. Deryagin himself held out for long enough that his eventual (and brief) capitulation to the demon of impurities, in August 1973, was able to pass almost unnoticed.

Short circuit

It is in the response to mistakes, of technique or of judgement, that one sees most clearly the human side of science's practitioners: a mix of pride, stubbornness, humility, honesty, dismay and regret that one could expect to encounter in any section of the population. There are the inevitable diehards, whose pride forces them to concoct ever more elaborate explanations for the facts in order to avoid confession of error. There are the crusaders, determined at all costs to expose the foolishness. And there are the 'told-you-so's', generally members of the scientific élite, who make little contribution while the debate is raging but are ready to heap scorn on the perpetrators afterwards for their 'obvious' mistakes. All of this, of course, is the antithesis of objective science, but follows as inevitably as night follows day from the fact that science is practised by people and not automatons.

To this extent, the trajectory of polywater and the lessons that can be learnt from it might be seen as generic to any controversial science. Felix Franks suggests, however, that one factor that played a major role in turning a controversy into a debacle was relatively new to the way science works: the involvement of the mass media. He argues that the readiness of newspapers to engage in wild speculation, to the extent of actively encouraging it from interviewees, as well as the capacity of the media for reporting new developments almost instantly and the realization by researchers that this was a way to raise their profiles and air their opinions, were all vastly destabilizing, short-circuiting the tempering influences of the conventional channels for scientific information exchange. The polywater affair saw the dawn of scientific disclosure by press release rather than by peer-reviewed publication. Needless to say, this kind of thing is self-amplifying, attracting self-publicists while the more sober-minded individuals stay aloof and silent. What was true in this regard in 1970 is more than ever the case today.

Polywater is clearly not 'normal science'. It exhibits pretty much all the characteristics of Irving Langmuir's pathological science: the experiments operated at the border of detectability, on quantities far too small for the claims of freedom from impurities to be credible, and they required a near-fantastic revision of theory: that water's known structure was not the most stable one. Criticisms were dismissed out of hand (Deryagin being particularly culpable), and the rise and fall were sudden and nearly absolute – polywater came out of nowhere and vanished into oblivion within a few years. Perhaps we should view episodes like this as a kind of 'epidemic science', an affliction to which the very nature of the scientific process makes it susceptible, and during which behaviour is abnormal, activity is feverish, and the 'immune system' is overwhelmed.

But that is still not quite enough. For it is surely the remarkable nature of water itself that made the whole business possible. Water is *already* anomalous, and so there is only a low hurdle to the idea that it can behave even more strangely. There were certainly enough mysteries about its hydrogen-bonded structure in the late 1960s to permit pictures of water polymers like Lippincott's to appear without ridicule. And the behaviour of water at surfaces is in many ways as mysterious as ever, as we saw in the previous chapter. One repercussion of polywater was that any discussion for some time after of 'modified water' at surfaces was regarded as tainted, even though we can be fairly sure that water's structure is indeed modified in some way close to a surface.

The implications for a new form of water stretch wider than they would if it had been just about any other substance that the Soviet scientists had studied in capillary tubes – after all, Fedyakin's similar observations for other liquids were soon forgotten. There is no doubt that media interest would have been just a fraction of what it was had another material, less central to life and to the planetary environment, been at the focus of the debate. All of this forces one to wonder whether the science of water is not abnormally sensitive to epidemics. In the following chapter I suggest that this feature of water might have deep and emotive roots that science may never excise.

11 A Drop of Something Stronger

Water as saviour

We attribute to water virtues that are antithetic to the ills of a sick person. Man projects his desire to be cured and dreams of a compassionate substance.

Gaston Bachelard, *Water and Dreams*

Should we speculate why water can remember something on some occasions and forget it on others?

Sergio Bonini *et al.* (1988), Report on a study into the 'memory of water'

In 1996 an Indian chemist named Ramar Pillai claimed to have found a way to turn water into 'petrol' – a yellow inflammable liquid – by brewing it with a local herb. Pillai showed that this liquid would power a scooter at a cost of about one rupee (three cents) a litre. Scientists at the Indian Institute of Technology in New Delhi and the National Chemical Laboratory in Pune were duly impressed, and Pillai was granted twenty acres of land by the state government of Tamil Nadu to scale up production of 'herbal petrol'. All this while the most rudimentary calculation showed that there was not nearly enough carbon in the added herbs to account for the amounts of kerosene-like fuel produced.

Within a month, the plans for mass production of the new fuel had evaporated, as Pillai proved unable to reproduce it in subsequent experiments requested by India's Department of Science and Technology.

Petrol from water: it is too irresistible, especially to a region as poor as Tamil Nadu. I find myself being asked from time to time if I have heard about the water-powered car. I have never been able to track down the source of these rumours, but am struck by their apparent tenacity in the face of chemical good sense. A car fuelled by water is like a car fuelled by exhaust fumes. Water is not a fuel – at least, not a chemical fuel, one that can be burned like petrol. Rather, it is the exhaust of a fuel: what you're left with when you burn hydrogen.

Water is spent fuel. Yet when water was found on the Moon in 1998, you could be forgiven for thinking, on the basis of media reports, that NASA had unearthed a reservoir of rocket fuel.

The idea of transforming water into something more potent – wine was once more tempting than petrol – has always captivated. Aptly or not, water is a symbol of abundance, and so the idea of extracting something of real value, like energy, from water smacks of getting something for nothing. 'Energy too cheap to meter' was the catch-phrase of the nascent nuclear industry; and when in 1919 Francis Aston made the discovery that started it all, revealing the vast amounts of energy locked in atomic nuclei which can be released by nuclear fusion, he immediately linked it with the powerful image of water as fuel:

> . . . to change the hydrogen in a glass of water into helium would release enough energy to drive the 'Queen Mary' across the Atlantic and back at full speed.[1]

Stephen Poliakoff captured this abiding dream in his play *Blinded by the Sun*. A chemist named Christopher at a provincial university in northern England makes a breakthrough in energy generation that he chooses to announce by press conference before publishing his results in a scientific journal. Christopher has succeeded in turning water cheaply and efficiently into hydrogen. His colleague Al, nervously occupying a new position as head of the department, stumbles across evidence suggesting that the 'discovery' is in fact a rather crude fraud, perpetrated perhaps to secure priority while Christopher works out the final steps in achieving the result for real. Whether or not fraud was indeed involved is never quite made clear; but in any event, the suspicion of it is enough to send Christopher spiralling towards ignomy and disrepute.

The idea of 'water as fuel' is one facet of what I am here calling the myth of *water as saviour*. We accord water the power to rescue us; we look to it for a panacea. After the OPEC oil scare of the 1970s, and with global warming looming and Chernobyl's lethal legacy scattered over the globe, we feel the need for rescue from the unpredictability of our fuel reserves – we yearn for a clean, plentiful juice to power civilization. The first of my two tales in this chapter resonates loudly with that yearning.

But older than this is the idea of water as the saviour from suffering. Water has the power to cleanse, and we imbue it with something more – the power to heal. Sacred waters, drawn from the Nile or blessed by priests, are considered more potent still, infused with a spiritual strength that can raise the dead. Gaston Bachelard gives us a marvellous encapsulation of this impulse in the passage that opens this chapter. Such dreams float beneath the surface of my second story, in which water becomes a fantastic, universal medicine.

To the extent that these cultural emblems of water have led to what I think can be reasonably described as bad science, it might seem that I am suggesting they are a Bad Thing. But you might as well suggest that the archetype of a hero is a bad thing, or the existence of religions. They are simply here with us, a part of human experience. By recognizing them, we may acquire some guidance for our judgements about the supposedly objective findings of science. Water is not – can never be – a value-free subject of study.

Water power

Poliakoff's play was inspired by the supposed discovery, in 1989, of a way to use water as the fuel for cheap, safe nuclear fusion. That story follows shortly. But the science behind his version of 'water as fuel' was rather different, and wholly sound – rather than a nuclear process, it involved the conversion of water to a chemical fuel, something that releases energy when burned in air: hydrogen. Christopher's putative breakthrough was the identification of a way to use sunlight to split water into its constituent elements hydrogen and oxygen. Al puts it this way:

> Water contains hydrogen. But how to get it out? Some chemical reactions are caused by shining a light. Find the right chemical to act as a catalyst – shine a light, a beam, above all the sun – and you can create hydrogen out of sunlight and water. Hydrogen, which will run planes, cars, anything you want. And when you burn it, it will turn back to water. Polluting nothing.[2]

Indeed, how to get it out? William Nicholson and Anthony Carlisle found one answer in 1800, when they split water using Alessandro Volta's primitive battery – a process named electrolysis by Michael Faraday in 1832. Breaking the chemical bonds that bind hydrogen and oxygen atoms in a water molecule requires energy: water splitting consumes power from the battery. And there's the rub. Today, electrolytically generated hydrogen is around ten times more costly than the amount of natural gas required to release the equivalent energy, and three times more costly than the equivalent in petroleum. In fact, most commercial hydrogen is not created electrolytically at all, but by a chemical process called steam reforming in which natural gas (methane) is reacted with steam at high temperature to generate hydrogen, carbon monoxide and carbon dioxide. The carbon monoxide can then be reacted further with water to convert it to carbon dioxide and more hydrogen. Again, you need to put a lot of energy in to get the hydrogen fuel out. This is the current Catch 22 of the 'hydrogen economy'.

But there is a better way. The sunlight streaming over the Earth provides, every day, ten thousand times the energy needed to power all human endeavours, and will continue to grant us this boon for billions of years. The trick is in finding out how to harvest it and put it to use. Conventional photovoltaic solar cells are not able to turn sunlight into electricity at a low enough cost to make them a viable power source for electrolytic hydrogen generation. Christopher's solution in *Blinded by the Sun* – to find a catalyst that captures light and uses the energy to break water's bonds – is 'the Holy Grail of photoelectric chemists', according to a researcher at the National Renewable Energy Laboratory (NREL) in Golden, Colorado. Christopher succeeds in finding this magic ingredient – we are never told quite what it is – and creates the 'Sun Battery'.

In fact, many versions of the Sun Battery exist already. Those developed at NREL use semiconducting materials with exotic names – gallium arsenide, gallium indium phosphide, amorphous silicon carbide – peppered with tiny particles of platinum and ruthenium dioxide as composite 'photocatalysts', and can convert up to a quarter of the energy in the sunlight that falls on them into chemical energy stored in hydrogen gas. Research on photocatalytic water splitting is pursued enthusiastically elsewhere too, particularly in the USA and Japan, and employs other magic ingredients – titanium dioxide (hinted to be involved in Poliakoff's Sun Battery), indium phosphide, complicated compounds of rubidium, tantalum and niobium. But no one is selling Sun Batteries by the truckload, because they are all too expensive. The materials are costly, and the devices just aren't economically viable.

Nature has its own way to split water – photosynthesis. As I explained in Chapter 8, photosynthetic organisms use sunlight to strip electrons from water, generating oxygen gas. The electrons are normally used to convert carbon dioxide to carbohydrates. But if there is not much oxygen in the vicinity, many organisms will instead use the electrical bounty to pull hydrogen from water. One such is the green alga *Chlamydomonas*, which some researchers hope to cultivate as a living source of hydrogen fuel. The algae can be genetically engineered to channel absorbed sunlight into the generation of oxygen and hydrogen rather than into carbohydrate formation. Alternatively, some strains of a type of bacteria called cyanobacteria can reshuffle carbon monoxide and water into carbon dioxide and hydrogen – the very same reaction as the second stage of steam reforming. A further possibility is to extract the various enzymes involved in water-splitting from cells and let them do the job in isolation. This allows oxygen and hydrogen production to be conducted in separate compartments – an advantage because the enzymes that are responsible for making hydrogen from water are typically inhibited by oxygen.

These are not by any means the only avenues being explored for making hydrogen fuel from water. One extreme approach is to subject water to temperatures so high (around 3000°C) that the molecules simply fall apart. But as yet, the vision of a bright, clean hydrogen-powered future remains out of reach.

Water batteries

Suppose that someone hits on Poliakoff's solution, and the Sun Battery becomes a commercial reality. How then do we make effective use of hydrogen for power generation?

The impulse behind hydrogen fuels is easy to appreciate: when it is burnt in air, a tremendous amount of energy is released. You need only think of the Hindenburg airship disaster of 1937 to realize that: using huge hydrogen-filled balloons for passenger transport was always inviting calamity.[3] You can use electricity to split water into its elements – but by goodness, they are keen to be reunited! The formation of water from hydrogen and oxygen is explosive, as anyone who has performed the classroom test for hydrogen will know: hold a match to the mouth of a test tube full of hydrogen, and hear the pop as it detonates.

This is both good and bad for a hydrogen-based power supply. It means that there is a lot of energy to be had from making water from its elements. But it also means that, unlike burning coal, you would be unwise to try to extract it just by setting your fuel alight. That will do for space rockets – the US Space Shuttles are powered by hydrogen – but hardly for a car.

The most promising solution has, however, been recognized for almost as long as electrolysis itself. It is simply to perform the electrolysis reaction in reverse: to get the energy back out as electricity, not heat. The electrochemical recombination of hydrogen and oxygen to make water was first demonstrated by William Grove, an English barrister, in 1830 before London's Royal Society. He bubbled the gases over two metal electrodes – and the electrons flowed.

A device like this is called a fuel cell, and it is a kind of battery that runs on gas. Rather than being charged up and then allowed gradually to discharge like a normal battery, a fuel cell requires a continuous supply of fuel. It is therefore really like a kind of hybrid of an internal combustion engine and an electrochemical device, producing electricity directly from a combustion process.

Despite their early start, hydrogen–oxygen fuel cells did not become useful devices until the 1960s, when technologies had developed sufficiently to enable them to provide appreciable and reliable amounts of power. All the same, their cost remained high relative to other forms of power generation, and so they became practical only in applications where factors such as their light weight were critical. Their first real application was as power supplies for instruments on board spacecraft, which carried the hydrogen and oxygen fuel with them in any case to power the rocket engines. The water exhaust from the cells can be used as drinking water for the spacecraft crew. Today hydrogen fuel cells remain a rather specialized form of power supply, although concerns over exhaust pollution are stimulating motor companies such as Daimler-Benz to explore hydrogen-powered vehicles. Ballard Power in British Columbia has developed a prototype bus powered by a hydrogen fuel cell, and several of these are already plying the streets of Vancouver.

Whether hydrogen fuel cells will ever be the answer to large-scale energy production is more a matter of economics than of science – the question is whether they can be fuelled and operated cheaply and efficiently enough to compete with other means of energy generation. My bet is that, sooner or later, the need to reduce pollution and greenhouse-gas emission will place hydrogen-powered vehicles on the road, even before dwindling supplies of oil and petroleum become a limiting factor. Large-scale energy generation might never rely on fuel cells – on the direct conversion of hydrogen burning to electricity – but there is no reason why the combustion of hydrogen should not compete with the burning of gas, coal and oil, if an efficient way is found to unlock the stuff from the bounty of the oceans. Big obstacles remain, amongst them issues of how to store and transport hydrogen safely. But partly it is simply a question of attitude, of believing that hydrogen is worth investing in. Says one researcher at NREL, 'If we really decided that we wanted a clean hydrogen economy, we could have it by 2010.'

More heat than light?

When Stephen Poliakoff's Christopher reveals his Sun Battery, Al gives us an aside that would bring smiles to the faces of the cognoscenti: 'I heard on the grapevine Utah were making progress.' For out of Utah came the play's inspiration: chemists Martin Fleischmann and Stanley Pons of the University of Utah announced at a press conference in March 1989 that they had performed nuclear fusion by electrolysing water in a test tube. This was 'power from water' with a vengeance, and the result was pandemonium.

The Utah chemists used heavy water D_2O, in which the hydrogen atoms were replaced by hydrogen's heavier isotope deuterium (see page 8).[4] They claimed that the deuterium atoms, stripped from the heavy-water molecules by electrolysis, fused in pairs to form a new element, helium, with the concomitant release of nuclear energy. By dunking two palladium electrodes into the heavy water and passing a current through them, Pons and Fleischmann asserted that they had recreated the process that powers the Sun. It was the kind of experiment a schoolchild could perform – and in a matter of days, many were doing so.

Hot stuff

Although it releases energy and so makes the nuclear particles more stable overall, this fusion process first has to overcome an energetic barrier, because of the electrostatic repulsion between the positively charged nuclei. So immense temperatures and pressures are generally needed to fire up a fusion reaction. In the centre of the Sun, the conditions are just adequate to keep the hydrogen fusing, but not so rapidly that the process runs away with itself to produce an unfettered thermonuclear explosion.

Fusing two hydrogen atoms requires such extreme conditions that it is impractical even to consider achieving it on Earth. Rather, efforts to harness fusion technologically focus on the use of deuterium, and hydrogen's still heavier radioactive isotope tritium (hydrogen-3), as the fuel. Thermonuclear 'hydrogen bombs' are really 'deuterium–tritium' bombs. Nuclear fusion for peaceful purposes has proved more elusive. While nuclear fission – the splitting of very heavy uranium nuclei, which also releases nuclear energy – now provides significant proportions of the power used by nations such as France and Japan, nuclear fusion is still not a practical energy-generation technology. Yet its potential payoffs are enticing. Not only is the energy release greater per nuclear event, but there should be far fewer problems of radioactive waste to contend with – the products of uranium fission are themselves radioactive and potentially lethal.

The problem is that, unlike nuclear fission, fusion requires extremely high temperatures and densities to get it going. To enable more energy to be extracted from a controlled fusion process than is put in to sustain it, the temperature must be even higher than that at the Sun's centre: hundreds of millions of degrees. There is no substance on Earth capable of containing a plasma of deuterium nuclei this hot. And to make matters worse, the deuterium has to be highly compressed before the fusion process will ignite. Fusion scientists are today attempting to confine deuterium within very strong magnetic fields – containers that have no material walls – or to fuse deuterium and tritium with a carefully orchestrated blast of laser beams. But the difficulties have so far defeated attempts to extract from the process more energy than goes into igniting it, let alone keeping the hot plasma confined for sustained periods of time. It is like trying to grasp jelly in your fist – it always manages to squirt out. Because of these difficulties, fusion research for power generation is a hugely expensive operation, requiring large and expensive apparatus and tremendous amounts of power.

Small wonder, then, that the announcement of fusion in a test tube at room temperature – cold fusion – sent the scientific community into a spin.

Storm in a teacup

Martin Fleischmann was a leading figure in electrochemistry, a Fellow of Britain's Royal Society with a distinguished career behind him. Together with Stanley Pons, chairman of the University of Utah's chemistry department, he convened the press conference on 23 March. The researchers showed the cameras an innocent-looking glass cell, a few inches in diameter, in which they had electrolysed heavy water (actually a solution of deuterated lithium hydroxide in heavy water) using palladium electrodes.

Palladium has been long known as a 'hydrogen sponge': it can absorb huge quantities of hydrogen atoms, which are small enough to slip into the gaps between the metal atoms. What Pons and Fleischmann claimed was that the deuterium released by the electrolysis of heavy water must have become concentrated so densely within the palladium lattice that it had undergone fusion – at room temperature.

The primary basis for these claims was that the electrochemical cell had produced a great deal of heat – several times more than they were feeding in electrically, and more than could be accounted for by the energy released in any chemical process that might be occurring. This 'excess heat', which the Utah chemists claimed to have measured very carefully using the standard techniques of calorimetry, suggested that something more than chemistry was required to explain the cell's energy output. They said that during one experimental run the temperature had suddenly soared so high – to an estimated 5000°C – that the palladium electrode was vaporized, and the fume cupboard containing the cell and laboratory floor around it were damaged.[5]

To back up this conclusion, they claimed to have detected the emission of neutrons from the cell. When two deuterium nuclei fuse, their constituent subatomic particles can be rearranged into a helium-3 nucleus and a neutron, or a hydrogen-3 (tritium) nucleus and a proton. So if fusion was indeed occurring, helium-3 and tritium should have been generated, and neutrons emitted. The Utah chemists claimed to have seen all three. They could conceive of no explanation for their results other than that 'we have sustained controlled nuclear fusion reaction by means which are considerably simpler than conventional'.[6]

All this was extraordinary enough and sure to provoke controversy.

But what set the story off on its spectacular and erratic trajectory was the decision to announce the findings by press conference – before any other specialists had had a chance to assess the experiments for themselves. Press conferences in advance of a proper scientific publication are almost universally frowned on by scientists, who feel that all results, and particularly shocking ones, should receive careful scrutiny by the scientific community before being passed on to the wider world. That way, outrageous and unjustified claims can be smoked out and either corrected or relegated to obscurity before they come to the attention of a mass media that lacks the training to evaluate them properly. But Pons and Fleischmann claimed that they had been forced into their premature disclosure by the fact that the incendiary story was on the verge of leaking out to the press anyway.

In fact, they did have a scientific paper written and accepted for publication when they called their press conference – it was to appear the following month in the *Journal of the Electrochemical Society*. But next to no one had set eyes on this paper when the Utah chemists made their revelation, and so all the scientific community had to go on was what the chemists told the press – which was an astonishing claim backed up by next to no details about the experiments. And when the paper finally appeared, it was distressingly short on crucial details. It even neglected to list all the authors: Pons's graduate student Marvin Hawkins, who had actually made most of the experimental measurements, was added only in a later correction.

This kind of sketchiness set many people's teeth on edge. Some researchers were relegated to trying to work out what the Utah team had done based on press photos of their electrochemical cell. Unconcerned with such minutiae, the press were glad to trumpet in bold headlines that a pair of chemists, using lab equipment available to most high schools, had cracked a problem in energy generation that had defeated highly funded physicists for decades. 'For £90', said the British *Daily Telegraph*, 'someone could attempt their own fusion experiment and rival the conventional approach . . . which costs £75 million a year.'[7]

But Pons and Fleischmann were not alone. Three weeks after their paper appeared, a report in *Nature* by a group at Brigham Young University in Provo, Utah, led by Steven Jones, also made tentative claims of 'cold fusion'. Jones had a solid background in unusual ways to harness nuclear fusion at low temperatures. He had for many years worked on a process called muon-catalysed fusion, an idea that originated in the 1950s and involves an exotic form of hydrogen in which the electron orbiting the atom's nucleus is replaced by a heavier, unstable version called a muon. Because the muon is 207 times heavier than the electron, its orbit is correspondingly times smaller, and so the hydrogen atoms can be brought closer together, helping them to overcome the electrostatic barrier to fusion of the nuclei. Muon-catalysed fusion has been observed in the laboratory, but it remains unclear that it could ever be developed into a practical fusion technology. Jones began work on muon-catalysed cold fusion in the early 1980s, but his studies had gradually led him towards essentially the same electrochemical route to deuterium fusion that Pons and Fleischmann had hit upon.

Jones too had come to realize that the hydrogen sponge of palladium metal might be the key to success. In September of 1988 he contacted the University of Utah group to suggest collaboration rather than competition. To avoid the iniquities of a publication race, in early 1989 they agreed to report their results in papers that they would submit simultaneously to *Nature*.

The collaboration, always strained at best, collapsed with the March press conference. Hearing of the paper that Pons and Fleischmann had already submitted to the *Journal of the Electrochemical Society*, Jones assumed that their agreement had been breached, and he sent his paper to *Nature* on the same day. Pons and Fleischmann, meanwhile, were under the impression that the agreement for simultaneous submission to *Nature* still stood, and were chagrined to learn that Jones had gone ahead without them. They duly submitted their paper in any case.

Nature sent both papers to referees for review, and they both drew a raft of criticisms. Jones and his colleagues answered the questions to the satisfaction of the referees; but Pons and Fleischmann chose not to, saying that they were too busy with further experiments and withdrawing the paper on 22 April.

The paper by Jones and colleagues reported no excess-heat measurements, but instead focused on the emission of neutrons – in principle a far more exacting test of whether fusion was occurring, since neutrons produced in fusion should have a well-defined energy of 2.5 million electron volts (denoted MeV; electron volts are conventional units of energy). Jones's team could not only detect emitted neutrons but measure their energies. Yet their observation of

2.5-MeV neutrons lay at the threshold of detectability, rising marginally above the background level to a degree that varied on each experimental run. They were careful to emphasize the very low rate of fusion that their results implied.

The plug is pulled

Cold fusion enjoyed just four months or so of pyrotechnic fame, before collapsing in a welter of calumny, bad blood and libel threats. Ultimately its fate was sealed for the same reason that all bad science eventually meets its demise: irreproducibility. Try as they might, no one could duplicate the findings clearly and consistently. Claims of successful cold fusion were always too sporadic, poorly documented, or on the verge of perceptibility, for the phenomenon to be credited.

The experiment was simple in principle, provided that one could obtain some palladium foil (stock prices in this metal rocketed after cold fusion made headline news, and even major laboratories encountered difficulties in obtaining it). Within weeks, everyone was at it, from the world's most sophisticated electrochemistry laboratories to schools and colleges. Inevitably the results were diverse, in all probability being somewhat a function of what their proponents expected (or hoped) to find. Many teams reported excess heat, and at Texas A&M University electrochemist John Bockris claimed also to see enhanced levels of tritium in his cells. These claims were to become particularly notorious, culminating in accusations that Bockris's experiments were being spiked with tritium by persons unknown.

Teams at Los Alamos National Laboratory, the Georgia Institute of Technology and the University of Washington also initially reported 'positive' results: heat, neutrons or tritium, inspiring NBC News to announce in April that 'an increasing number of scientists now agree that this experiment does produce at least small amounts of nuclear fusion'. And there was further confirmation worldwide: in India, Italy, China and Japan.

But other groups looked hard and found nothing. One of the most extensive attempts to verify the Utah claims was made at Harwell by a team led by electrochemist David Williams. His group assembled a vast array of different designs and compositions, and proceeded to test each one in turn for neutron production – with not a hint of 'success'. Another particularly thorough search was conducted in the chemistry department of the California Institute of Technology, led by respected electrochemist Nathan Lewis and nuclear physicist Charlie Barnes. They measured the heat output from their legion of electrochemical cells, as well as deploying a battery of neutron detectors and seeking the fingerprint of fusion in gamma rays, helium-4 and tritium too. At the beginning of May Lewis told the American Physical Society that they had seen no signature of fusion whatsoever. If cold fusion was going to solve the energy crisis, it was hard to see how such an extensive study in such a renowned laboratory would have missed it entirely.

By June 1989 cold fusion looked doomed – although that did not stop the state of Utah investing five million dollars in a centre to study cold fusion at the university. This was finally and sheepishly shut down in 1998.

One of the most damning studies was performed in the University of Utah physics department, where Michael Salamon and coworkers had persuaded Pons and Fleischmann to lend them their electrochemical cells – not, it seems, without considerable reluctance on the part of the chemists. This was the first independent test of cold fusion using the very same cells as the Utah chemists, so there could be no recourse to the explanation that 'you didn't design your cells right'. In March of 1990, Salamon's team recorded that they had seen nothing out of the ordinary.

Irreproducibility does not by itself indict cold fusion as bad science. I make that accusation because of the remarkably sloppy way in which critical questions were treated by Pons and Fleischmann. (Steve Jones was considerably more careful, and eventually became a cold-fusion sceptic.) This trail of evasion and fuzziness is admirably followed by physicist Frank Close in his book *Too Hot to Handle*.

After the event

Cold fusion fits Langmuir's criteria for pathological science with room to spare. Glance back at these on page 257 and you will see that the list could almost have been designed on cold fusion's model:

i. The phenomena were never more than barely detectable. Steven Jones's results made this clear, and to his enduring credit, he never implied otherwise. On one occasion he indicated that he would be surprised if the supposed fusion process could be prolific enough to power a flashlight. Pons and Fleischmann made much of their claimed meltdown in one experimental run, but only later confessed that no one had ever witnessed it. None of the claims for significant and reproducible excess-heat production ever stood up to scrutiny when assessed independently by reliable methods.

ii. The 'cause' of the excess heat – nuclear fusion of deuterium – never produced the clear neutron and gamma-ray signatures that would be expected of it, let alone producing them in sufficient abundance to match the claimed heat output.

iii. Accuracy in measurement has seldom been treated with less dignity. Says Frank Close of Pons and Fleischmann's original paper, 'Numbers were quoted to several significant places without details of where they had come from.' Some of the arguments centred on the intricacies of neutron detectors, as though these devices used for years in physics experiments had suddenly become questionable items for the task at hand. In fact, the low levels of neutron emission that were being measured were pushing the detectors to the limits of their useful range. The group at the Georgia Institute of Technology were misled into 'seeing' fusion neutrons in April 1989 by failing to account for the subtle effects of temperature changes on the neutron detector when only low fluxes of neutrons are being registered. One alarming conclusion that emerged from the affair was just how much poor calorimetry was seemingly performed routinely in laboratories around the world. And the gamma-ray measurements made by the Utah team became notorious for their fluidity.

iv. It takes relatively little theory to show that fusion of deuterium absorbed in palladium should be utterly out of the question. The pressures developed inside the metal are enormous by everyday standards, but pitiful compared with what is known to be needed in conventional fusion devices. It is healthy that experimentalists should investigate 'theoretical impossibilities' from time to time; but the willingness on the part of cold-fusion protagonists to disregard theoretical realities was occasionally staggering.

v. *Ad hoc* excuses were the norm, from vague 'recalibration' of spectra to the idea that some electrochemical cells were inherently 'dead' while others were 'live'.

vi. It is hard to evaluate the proportion of supporters at the height of cold fusion – but it is safe to say that at one stage hundreds of researchers were looking into some aspect of the experiment or theory, whether they believed the claims or not. One can generalize to say that many chemists were sympathetic, and most physicists antagonistic. And the proportion of supporters today is certainly as near zero as makes no difference.

Pointedly, the bulletin journal of the American Physical Society *Physics Today* reprinted Langmuir's transcribed 1953 lecture in October 1989. The journal made no reference to cold fusion – there was no need to labour the point.

It is clear that the involvement of the popular media played an even bigger part in raising the profile, and the temperature, of the cold-fusion affair than it did for polywater. It was television, radio and the newspapers that brought the claims to the attention of the scientific community, not the scientific journals. 'Taming H-Bombs?' was the emotive initial headline of the *Wall Street Journal* report that introduced cold fusion to the USA. For some time, these media were the only sources of information for frustrated scientists: on one occasion, a pivotal analysis of Pons and Fleischmann's gamma-ray measurements by physicists at Princeton University was based on a graph flashed on the screen during a televised interview. Reporters besieged laboratories, and as ever it was those players with the bravado that the media loves who came to steal the limelight. It is, then, more than ever relevant to ask which factors of the story stimulated this extraordinary media attention.

The 'David and Goliath' dimension was obviously an attraction: two chemists from Utah outwit the might and million-dollar resources of the fusion physics community. 'For the media and the public,' says Peter Bond of the physics department at Brookhaven National Laboratory, 'this was the US dream; the little guy beats the big battalions.'[8] But at least equally crucial was the fact that this was not a scientific story about discovering how the world works – it was about energy, and therefore about money and power.

Frank Close suggests that the oil spillage by the *Exxon Valdez* tanker in Alaska the day before the Utah press conference, as well as the Chernobyl accident three years earlier, primed the world to be receptive to a new, cheap and clean source of energy. There is surely something in this, but I suspect that the Utah claims were also tapping into a more abiding myth: that of turning water into precious fuel.

The fact that it was heavy water that was implicated in cold fusion detracted not at all from the allure of this myth. There is plenty of heavy water in the sea, and it is not so difficult to extract. In 1989 you could purchase enough heavy water for a Pons/Fleischmann cell for around thirty dollars. Indeed, to the lay person a nuclear process involving heavy water sounds all the more feasible, given its associations with today's nuclear power – wherein, however, it is used as a moderator to slow down fission neutrons, not as a fuel. Of course, at root there is nothing more to this claim than to that of conventional fusion research, which also aims to stimulate fusion of deuterium extracted from sea water. But the 'power from water' case was far more vivid and immediate in cold fusion, since the deuterium had not been extracted in pure form by some high-tech process before being heated and compressed to unearthly conditions; the water was right there in the glass cell, sloshing about for all to see.

Bad memory

Commenting on the announcement of cold fusion in advance of peer review, *Nature*'s editor John Maddox proclaimed in April 1989 that 'It is no disrespect to any of those concerned to compare the dilemma created for *Nature* by these events to that occasioned a year ago by the article in which Professor Jacques Benveniste and colleagues claimed that infinitely diluted reagents retain their biological effectiveness.'[9]

Maddox refers to a paper that was published just weeks before I joined *Nature*'s editorial staff, and I remember all too uncomfortably how my PhD supervisor, a physicist widely respected for his work on the liquid state, remarked with a grin 'So, you're going to work for a journal that believes water has a memory?' What was going on?

In Clamart in the western suburbs of Paris, Unit 200 of the French medical research organization INSERM, the equivalent of the National Institutes of Health in the USA, is devoted to the study of immune response and allergy. Jacques Benveniste became director of the unit in 1981, and in the following seven years his laboratory acquired a respectable reputation for its work. In 1987, his team discovered something peculiar while studying the response of a certain type of human white blood cell, called basophils, to antibodies.

Basophils are part of the body's immune system. Their job is to patrol the bloodstream for foreign particles that originate in other organisms. Basophils are triggered into activity by the presence of allergy-inducing substances denoted allergens. When basophils meet allergens, they release from their interior tiny parcels, or 'granules', containing the amino acid histamine. This response is called degranulation. It is part of the body's automatic defence mechanism, but it provokes the unpleasant symptoms associated with allergies such as hay fever and asthma: irritation and inflammation. That is why anti-allergy medications contain an antihistamine ingredient that mops up histamine.

The interaction of basophils with allergens is mediated by antibodies of a class called immunoglobulin E (IgE), which are protein molecules that the body produces. These antibodies become attached to sites at the surface of the basophil cells, and it is the interaction of the anchored IgE molecules with the allergens that promotes degranulation. Benveniste's group were studying this allergenic response – degranulation of basophils – in test tubes rather than in the body. Instead of employing typical allergens to stimulate this response, they found it more convenient to use another kind of antibody as a 'fake allergen'. Such 'imposter' antibodies, which recognize and bind to human IgE molecules, can be produced by injecting IgE into another animal. In a non-human organism the IgE molecules – part of the immune defence system in humans – are just the kind of invaders that in humans they help to destroy, and the animal's immune system produces antibodies that recognize them and mark them out for destruction by white blood cells. Antibodies geared to bind to human IgE, called anti-IgE antibodies, can be generated in this way. They induce degranulation when added to a solution containing IgE-studded basophils. Benveniste's team used commercially available anti-IgE from goats in their experiments.

They noticed that sometimes degranulation was triggered by concentrations of anti-IgE that should be too low to have any effect. One expects the extent of basophil degranulation to depend on the amount of anti-IgE added: the less you add, the less there is to bind to IgE and trigger degranulation. But occasionally the French researchers saw extensive degranulation even when extremely small amounts of anti-IgE were added to the basophil suspension in water. The extent of degranulation was monitored by adding to the solution a dye called toluidine blue. This stains intact basophils red, but does not stain degranulated basophils. So the degree of degranulation can be measured by counting, under a microscope, how many cells have become stained and how many have not.

Thus anti-IgE would sometimes apparently produce a response out of all proportion to its abundance in the solution. Intrigued, the researchers decided to look systematically at how the degranulation response varied as the anti-IgE was diluted. Again, the expectation is that the degree of degranulation will fall smoothly to zero with increasing dilution. But what the French researchers found was something else entirely. As they diluted the anti-IgE, degranulation would first fall but then rise again – and keep falling and rising repeatedly as the dilution increased (Fig. 11.1). This rise and fall seemed to go on inexorably, no matter how great the degree of dilution. It continued, in fact, well beyond the point at which the dilution was so great that no anti-IgE molecules were left in the water at all! It was as though the anti-IgE molecules had left 'ghosts' in the progressively diluted solution that were equally adept at carrying out their biological role. The water, it seemed, retained a memory of the molecules' presence.

A spoonful of sugar

If you add a spoonful of sugar to a glass of water, you are putting into solution an awful lot of molecules. A teaspoon of sugar contains around 10^{22} molecules – that's one followed by twenty-two zeros. It's a number larger than we can ever picture, and you might think that it would take a tremendous amount of diluting to end up with a

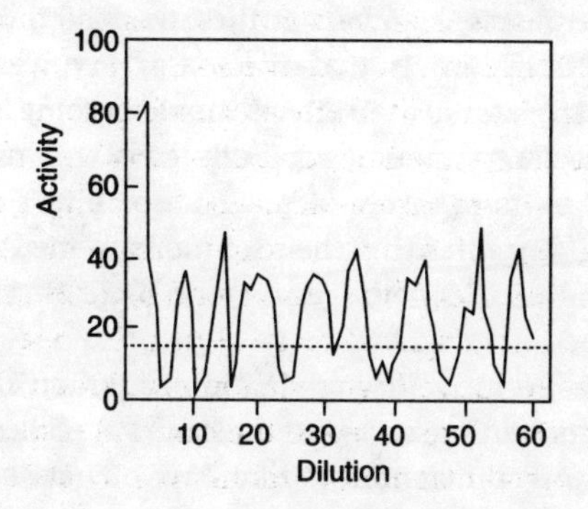

Fig. 11.1 Does water have a memory? In the experiments of Jacques Benveniste and colleagues in 1987, the biological activity of protein molecules involved in the immune response seemed to persist even when their solutions were so diluted that not a single protein molecule remained in the water. (This point is reached at least halfway along the dilution scale here.)

sugar solution containing just a handful of molecules in it. But that's really not so hard to achieve. If you dissolve a teaspoon of sugar in a litre of water, pour out one tenth of it and make that extract up to a full litre again with water, you'll have diluted the initial solution by a factor of ten, assuming that the sugar was well dispersed. In other words, the new solution should have around 10^{21} sugar molecules in it. Now repeat that twenty times, and you'll have a solution containing something of the order of ten sugar molecules in it. All the others remain in the twenty-one solutions from which you've decanted an extract. Do the same once more, and you have a solution with just one sugar molecule in it – a high dilution indeed.

How do you know you have just one molecule in the solution, not two or three, or none? This would in fact be exceedingly hard – indeed all but impossible with current techniques – to prove. If all the molecules are dispersed at random, then as their numbers get very small the chances of picking out slightly more or less than a tenth of them with a tenth of the liquid get greater. With just ten molecules in a litre of water, it is not at all unlikely that you'll happen to end up with two or three or none in an extract of one tenth of a litre. It's very unlikely that such a sample would happen to capture all ten, however. So when you get down to these extremely small concentrations, you can't be sure exactly what the dilution procedure is giving you. For all practical purposes, however, the uncertainties in precise numbers at these very high dilutions don't matter. What is certain is that there are very, very few of the molecules present, and so their effects should become negligible. You would not expect a solution of sugar diluted twenty or so times by a factor of ten to taste sweet – and it doesn't. The surprise of Benveniste's findings was that their solutions seemed to get episodically 'sweet' and then 'tasteless' again with increasing dilution. *That* you don't expect.

What is more, for a sufficient degree of dilution the chances of your solution containing any sugar molecules at all becomes negligible – insignificantly different from zero. Say you have diluted twenty-one times, down to a concentration of around ten molecules in a litre. If by good fortune you capture as many as three of these in the next extract you pour off, well, lucky for you: you have three molecules per litre in the next dilution. What are the chances of catching all three in the next dilution, leaving the remainder of the solution devoid of molecules? Pretty slim – the chances are that you won't catch any. But even if you get one, giving you a solution of one molecule in a litre of water, you'd have to be very lucky to get it on the next round too – and the chances of doing that for one round further are tiny. So by the time you have performed the dilution, let's say, thirty times, you can probably see that the chance of having any molecules at all left in your 'solution' is utterly negligible.

Now, Benveniste's team progressed to serial anti-IgE dilutions of 120 orders of magnitude – they repeated a procedure like this 120 times – and still saw the apparent effect of anti-IgE on degranulation. The signature of the molecules just would not go away. And significantly for what was to follow, the INSERM team reported that vigorous mixing – 'vortexing' – of the solutions was essential for this effect to be revealed. You can see, I hope, why John Maddox felt obliged, when he published Benveniste's results in June 1988, to preface them with an editorial comment stating that 'There is no objective explanation of these observations.'[10]

You can see also, perhaps, where they are leading. For retention of biological activity in solutions of essentially 'infinite' dilution is precisely the claim made for homeopathic remedies. Here, it seemed, was scientific vindication, in the world's premier scientific journal, of the assertions that homeopaths have made for two hundred years: that solutions diluted to the point where they can contain no molecules of the active ingredients nevertheless have a medical effect.

Alternatives

Homeopathy surely poses one of the boldest and most controversial challenges to our understanding of what water is and what it does. In recent years, there have been modest efforts to conduct truly scientific investigations of alternative or complementary medicines, notably at the National Institutes of Health in the USA. That these have provoked fury in some quarters of the medical establishment[11] is a great shame, although such a response is understandable in a country beleaguered more than most by powerful anti-scientific sentiments. The vitriol heaped on homeopathy by many scientists seems to me to be symptomatic of a frustration at the broader cultural flight from rationality. For all that this frustration is understandable and to some degree warranted, science's champions need to recognize it for what it is. That the medical community should devote a tiny fraction of its resources to investigating treatments that appear to have no clear scientific basis but which huge numbers of people use routinely does not strike me as a cause for complaint – indeed, it is long overdue. In my view, the medical community would be extremely short-sighted to try to merely ignore an industry as substantial as homeopathy. Part of this community's antagonism, however, no doubt originates from a sense that the results of such an investigation are unlikely to change anyone's opinion, particularly if they are negative: 'disproving' homeopathy is not likely to cut much ice with its proponents or adherents. But that is no reason to decide that we do not even need to 'know' whether it is effective or not.

An empirical approach simply asks whether the treatment works. But measuring effectiveness is no simple matter. Homeopaths take a wide view of health: they take account of the patient's character and individuality, and they afford time to explore the emotional side of illness as well as related issues such as diet. This, one might imagine, not only is a commendable thing but holds out the prospect of offering some real therapeutic benefits. Where the real conflict arises is over the question of whether the diluted homeopathic remedies have any direct physiological effects in themselves. Thorniest of all are the existing attempts to explain the influence of these remedies 'scientifically': to formulate 'theories' of how plain water can act as a universal medicine, its specific effects determined by its manner of preparation.

In this respect, it troubles me that homeopathy seems able to offer little more by way of conceptual vindication than the doctrine of an eighteenth-century physician. Conventional medicine would be in a dreadful state if the same were true, since we know that physicians of the eighteenth century seldom got anything right beyond basic anatomy and a smattering of empirical wisdom.

Homeopathy's founding father Samuel Hahnemann was born in Dresden in 1755, and studied medicine at Leipzig and Vienna. His subsequent experiences in the hospitals of Germany led to a considerable disenchantment with the methods and beliefs of his contemporaries, and not without good reason: these were the days when bloodletting was still common practice, when prescribed medicines were as likely to cause as to relieve ailments, when mentally disturbed people were locked away, regarded as barely human.

Hahnemann left medicine to become a translator, but was drawn back to it when he came across a prescription for quinine as a treatment for malaria. Experimenting on himself, he found that quinine could induce the very symptoms that it was held to cure in malaria sufferers.[12] He was led to propose a principle that homeopaths now call the law of similars, which states that the cure lies in the disease. Substances that provoke certain symptoms when taken by healthy people would, he said, alleviate those symptoms in people suffering from a disease. *Similia similibus curentur*: like cures like; hence homeopathy, meaning 'similar suffering'. Hahnemann believed that the body could not support two diseases at the same time, and that a 'disease' introduced artificially by the administering of weak doses of some agent could expel one already present – but without replacing it with an equally distressing illness, for the weak dose prevented the administered disease from taking hold.

Hahnemann outlined his homeopathic theories in his treatise *An Organon of Rational Healing* (1810). But their rejection by the medical community meant that he was restricted in his ability to put them into practice. In the end, however, Hahnemann may have been an effective doctor nonetheless, since he rejected some of the unhelpful practices of his time and focused on the benefits of a good diet and exercise, and on good emotional health. And those homeopathic treatments he did administer could scarcely have done any harm, since he soon began to dilute the ingredients to an extreme degree. Some of the 'cures' with which he experimented, which were in widespread use by traditional physicians, were toxic substances like mercury, arsenic and belladonna, which could be severely deleterious to a patient's health even in small doses. So Hahnemann experimented with dilution to reduce the harmful effects, whereupon he made his second key discovery: dilution, he said, could increase the efficacy of the medicine while reducing its toxicity. But these benefits were conveyed only if dilution were carried out in the right way: the remedy had to be mixed by vigorous shaking, called succussion, which involved repeatedly banging the container on a hard surface. According to Hahnemann, succussion releases the potency of the diluted substance.

Hahnemann came eventually to use dilutions that contained negligible amounts of the active ingredient, to the point of their being absent entirely. Homeopathic remedies today are classified according to their degree of dilution, with the most dilute regarded as the most potent. Successive dilutions are usually by a factor of 100: one drop of the mixture is mixed with ninety-nine drops of the solvent. The potency is thus denoted on a centesimal scale: '6c', which is recommended for most temporary (acute) conditions, indicates six sequential 100-fold dilutions, while the more 'potent' 30c remedies for chronic conditions represent thirty 100-fold dilutions, which is certainly sufficient to remove any trace of the active ingredient. In Hahnemann's defence, one must point out that he almost certainly did not know this was so. For modern-day homeopaths, however, there can be no question that most of their remedies contain no 'medicine' at all in the conventional sense.

This, more or less, is the form of homeopathy that operates today. It may be summarized in three concepts:

i. Like cures like.
ii. High ('infinite') dilutions have physiological effects, and the greater the dilution, the greater the potency of the treatment.
iii. Succussion is essential to preparation of the cure.

It is, I hope, apparent already why the second principle, at least, makes no sense whatsoever in terms of conventional science. Hahnemann explained his observations on the basis that the body contains a 'vital force' that maintains the body's health. In sickness, the vital force is depressed or out of balance. Homeopathic remedies somehow re-energize this extraordinarily sensitive life-giving agency. It is not clear, however, exactly what Hahnemann had in mind here. His term was *vis mediatrix naturae* – literally, the 'natural healing force'. On one level you might equate this with the evident tendency of the body to heal its ailments – something that, even if unknown, might conceivably be biochemical in its modus operandi. This would then offer no affront to either common sense or contemporary knowledge – but neither would it represent the basis of a theory in any real sense: replacing 'vital force' with 'healing potential' leaves you with a tautology.

Yet I think there are very good reasons to suppose that Hahnemann had in mind something more: a kind of animating spirit of non-corporeal origin. Even if so, Hahnemann was proposing nothing outrageous or even controversial according to the standards of his time. It was a common belief amongst conventional scientists and physicians of the late eighteenth century that living matter was imbued with some vital ingredient that distinguished it qualitatively from inorganic stuff. Well into the first half of the nineteenth century, many chemists believed that organic compounds, the carbon-based molecules found in cells, could not be put together from inorganic ingredients but only through 'life processes' such as fermentation.

Today this concept of vitalism serves no useful purpose in the life sciences. To the non-scientist, however, I can understand why there might seem nothing objectionable in it. Surely living organisms are so remarkable that it is not too much to hypothesize the influence of a non-material cause? But the point is that this is fundamentally a non-scientific belief, an act of faith. We can explain a vast amount of biology in a formal, rational, quantifiable and testable way – part of it through molecular 'reductionism', part relying on ideas about emergent properties in complex, dynamic systems of many interacting components. That there is a still greater amount that we do not understand is disputed by no one. It is not, however, a scientific way of proceeding to attribute all this murky stuff to the influence of a vital force, any more than it is to attribute it to divine providence. That merely adds a hypothesis for which there is no strict requirement, and does not 'explain' anything in the scientific sense of providing a predictive, testable theory. It is to introduce a 'God of the gaps'. While I certainly do not believe that scientists need regard such beliefs as inherently fit for ridicule, their adherents need to be aware of this fundamental, qualitative difference from scientific thought. Any belief in vitalism is not an alternative 'scientific' theory that can be judged against the criteria of orthodox science.

While the evocation of some 'natural healing force' closely allied to eighteenth-century vitalism is common amongst homeopaths today, it is by no means universal. Some later homeopaths regarded it as transparently unscientific, and in the late nineteenth century the British homeopath Richard Hughes initiated a school of thought that attempted to establish a more rational, scientific basis for his practice. None, however, has yet emerged.

Other homeopaths say that there is no point yet in trying to concoct a scientific explanation for the effectiveness of their treatments; rather, the primary task is to show that they work. This is commendably realistic. Says Anthony Campbell, a consultant at the Royal London Homeopathic Hospital,

> Modern molecular theory is much too well established to be easily overthrown, and there is no real doubt that homeopathic medicines above the 12th centesimal dilution can contain only water. If,

> therefore, they do have a measurable effect, we have to conclude that in some unexplained way the substance they originally contained has somehow impressed itself on the water . . . Unfortunately it is very difficult to imagine how water could preserve traces of the original substance in the way I have suggested. Theories of this kind have been proposed but they depend on ideas about the nature and structure of water that are not universally accepted among physicists.[13]

Clearly Campbell understands the problems his practice faces, even if he does them the favour of occasional understatement. But we live in an age in which people have come to look to science for explanations, even if they do not always like and trust what it tells them. The efficacy of antibiotics, and the drawbacks too, can be rationalized in clear scientific terms. The tools of science can be used to design drugs that will combat viruses, and to show that they work. Or if they do not, this can generally be explained too, with reproducible experiments that test clearly stated hypotheses. There is some perception – as much within the homeopathic community as without, it seems – that if homeopathy is to compete with the demonstrable successes of modern medicine, it must explain itself in a similar language.

Which is why, when Jacques Benveniste and his colleagues claimed that 'highly dilute'[14] solutions are biochemically active, and that something akin to succussion is required to realize this effect, homeopaths believed they had been vindicated at last. 'Homeopathy finds scientific support', proclaimed *Newsweek*; 'Homeopathic enthusiasts are rejoicing while scientists scratch their heads', *New Scientist* told us. Had Benveniste made homeopathy scientifically respectable?

Suspension of disbelief

It was not the first such claim to appear in the scientific literature. In 1986, David Taylor Reilly and coworkers from the Glasgow Homeopathic Hospital reported in the respected medical journal the *Lancet* that grass pollen administered to hay-fever sufferers in homeopathic doses, which means diluted to the extent of removing all the pollen, reduced their symptoms. The study was conducted with all of the control experiments that would be expected to identify placebo effects.[15] Reilly and colleagues adopted the full trappings of homeopathy, commenting on the importance of 'succussion' for a successful result. It beggars belief, however, that their explanation of the effect could appear in a scientific journal – not so much because it is flawed as because it is incoherent:

> . . . succussion produces energy storage in the bonds of the diluent in the infrared spectrum which 'downloads' in contact with the water in living systems. Perhaps this information then spreads like a 'liquid crystal' through the body water, modifying receptor sites or enzyme action.[16]

'Infrared spectrum' and 'liquid crystal' sound impressive, but this explanation could have come from Hahnemann himself if he'd known the jargon. What is more, to the extent that one can make sense of it at all, it is relatively easy to test. Why not measure the infrared spectrum of the water and see if succussion changes it?

But these results passed without note until Jacques Benveniste and his colleagues at INSERM 200 submitted their report of basophil degranulation at 'high dilution' to *Nature* in August 1987. At face value, these high-dilution results seemed to be scientific heresy, for they challenge a well-established chemical principle called the law of mass action. This says in essence that the outcome of a chemical process is determined by the amounts of substances present. Say that two chemical reagents, A and B, come together and react to form two others, C and D. If conducted in a sealed environment, so that it can come to equilibrium, such a reaction never consumes every last scrap of A and B, although it might often seem that way. Rather, the equilibrium point, at which no further reaction occurs, corresponds to a certain ratio of reagents to products – of A and B to C and D. What this implies is that the reaction can go in either direction: if there is more C and D present than there would be at equilibrium, then C and D react to form A and B rather than vice versa. The reaction goes whichever way it needs to in order to reach the equilibrium state. This is the law of mass action, and was discovered in the early 1800s by French chemist Claude Berthollet. It implies that a reagent can't go on and on producing the same amount of product as you apply less and less of it *ad nauseam*. Yet this is in effect what Benveniste's findings seemed to assert.

The law of mass action may seem like an obscure and rather technical law to be challenged by Benveniste's results; but in essence it amounts to saying no more than that molecules cannot have an effect in their absence – which sounds like common sense. The paper by Benveniste and colleagues challenged the common sense of chemistry.

In science, that is no crime. Indeed, if there is a single lesson to have emerged from science in the twentieth century, it is that common sense is not always a good guide to our intuition about how the world works.[17] There is much in quantum theory, as in Einstein's theory of special relativity, that challenges our common-sense notions about such familiar entities as time, space and matter. For this reason, John Maddox comments that in science the concept of heresy may in fact be meaningless. One definition of heresy is 'an opinion contrary to generally accepted beliefs' – and scientific enquiry cannot survive without such things.

Science elects to embrace some such contrary opinions, yet to reject and even, perhaps regrettably, to ridicule others. I hope I have already given you some inkling of why this is so, and why the choice is not arbitrary. New discoveries can be startling and challenging without demanding that we blank out everything that we have previously believed to be true and have validated with experiment. That was so with Einstein's theory of relativity, which did not require us to believe that the apparent success of Newtonian mechanics for two centuries previously was the result of some collective delusion. Relativity is entirely consistent with Newtonian mechanics except in some extreme circumstances where new rules apply. But as Maddox said, 'Benveniste's observations, on the other hand, are startling not merely because they point to a new phenomenon, but because they strike at the roots of two centuries of observation and rationalization of physical phenomena.'[18]

Even that is not cause to dismiss them out of hand; but it is cause to demand that the evidence for them be unusually compelling. 'Extreme claims require extreme evidence', as Carl Sagan has put it. If this seems like shifting the goal posts, bear in mind that we do it in life all the time; otherwise the business of living would be an impossibly troublesome affair. It is simply the blue-apple response. But Benveniste emphatically rejected this standard position of the scientific community: at suggestions that 'a different editorial standard is required' for such claims, he complained that 'changing the rules of experimental science will first kill fragile data critical to fringe advances and then science as a whole'.[19] On the contrary, unless such criteria are applied, science will forever be seeing unicorns, telling us that there is infinite energy available for free.[20]

Nothing's perfect

Benveniste's paper appeared free from glaring errors. The experiments had been performed in a competent laboratory, the procedures were documented clearly enough, and potential sources of error seemed to have been taken into account. But I need here to say a word about errors.

No experiment is perfect, in the sense that the numbers that come out can be treated as 'absolute' measurements. Rather, every quantitative measurement must be afforded a kind of haziness owing to the limitations on measurement technique. If I ask for ten lengths of six-foot planking from my timber merchants, I don't expect that they will all be exactly the same length, even if the timber workers are conscientious and skilled. Indeed, it's not even clear what one can mean by 'exact' length, since the end of a sawn piece of wood has a roughness imparted by the action of the saw across the grain – if we're talking in terms of tenths of a millimetre, there is no unique length to a plank at all. On top of this roughness, there are limitations in how accurately the saw can be placed on a ruled six-foot mark, and also on how accurately that mark can be inscribed by eye, especially with the implements and time constraints of a busy timber shop. So I expect that the shop will supply me with six-foot lengths within an error margin of, say, an eighth of an inch. Measurements in scientific experiments are like this: there is only so much accuracy one can expect of them, imposed by the limitations of the apparatus and of the researchers themselves. Any kind of scientific data is therefore essentially meaningless without some estimate of the errors involved.

Assessing degranulation of basophils is a tricky business. It involves inspecting, by eye with the aid of a microscope, samples of the suspension of white blood cells stained with the toluidine-blue dye, and counting the number of cells that have turned red (indicating degranulation) relative to the number that have not. This kind of counting by eye is particularly prone to error, since it depends on the care and tenacity of the experimenter. Benveniste and colleagues were aware of this source of 'sampling error', and in the early 1980s they had investigated the kind of uncertainties that it introduces into the data. They had concluded that the measured degree of degranulation should be trusted to within about twenty per cent or so. One way of expressing this is to say that degranulation of less than twenty per cent in any sample is not statistically distinguishable from no degranulation at all.

In their experiments, however, they reported degrees of degranulation way outside this error boundary: the peaks in degranulation at successive dilutions sometimes reached as high as sixty per cent. So to this extent, it seemed that the experiments had taken proper account of errors, and that such errors would not suffice to explain the results.

A second important check on the results was to run control experiments. These generally comprise 'blank' experiments run without the supposed causative agent: they ensure that any effect observed can be correlated with the presence of the agent that is thought to be causing it, and does not happen in its absence. In this case, the appropriate control experiment was to look for degranulation of basophils in solutions that were serially diluted but without any anti-IgE present to begin with. The researchers saw degranulation at 'high dilution' only in the solutions that initially contained anti-IgE, not in the control experiments. It seemed that anti-IgE really was responsible for the effect.

And to be absolutely sure that there was no trickery or self-deception going on within the French team, they conducted some experimental runs 'blind'. Several samples – some at high concentration of anti-IgE, some at very high dilutions, and some that were just controls with no anti-IgE added at any point – were prepared and coded by one team of researchers, and given to another team, who did not know the code, to analyse for degranulation. They saw it in the low- and high-dilution samples, but not in the controls. And they confirmed that no anti-IgE at all could be detected in the high-dilution samples, even though degranulation was observed.

I have explained these tests in some detail to make it clear that the INSERM 200 team went to some pains to show that their results were not the consequence of poor technique or self-delusion. Yet even then, the referees for *Nature* could not accept the results at face value, and they insisted on the very unusual measure of having the experiments repeated in different laboratories. This should have provided a very stringent test of the claims, for it is hard to see how any idiosyncracy in Benveniste's laboratory that could have generated spurious results would be duplicated elsewhere. Benveniste arranged for three different groups in Israel, Canada and Italy to perform similar experiments, and they duly reported evidence for the high-dilution activity of anti-IgE.

The case for publication of these seemingly unbelievable results was thus mounting. And the urgency was highlighted when the French press caught wind of Benveniste's work, and reports of it appeared in *Le Monde*. Almost one year after the paper had been submitted to *Nature*, Maddox and *Nature*'s referees were forced to conclude that they had 'run out of hurdles to set' – and the paper was published, accompanied not only by a leading editorial comment ('When to believe the unbelievable') but by an 'Editorial reservation' at the foot of the paper itself which read like a disclaimer, confessing that 'Readers of this article may share the incredulity of the many referees who have commented on several versions of it during the past several months.'[21]

The memory of water

Amidst all of this, what passed almost without comment was the explanation that the paper offered for the high-dilution effect. It was as though so much effort had gone into ensuring that the results themselves were secure that no one had bothered to worry about how the authors interpreted them. It is hard to understand how such vague and ill-formed notions as they offered could possibly have been considered to augment the paper:

> We propose that . . . specific information must have been transmitted during the dilution/shaking process. Water could act as a 'template' for the [anti-IgE] molecule, for example by an infinite hydrogen-bonded network or electric and magnetic fields.[22]

The authors cautioned that 'any such hypothesis is unsubstantiated at present' – but in reality there is no 'hypothesis' here at all. We are in the same territory where we find the 'scientific' explanations of many alternative medicines: words and concepts are mis-used with abandon, or evoked as if to accrue credibility by association. 'Water could act as a template . . . by electric and magnetic fields' has no objective meaning that I can think of, even if one replaces 'by' with 'through the influence of'. All chemical interactions are electrical in origin, so the second part of the phrase is redundant. We can assume, however, that the authors actually said in reverse what they intended to express: that the biomolecule might act as a template for structuring water. And it seems fair to assume that, by throwing in the 'infinite hydrogen-bonded network' – a term that they had clearly picked up from a physics paper on water structure – they were thinking that the existence of hydrogen bonding in water could enable such a structurally well-defined arrangement of water molecules to be assembled and maintained around the biomolecule. The idea, then, appears to be that the anti-IgE molecule interacts with water in such a way as to leave an imprint in the hydration shell (see page 238), which acts as a mimic of the biomolecule even after it is withdrawn. This, at least, is how Benveniste's vague 'explanation' was interpreted, and something of this sort continues to be cited from time to time in support of homeopathy.

Given, as I shall show, that this idea is so untenable, how can it have been suggested by otherwise serious scientists?

We have seen in Chapter 9 that there are very good reasons to believe that molecules in solution do impose some kind of structure on the water molecules that surround them. Whether this restructuring of water is profound or minor remains an open question; but it seems clear that it must occur to some extent, even for relatively simple molecules, and that it is influenced by the peculiar nature of the hydrogen bond. Moreover, we know that water trapped within or around crystalline proteins can be highly structured, the molecules occupying well-defined positions and orientations in space. There can be little doubt that biomolecules like proteins, of which anti-IgE is one, may induce significant reorganization of the infinite hydrogen-bonded network of liquid water.

But can this reorganization really amount to a complete imprint, so that the shape of the biomolecule is faithfully reproduced 'in negative' in the surrounding hydration shell, an imprint of the molecule that could act on its behalf? I'm afraid that such an idea has serious flaws not only scientifically but logically. To deal with the latter first: an object is not identical to its mould. A cast of a bust is not made by pouring molten metal over an identical bust. Similarly, an imprint of anti-IgE in water shouldn't act like anti-IgE itself, but like something that binds to it – like the IgE molecules on the basophil surface. Where then would the degranulation come from?

There's a further logical objection, which amounts to a *reductio ad absurdum*. If this kind of templating, or some other mechanism that makes water act like anti-IgE, takes place, why doesn't it happen all the time? Why isn't the cytoplasm of our cells for ever getting clogged up with the ghosts of the proteins they contain? How is it that the sea, which has been awash in biomolecules (and has been thoroughly vortexed!) since time immemorial, has not been transformed into the most potent bioactive soup in the world, full of imprints of just about every enzyme in existence and so as inimical to swimmers as our gastric juices are to a swallowed fly? How is it that we can ever dilute any solution, that orange juice does not keep tasting of oranges however much water we add to it because the flavoursome molecules have been templated? And why should such templating come and go as the dilution is increased – for, once gone, how does it come back again without templates to induce it? Setting aside this periodic appearance and disappearance, the templating explanation would require that each individual molecule leave an increasing number of imprints as dilution is increased, and that it do so in exactly the proportion needed to counterbalance the fact that there are fewer molecules around after each dilution.

And finally, why is it that factors of ten seem to be so important for this 'memory' process? Benveniste's laboratory apparently avoided dilutions by other factors, such as three or seven, on the grounds that these did not 'work'. Can there by any logical reason why two dilutions by a factor of three should be so significantly different to one by a factor of ten?

One could advance a number of more technical reasons why the 'templating' idea makes no sense, but probably the most damning is the indisputable fact that water is a liquid. Each hydrogen bond is broken and reformed, on average, once every billionth of a second. In other words, no specific hydrogen-bonded structure can persist in pure water for more than roughly this length of time: it gets randomized away by the breaking and reforming of bonds. Even if proteins can hold on to some water molecules in their hydration shells for longer than this – and measurements indicate that it's only a little longer, at best – there is simply no good reason why any water molecule should be less mobile and so less readily randomized in the absence of such a perturbing influence, any more than a tub of water should retain the imprint of your hand after you withdraw it.

To anyone who wants to believe that liquid water has a memory, all this is splitting hairs. I'll simply ask that you make up your own mind.

Only the smile is left

When cold fusion was first reported, it prompted a mad dash to laboratories all over the world in attempts to reproduce the results. Although polywater had a slower start, it too provoked furious research activity once the story broke. Yet Benveniste's stunning claims were met with, for the most part, a stony silence. A year after the paper was published, *Nature* had received only one other paper reporting similar observations (which, after peer review, it did not publish). Given the extent to which the INSERM 200 team's claims otherwise conformed to the model of pathological science, how can we explain this difference in reception?

Ultimately the reason is most probably that Benveniste's claims were just too far beyond the pale. Cold fusion required a suspension of disbelief, but only in quantitative terms: the basic phenomenon claimed, nuclear fusion, was known to occur, but was not expected to happen to any significant degree under the conditions reported. Polywater played on the well-known ability of water to become structured near surfaces, as well as the abiding mystery about the fine details of that structure. They threatened no fundamental principles. Yet Benveniste's results could be seen at a glance to do so, and it is likely that most researchers chose straight away to regard them as either an obvious example of poor experimental technique or, at worst, fabrication.

Not everyone, however, decided that the results were just too implausible to be even worth testing. Henry Metzger and Stephen Dreskin from the National Institutes of Health in Bethesda, Maryland, looked at the response of rat basophilic leukaemia cells to the presence of anti-IgE produced in rabbits. In blind tests, so that Dreskin analysed the samples in ignorance of how Metzger had coded the dilutions, they saw nothing but the expected sharp drop-off of basophil response with increasing anti-IgE dilution. 'It's a shame really,' they concluded – 'It still takes a full teaspoon of sugar to sweeten our tea.'[23] JeanClare Seagrave from the University of New Mexico School of Medicine conducted similar experiments and saw similar results. An Italian team, Sergio Bonini and colleagues from the University of Rome 'La Sapienza', looked for histamine release from degranulated basophils rather than using staining to reveal degranulation itself. As I explained earlier, histamine should be an inevitable consequence of degranulation, and indeed the Italians saw it generated when high concentrations of anti-IgE were used – but not as the dilution increased. 'Should we speculate why water can remember something on some occasions and forget it on others?' they asked.[24]

Benveniste was unimpressed: 'our exact experimental design was not reproduced', echoing the rebuttals of Deryagin, Pons and Fleischmann. In science, this broad-brush argument will not wash – it is the *ad hoc* self-justification of pathological science. In Michel Schiff's book *The Memory of Water*, the author – a physicist and sometime collaborator with Benveniste – charges that all of the attempts to duplicate Benveniste's findings were launched in the hope that no such verification would be found, and that the experiments were designed accordingly. This accusation seems hard to sustain when one remembers how cold fusion, a comparably challenging notion, was apparently duplicated very quickly in laboratories all around the world. There is never any shortage of scientists eager to jump aboard a controversial bandwagon.

But by the time these counter-experiments were reported, readers of *Nature* eager to learn more about water's memory had already been treated to a further colourful addition to the tale. In his editorial footnote to Benveniste's paper, John Maddox advertised that 'With the kind collaboration of Professor Benveniste, *Nature* has... arranged for independent investigators to observe repetitions of the experiments. A report of this investigation will appear shortly.'[25] He was as good as his word, and the report appeared a month later.[26] It was extraordinary.

Those sympathetic to Benveniste's message, and even some who were not, were quick to accuse John Maddox of a set-up job, of exacting revenge for being 'forced' to publish Benveniste's paper. Anyone who knows Maddox will be aware that this just doesn't ring true. The fact is that he is simply a man of insatiable curiosity, particularly for the offbeat. I am sure he just wanted to discover what on earth was going on in INSERM 200. To assist in this endeavour, he put together 'an oddly constituted group': biologist Walter Stewart of the National Institutes of Health, who had set himself up as an arbiter of misconduct in science, and James Randi, a professional magician. Concerned about the increasing incidence of fraud in scientific (particularly biological) research in recent years, Stewart had become involved in several celebrated cases, most notably that involving a researcher from the laboratory of biology Nobel laureate David Baltimore.[27] Some regard Stewart as playing an important role in policing the scientific community; others feel that he does so with no authority and with the persecutory zeal of a Grand Inquisitor. Randi is most widely known for his masterful debunking of the antics of psychic Uri Geller, whose feats he is able to reproduce through standard yet highly accomplished techniques of stage magic. I can testify, following Randi's materialization in the *Nature* office one day, that his dexterity in this respect is stunning.

This trio made their way to INSERM Unit 200 in Clamart, prompting one commentator to remark, 'So now at last confirmation of what I have always suspected. Papers for publication in *Nature* are refereed by the Editor, a magician and his rabbit.'[28]

The investigation was described as 'tense', which probably reflects Maddox's penchant for understatement. Benveniste's team clearly bristled with distrust and believed themselves to be the victims of a 'Salem witchhunt or McCarthy-like prosecution'. The mere inclusion of James Randi amongst the investigators carried the unspoken accusation of fraud; and indeed, the team justified his presence 'in case the remarkable results reported had been produced by trickery'. It seemed clear that the French researchers were as concerned that their investigators might be out to use sleight of hand to dupe them as the latter were that the results might be generated in this manner: Benveniste insisted that Randi be kept away from the samples when blind tests were being set up. 'A tornado of intense and constant suspicion, fear and psychological and intellectual pressure unfit for scientific work swept our lab,' he subsequently lamented.[29]

The report of this strange and, as far as I know, unprecedented extension of the scientific process is illuminating, entertaining and confusing in roughly equal measure. The investigators spent a week in Benveniste's laboratory, during which time they conducted or supervised many more experiments than the French team would normally have managed in that space of time. Four runs of the dilution procedure were first conducted by the INSERM team according to the procedures outlined in their paper, and all were considered by Benveniste to be 'positive' – that is, to show the high-dilution effect. The first three runs showed statistically significant degranulation at high dilution, but the peaks showed no sign of being either periodic or reproducible – they appeared in different places each time. It seemed that this was typical: strictly speaking, the 'repetitive, reproducible waves' claimed in Benveniste's paper justified neither adjective.

The fourth experiment was different. Oddly, this was one in which Walter Stewart coded the samples for Benveniste's colleague Elisabeth Davenas ('in whose hands the experiment most often "works" '[30]) to read blind. It showed very strong degranulation peaks (up to about seventy per cent) at roughly repeating dilutions of 10,000. Nothing in the report sheds much light on this strange result.

Unenlightened by these experimental runs, and concerned about the way the group were handling sampling errors, Maddox and colleagues proposed to devise a procedure for running three more tests that should expose any experimental or analytical flaws. They concocted an elaborate coding procedure for blind testing, recording the proceedings on video tape and, in a gesture that later provoked much hilarity, taping the code, inside an envelope, to the ceiling of the lab. Randi explains that this was not so much to protect the code as to offer a bait to any who would like to fix the results.

The results of these blind tests were unambiguous: 'The anti-IgE at conventional dilutions caused degranulation, but at "high dilution" there was no effect.' The effects reported by Benveniste's team were, in the investigators' eyes, a delusion.

The systematic list of conclusions that Maddox's team drew up is fully transferable in remarkable degree to both polywater and cold fusion, and is no doubt equally so to most other examples of pathological science:

i. The care with which the experiments reported have been carried out does not match the extraordinary character of the claims made in their interpretation.
ii. The phenomena described are not reproducible, but there has been no serious investigation of the reasons.
iii. The data lack errors of the magnitude that would be expected, and which are unavoidable.
iv. No serious attempt has been made to eliminate systematic errors, including observer bias.
v. The climate of the laboratory is inimical to an objective evaluation of the exceptional data.

The last of these is particularly revealing: Benveniste's team had become accustomed to speak in terms of experiments as successes or failures, as 'working' or not. They are undoubtedly not alone amongst scientists in adopting such language, but in truth an experiment only does not 'work' if, say, the refrigerant leaks or the power fails – not if it simply doesn't produce the results you want to see. Experimenters almost always have a preference for the outcome of an experiment; but nature always answers honestly. Benveniste did not seem to appreciate this point – in his rebuttal of the *Nature* report, he says of the last three experiments that they 'worked poorly'.[31]

This rebuttal did him little credit in other ways. The excuses were based mainly on attempts to discredit the investigators' competence and the reliability of the methods used, with no specific indication of where their procedures might have erred. But one *cri de coeur* probably resonated with other scientists: 'never let these people get in your lab'. For all that Benveniste took to donning the mantle of victim, it isn't clear that the *Nature* investigation gave much enlightenment, or comfort, to those following the affair. It would 'dismay serious scientists' and added to the 'circus atmosphere', in the words of one critic.[32] Despite my own belief that it was motivated by John Maddox's curiosity rather than by any desire to blacken Benveniste's name, this sort of investigation is simply not needed in science. As the earlier cases have demonstrated, the standard procedures of the scientific community are quite sufficient to sort the wheat from the chaff. Questionable work will not stand for long in the sun – if it cannot be reproduced, that is enough to spell its doom. This process might take months, even years,[33] it might divert money and time – but that is how it must be.

Curiously, for all his bluster Benveniste emerges from this episode with more credit than Pons and Fleischmann. His dismissal of critics was often shallow and cavalier, but it is hard to avoid the impression that his belief in his results was sincere, that he and his colleagues made every attempt to perform what they believed were careful and well-documented experiments, and that he made some efforts to cooperate with sceptics rather than imposing secrecy. 'It may be that all of us are wrong in good faith,' he said. 'This is no crime but science as usual.' With this at least, John Maddox agrees: 'Nobody will dispute people's right to be wrong. It is also proper that people should persist with their beliefs in the face of scepticism.'[34]

For these reasons alone, it seems only right that, once the claims made for the high-dilution experiments had clearly fallen into disrepute, the response of INSERM was not – as some might have anticipated – to dismiss Benveniste from his post. Following a routine evaluation of Unit 200 in 1989, he was put on a six-month probationary period during which he was expected to establish 'the full guarantee of his peers' and to adopt the 'critical and reserved attitude' expected of a scientist. The evaluation committee recommended that he stop the work on high dilutions, which they considered Benveniste had interpreted 'out of proportion with the facts'. Philippe Lazar, the director-general of INSERM, chose not to endorse this particular recommendation, however, feeling that it would be inappropriately compromising to the freedom that a laboratory director should enjoy.

High-dilution experiments are by no means moribund, and neither do I think they should be. As a paper in the *Lancet* from September 1997 declares, 'Homeopathy seems scientifically implausible, but has widespread use.'[35] This study, by academics from Germany and the USA, concluded that the results of eighty-nine previous trials on medical effects of high-dilution remedies 'are not compatible with the hypothesis that the clinical effects of homeopathy are completely due to placebo'. Yet at the same time, the researchers stated that 'We found insufficient evidence from these studies that homeopathy is clearly efficacious for any single clinical condition.' Clearly, one could interpret this study whichever way one pleases; personally, it does not give me confidence that homeopathic remedies will cure me of any ailments. It indicated, however, that two thirds of the trials included in the analysis were 'methodologically poor', while not strictly invalid – often an outcome of their low-budget nature and the fact that they were done by 'advocates with high enthusiasm'. That 'a serious effort to research homeopathy is clearly warranted despite its implausibility' seems to me to be a fair deduction. As to whether or not water has a memory, the study is wisely silent.

Benveniste's case is put sympathetically in Michel Schiff's book, which charges the scientific community with complacency and censorship at several levels. I do not wish to be unduly harsh about Schiff's portrayal, but neither would I like to feel that the reader will be misled by it. I have said enough in earlier chapters, I hope, to demonstrate that Schiff's characterization of the state of knowledge about liquids – 'like the state of knowledge of astronomy before Galileo, Kepler and Newton' – is woefully inadequate. It is particularly worrying to see water's structure described without any reference to hydrogen bonds. Schiff suggests that 'the memory of water' need not revoke any fundamental principles of physics after all, but might rely instead on a 'theory of coherent domains' in liquids, published in respectable journals by two Italian physicists.[36] It is left rather vague exactly how the theory accounts for Benveniste's results, but in any event this theory has received robust criticism from leading physicists in a quite independent context and I think it is fair to say that it is not a mainstream view.

Benveniste himself is unrepentant about his ephemeral notoriety. 'I did not then know that physicists who deal with infinity have the right to dream, but that soft savants like biologists do not! Now I know.' And it seems he is dreaming still, claiming that he has now mastered the electromagnetic basis of water's memory sufficiently to be able to programme biological activity into pure water with data sent down a telephone line. The 'molecular signal,' says Benveniste, 'consists of waves in the kilohertz range':

> In the course of several thousand experiments, we have led receptors (specific to simple or complex molecules) to 'believe' that they are in the presence of their favourite molecules by playing the recorded frequencies of those molecules. In order to arrive at this result, two operations are necessary: 1) record the activity of the substance on a computer; 2) 'replay' it to a biological system.[37]

Water, says Benveniste, acts as the 'vehicle for information': the biomolecular effect is transmitted by 'perimolecular water which relays and possibly amplifies the signal'. He predicts a 'digital biology', in which the electromagnetic signature of antibodies, bacteria and viruses will be digitally recorded and programmed into water from a distance. Benveniste is now the director of a private company, DigiBio, formed to research and promote the phenomenon. While he has published some of his results on digital biology in peer-reviewed journals, the road he is taking appears ever more fantastic – a fantasy such as only today's digital age could have manufactured.

Perchance to dream

I should be most surprised if water does not continue to inspire extravagant fantasies. And just as surely, they will resonate sufficiently with the scientific mysteries that still persist about water, or with the psychological significance that it holds for us, that these flights of imagination will secure adherents. In the end, it seems to me that the 'magical' properties attributed to water – an inexhaustible fuel, a universal remedy – derive from these psychological and emotional correspondences.

The same can be said for the work of Theodor Schwenk and Viktor Schauberger on 'flow forms' – work that often crops up when water's mystery is discussed outside of scientific circles. Schwenk, who was profoundly guided by the spiritual teachings of Rudolf Steiner, believed that flowing water acts as a kind of sensory organ through which celestial influences enter into the world. His is a teleological Universe in which water is an 'appropriate medium' to 'receive formative impulses from the spiritual world'.[38] Because Schwenk's poetic imagery borrows occasionally from the scientific lexicon, it is unfortunately sometimes mistaken as an expression of a scientific theory. Possibly Schwenk believed this himself, but the idea is untenable. Schauberger, an Austrian forest warden who lived from 1885 to 1958, voiced similar opinions, clinging to the notion of water as an element and awarding it a kind of non-physical vital force. His life story is extraordinary, and I recommend Callum Coats's biography *Living Energies*[39] for this reason, while cautioning most strongly against regarding Schauberger's ideas as in any way connected with science.

If nothing else, these romantic fancies remind us that water is inspirational. It infuses us with spirit. It quickens the heart to watch a clear, gurgling brook or the pounding of wild surf. The rhythmic lap of the waves on a beach soothes and reassures. So we bring to the study of water an emotional response that does not accompany the investigation of, say, bismuth. Such responses are not irrelevant in scientific work – I feel sure that something of the kind initially drew many a field ecologist to the study of nature in the wild, or an astronomer to contemplate the starry heavens.

This excitement can provide the fuel, the motivation for research. Scientists can retain this emotional commitment to their topic of study, if they do not become too jaded with funding struggles or the need for overspecialization, while still engaging in thoroughly objective work. But they will, on the whole, make a distinction between their emotional relation to their object of study and the object itself, at least to the extent of disconnecting the inspiration from the phenomena. That sounds perhaps like a brutal thing to do, but it is generally done unconsciously, and it really amounts to nothing more than accepting that the thrill we feel at the aroma of freshly baked bread arises within us and is not transmitted by the loaf. The aroma is the loaf's – the thrill is ours.

I think only a fool would deny the possibility that water might still hold unguessed secrets and wonders in its molecular make-up. But there is no earthly reason to believe that water should be so magical as to escape the laws of physics, nor can I really believe that we would be affording it due respect by imagining otherwise. Water is already abundant enough in its granting of marvels.

Epilogue

Blue Gold

Water as a resource

If there be sufficient authority to remove a putrescent pond from the neighbourhood of a few simple dwellings, surely the river which flows for so many miles through London ought not to be allowed to become a fermenting sewer . . . If we neglect this subject, we cannot expect to do so with impunity; nor ought we to be surprised if, ere many years are over, a hot season give us sad proof of the folly of our carelessness.

Michael Faraday (1855), Letter to *The Times*

Water is the true wealth in a dry land; without it, land is worthless or nearly so. And if you control water, you control the land that depends on it.

Wallace Stegner (1954)

I blame it all on the person who started destroying the land, the river. Now we do not have good drinking water, and a river that was once so strong and deep and that had provided for a lot of families is now all dried up and sad to look at.

Alice M., Cree Elder (1994)

When Coleridge's Ancient Mariner said 'Water, water, everywhere, nor any drop to drink', he gave a fair picture of the global situation. The 'drop to drink' is a hundredth of a per cent of the world's water: about one drop in every bucketful. The proportion of planetary water that is fresh is rather larger – around 3.5 per cent – but most is frozen in the ice caps and mountain glaciers. As sea water is corrosive and toxic to land-based animals and plants, nearly all of the water that we use must come from that precious one hundredth of a per cent. But unlike many other natural resources, water is renewable, continually replenished by the hydrological cycle.

Around thirty trillion gallons of fresh water are recycled from the sea to the land every day – a yearly throughput of 40,000 cubic kilometres, or one hundredth the volume of the Mediterranean Sea. But two thirds of this water returns to the sea as flood water, which cannot be captured for human use. Some 5000 cubic kilometres more drains away in uninhabited and inaccessible areas. In all, just a quarter to a third of the total amount of recycled fresh water is easily available to us.

All the same, that sounds like a lot to go round, and indeed it's thought that it could in principle sustain a world population of about twice the size projected for the end of the twenty-first century.

Yet it seems likely that water will become a scarce commodity, with potentially incendiary consequences, well before then. Between 1950 and 1990, global water use tripled, and by 1996 it was estimated that we were using over half of the available run-off. In other words, if, as some predict, water use doubles over the next thirty-five years, the taps will have run dry. We'll be looking for more than the skies can offer.

If the last two paragraphs seem contradictory, think again. There's a distinction between what we need and what we use. Sustainable use of the world's water supplies at a level commensurate with population growth requires that water be used with restraint and efficiency, and this simply does not happen on the whole. Moreover, it presupposes equitable distribution of the available resources, and this is not the case either.

This second consideration is largely a fact of nature: rainfall is not evenly distributed throughout the world (Fig. 12.1). In many lands the supply of water is dangerously sporadic and scanty, while sixty-five per cent of the world's natural water resources are located in just ten countries. Since water is emerging as a fundamental limit to economic growth, this means that we must come to terms with a world that knows nothing of equal opportunities. There has been little indication so far that our political and economic systems are well suited to redressing nature's iniquities.

Restrained and efficient water use should, however, be a more achievable objective. But in industrialized nations, better use of water resources means shedding some deeply ingrained habits. The Industrial Revolution paid scant heed to questions of water purity and conservation, nor to the water needs of aquatic ecosystems. Slowly this is changing, though legislation rather than individual or corporate self-restraint is commonly the agent of change.

In many less developed countries the problem of proper water use

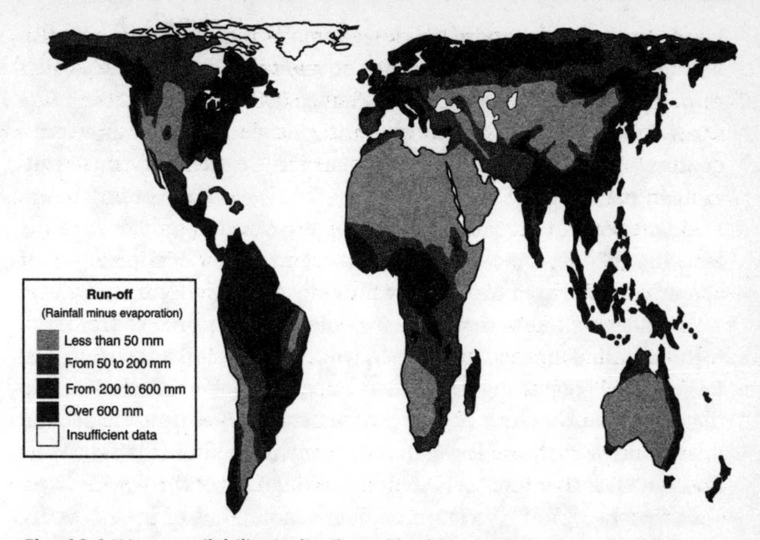

Fig. 12.1 Water availability is distributed highly unevenly across the globe. Here the Peters projection is used, which ascribes the true relative surface areas to the land masses.

is more stark. On the one hand there is the threat of scarcity; on the other, the threat of pollution. In 1977 the United Nations proclaimed that 'all peoples . . . have the right to have access to drinking water in quantities and of a quality equal to their basic needs',[1] and with this in mind, the International Decade of Drinking Water Supply and Sanitation was launched at the start of the 1980s. But today one in every three persons in the developing world – 1200 million people according to the estimate of the World Health Organization – lacks these two provisions. Largely because of the UN initiative, the number of people without access to safe drinking water decreased between 1980 and 1990. But this number is now rising again as population growth outstrips efforts to improve water quality and distribution.

Many millions of people worldwide live with water supplies that are a major health hazard. In India, for instance, water is relatively abundant, but rivers and urban supplies are highly polluted. Half the world's population does not have basic sanitation systems, and a quarter has no access to clean water. As a result, around eighty per cent of all diseases and one third of the deaths in developing countries are the result of contaminated water.

Amongst these diseases are dysentery, cholera, typhoid, hepatitis and guinea worm. Standing bodies of water can harbour the carriers of dengue fever and the larvae of mosquitoes. Many of these afflictions were common in Europe until the end of the nineteenth century – the direct dumping of human waste into rivers, which initiates so many of the health problems in developing countries, was standard practice in Britain, for example, until laws forbade it in the late 1800s. The link between disease and water quality was first clearly established in 1854 by the London physician John Snow, who traced the highest incidences of cholera during one of the city's recurrent epidemics to water drawn from a particular well. Yet knowledge is one thing, implementation another – and cholera epidemics remained frequent in Victorian London. The near-eradication of such diseases in the last hundred years in the industrialized West makes it clear that their persistence in other parts of the world is likewise a problem not of scientific understanding but of the necessary means to implement even elementary sanitary standards. We should not forget that it is economic considerations, not limitations of medical science, which hold back amelioration of the majority of suffering, affliction and fatality in the world.

It is hard to appreciate in the industrialized West what a lack of access to clean, safe water implies for the way one lives. Here the phrase 'on tap' is synonymous with 'freely available' – by which is meant 'as freely available as water'. In these parts of the world, it is a resource so taken for granted that we use it in profligate amounts that would be impossible and catastrophic elsewhere. An average US citizen consumes over one hundred times the amount of water as a citizen of Uganda or Burundi. In Western Europe people use an average of between 100 and 260 litres each per day. In these wealthy countries, you turn on the tap and out it comes, ready and safe (one imagines) for consumption. Prime-quality water is used not only for drinking, cooking and bathing but for washing cars and streets, and watering lawns and parks. In Europe, nearly eighty per cent of water withdrawn for domestic and municipal supplies ends as waste rather than being consumed, and this is before you consider the huge quantities – an estimated thirty per cent or so for England and Wales – squandered as leaks before it even reaches the tap. In Asia, this profligacy cannot be entertained: about sixty-two per cent of domestic and municipal water there is consumed.

With future increases in population, industry and agriculture, things will get worse. UNESCO predicts a serious worldwide shortage by around 2020, which is consistent with the figures quoted earlier. The implications for human health and for political stability are immense. Where the problems arise, what might be done about them, and what might happen if we ignore them – these are issues that lie outside a survey of the science of water. But it seems fitting, indeed essential, to conclude water's scientific biography with a coda that looks at the aspect of water that none of us can ignore: we all need it, yet it is getting scarcer, per capita, by the day.

Of course, one cannot separate a discussion of the physical nature of water resources from the broader cultural issues that surround them. I'd submit that the political and economic dimensions of water's human significance are linked to some of the features of our psychological relationship with the material aspects of water that I have highlighted in this book. The water of myth, says Gaston Bachelard, is not the abundant brine of the oceans but the precious, sweet fresh water that we need for drinking, for agriculture, for bathing. Legends tell of punishments inflicted on polluters of fountains; today we taint our fountains as never before, with impunity or, in developing countries, out of tragic economic necessity. In legend, the slightest impurity can place a curse on water; today millions live with that curse, and its attendant misery is very real.

Watering the fields

Water is essential to virtually every economic activity, but none more so than agriculture. Around seventy to eighty per cent of the water withdrawn globally is used for irrigation, and the proportion is higher still in less developed countries. Concerns over the impending water shortage are not so much about thirst as about hunger: irrigation is the key to our ability to feed the future world.

Irrigation is an ancient agricultural technique. It was practised in the Middle East at least 7000 years ago using the same principle that is most commonly exploited today: gravity-driven flows of run-off water in streams, rivers, lakes, reservoirs and canals are directed through basins, pipes and furrows onto the cultivated soil. Over half of the increase in crop production of developing countries between 1960 and 1980 was the result of irrigation, and the area of land rendered arable by this means is increasing at a rate of around eight per cent per year.[2]

But agriculture is a voracious consumer of water: it takes something like forty litres of water to produce one can of vegetables. And the capacity for wastefulness is tremendous. The global efficiency of irrigation methods is estimated at below forty per cent – that is, well over half of the water is wasted. Water supplies for agriculture are generally subsidized so that farmers tend to pay far less for water than is warranted by the regional supply and demand, and usually less than the cost of delivering it.[3] As a result, there is often little incentive to install costly water-saving schemes. New methods such as drip or trickle irrigation, in which water is directed through tubes directly onto the soil around individual plants, are more efficient but are unlikely to find wide application while it remains so much cheaper simply to let water go to waste.

Over-use of irrigation is widespread, and as well as squandering fresh water it brings other problems, the most serious of which is salinization. Soils contain a variety of soluble mineral salts which are dissolved and carried away by run-off water. If water is prevented from draining normally from the soil, these salts remain behind as the water is removed by evaporation. Their accumulation turns the soil alkaline and caustic, inhibiting plant growth. In extreme cases, the salt forms a hard crust at the soil surface.

Salinization can occur naturally whenever water evaporates more rapidly than it is delivered by rain, and so it is common in arid regions. Because groundwater is enriched with salts leached from the soil through which it percolates, salinization can be exacerbated by human practices that cause the water table to rise up into the cultivated soil. The drawing-up of groundwater can leave irrigation water one tenth as salty as sea water.

Centuries of over-irrigation in the Middle East and in the Indus valley of Pakistan have led to acute problems of salinization, in the latter case giving the region one of the lowest crop productivities in the world. Poor management of irrigation techniques in projects in the Sudan, Iraq and China have ruined soils and incurred huge debts with no pay-off in terms of actually feeding people – a tragically common result of schemes designed to render arid lands farmable.

Conversely, in coastal regions salinization can arise from *depletion* of fresh groundwater and consequent lowering of the water table. Here salty water from the sea percolates inland through the soil until it meets the freshwater aquifer. If the water table is lowered by the withdrawal of water through wells and boreholes for crop irrigation or other uses, the saline water is able to push its way further into the aquifer. Eventually the wells themselves become contaminated. It is very difficult to reverse this process, restoring the boundary between fresh and salty water to its previous position. This kind of salinization has become a serious problem all around the Mediterranean coast, from France to Libya, exacerbated by the increasing cultivation of water-hungry export crops such as citrus fruits in place of the more traditional cereals and olives.

About two million hectares of agricultural land become affected by salinization each year and the United Nations estimates that, through the expansion of irrigation schemes in the past decades, around a quarter of all irrigated land is now afflicted. This figure may be as high as fifty per cent in extensively irrigated Iraq. The Earth Report (1988) painted a still grimmer global picture, estimating that between thirty and eighty per cent of irrigated land suffers from salinization, excess alkalinity and/or waterlogging.

Salinization is a tough problem, because there isn't really any effective way of getting rid of all that salt except by flushing it out with water. Not only does this consume even more water without it being used directly for agriculture, but it just makes the problem someone else's: the salt gets washed away downstream to contaminate other fields. As a result, salinized land may be simply abandoned by farmers. Large areas of land in the Tigris and Euphrates valleys, which were irrigated since time immemorial, today lie barren because the cost of remediating the salinization is simply too great for the local farmers.

Most mineral salts are harmful only in considerable excess. But poor irrigation practices can also lead to the accumulation of more hazardous substances from the ground. An irrigation scheme in central California diverted water through soils rich in selenium, a relatively toxic element, into a reservoir in the San Joachim valley. The selenium-rich waters of the reservoir have caused death and deformity amongst wildlife in the nearby Kesterson National Wildlife Refuge.

Increased agriculture brings with it other problems for water supplies. Clearing of forested land for arable use, and overgrazing of livestock, denudes land previously covered in vegetation. This means that less rain water is retained, the risk of floods is greater, and soil is eroded more rapidly. Soil erosion is particularly acute in dry tropical regions such as equatorial Africa and China, leading to the silting-up of lakes and reservoirs. Livestock rearing places an additional stress on water consumption, requiring around sixty billion litres each day globally. The meat-rich diet of the Western world is fantastically wasteful of resources – there is simply not enough land and water to allow all the world to indulge such tastes.

Advance of the dunes

One of the most dramatic consequences of intensified agricultural activity is evident in the Sahel region of western North Africa, where desert dunes have been relentlessly creeping over land once used for pasture and crop growing. The causes of, future prospects for, and even the meaning of 'desertification', the expansion of deserts, have been the subject of much recent controversy amongst environmental scientists. But one thing seems sure: the process is tied up with the availability of water.

According to the definition prescribed by the United Nations in 1990, desertification is the spread of desert-like conditions outside existing desert boundaries as a result of 'adverse human impact' – which generally entails inappropriate agricultural practices. It's not yet clear to what extent desertification in the Sahel and elsewhere conforms to this prescription. Certainly, overgrazing, overcultivation and deforestation have the potential to reduce once-productive soils to barren dust. When vegetation is removed or is consumed or damaged by grazing animals, the bare soil dries out more rapidly and its complex, porous structure – maintained by biological growth – is broken down, reducing the soil's ability to retain moisture. Trampling by animals only makes this worse. The process of deterioration may be self-sustaining: reduced vegetation cover leads to a decline in air humidity, cloud formation and rainfall. In other words, the land becomes trapped on a gradual decline towards desert. Overcultivation can deplete soil nutrients, leading to declining productivity and so less vegetation cover; and deep ploughing also disrupts the structure and integrity of the soil. The result is a dust-bowl like that created by intensive agriculture in the mid-twentieth century in the Great Plains of the American Midwest.

But desertification like that seen in the Sahel since the 1960s could instead have meteorological causes. Rainfall here has been below the long-term average since the 1960s, possibly as a result of natural fluctuations to which arid regions are prone – or perhaps wrought by human-induced global warming. Either way, the changes arising from increased aridization and the encroachment of desert dunes have taken a terrible toll in human life through drought and famine. Desertification has also afflicted Kuwait and South America, while in China around thirty per cent of the desert land has appeared since 1920. One episode of desertification of unambiguously human origin has taken place in Central Asia (formerly part of the Soviet Union), where the inland Aral Sea has been drained almost to oblivion to irrigate cotton crops, one of the most water-thirsty of all crop types. The pitiful, salty remnant of this sea now supports no fish, and deserted fishing villages stand silent on the dried-up coastline. The Uzbek poet Mukhammed Salikh laments, 'You cannot fill the Aral with tears.'

Because the uncertainties in monitoring efforts remain huge, it is not clear whether desertification is still increasing globally. Nor is it really known how feasible remediation is – there have been successes in developed nations, but only at considerable cost. But the problem it poses is potentially enormous: the United Nations Environmental Program estimates that sixty per cent of agricultural land outside humid regions is affected to some degree by desertification, and other estimates suggest that it places at risk the livelihoods of between 200 and 700 million people. It surely looms large in any assessment of agricultural impacts on water availability.

Making a mess

Wherever water is held sacred, the ideal of purity manifests itself. Our social attitudes towards pollution and contamination harken more to the mythic associations of purity than to considerations of natural science. 'Pure' water does not, to most of us, mean water free from all substances that are not H_2O – you can buy such stuff, but only as an expensive fine chemical, not in clear plastic bottles at the supermarket. Rather, we cherish our 'pure' spring water for its very impurities, for the health-giving mineral salts proudly listed on the label. 'Natural' and 'pure', synonyms to the consumer, carry quite different connotations to the chemist.

I suspect that producers of bottled water understand the mythical dimension of water purity only vaguely, but nonetheless have a strong sense of its selling power. In the 1970s and 1980s US citizens experienced a string of scares over drinking-water quality related to toxic and possibly carcinogenic contaminants: DDT (a pesticide), carbon tetrachloride (an industrial solvent), trichloroethylene (a supposedly safer replacement for carbon tetrachloride). The introduction of the US Clean Water Act in 1972 did little to quell either the fears or the real problems associated with these compounds, while during the Reagan administration the Environmental Protection Agency had its teeth pulled by a government that was unwilling to impose controls on industry. As a result, US citizens turned in increasing numbers to bottled water. In effect, the supply of drinking water in the Western world became largely a private enterprise performed by the bottled-water companies.

DDT was the villain of Rachel Carson's pivotal book *Silent Spring* (1962), which surely derived much of its impact from Carson's ability to tap into the psychology of purity. Whether consciously or not, she invoked the ancient idea that befouled water is cursed. She spoke of 'chemists brewing' their 'elixirs of death' in industrial laboratories, of the 'chain of evil' initiated by these pollutants.

Only by recognizing these associations can we understand why it is that the true hazards posed by pollutants are not always correlated with the level of public concern that they provoke. DDT represented a danger of relatively limited proportions that proved rather easy to contain. Radioactivity too is a relatively minor issue as far as water pollution is concerned, although this is by no means to downplay the local hazards posed by leaks and effluent from nuclear power stations – nor indeed the damage wrought by airborne release from Chernobyl. Dumping of nuclear waste at sea is outrageous not because it threatens us all with radiation sickness but because it demonstrates a disrespect for the integrity of natural waters that, applied globally, would be disastrous.[4]

Nonetheless, the detrimental influence of DDT was very real at the time of Carson's book. She recounted how fish and birds in a mid-Western region of the USA were gradually killed off by a mysterious ailment that was eventually identified as a toxic dose of this synthetic compound. The toxicity of DDT is not, in fact, particularly great; but the crucial point is that it accumulates in body fat, and so was transferred up the food chain from plants to insects and ultimately to birds and fish that have enough body fat to acquire high doses. DDT is now banned in the USA, but has been cynically marketed to developing countries where no such prohibition applies.

DDT is an 'organochlorine' compound, as are the related pesticides dieldrin and aldrin. These too have caused deaths amongst birds of prey. They are all very insoluble substances, but can nevertheless be transported in streams and groundwater on the surfaces of suspended particles of organic matter.

Perhaps more than anything else, *Silent Spring* highlighted the 'chemicalization' of the agricultural industry. Says one Californian farmer of pesticides, 'It's got to the point where we can't farm without them. Kind of like a dope addict, we're hooked.'[5] The increasing reliance on synthetic chemicals has two main driving forces: to stimulate crop growth with artificial fertilizers, and to suppress organisms that impair it, by deploying a veritable armoury of toxins. Herbicides are targeted at weeds, which compete with crops for space, water and nutrients. Fungicides, insecticides, rodent and slug poisons are amongst the chemicals deployed to ward off crop attack. As the main objective of these chemicals is to prevent the ecosystem from reaching a dynamic equilibrium by selective extermination, it is hardly surprising that many constitute or give rise to health-threatening contaminants when they find their way into natural waters.

Also chemically related to DDT are the polychlorinated biphenyls (PCBs), a large group of synthetic chemicals used by industry in the manufacture of paints, plastics, adhesives and electrical components, as well as being employed as hydraulic and cooling fluids. Most release of PCBs into the environment has been unintentional, through spills and improper disposal. Because they are biodegraded only very slowly, they have become widely dispersed in the environment, and have been found in high concentrations in the body tissue of several seafaring vertebrates, including sea birds, seals, dolphins, porpoises and whales. Their toxicity has been implicated in the deaths of many of these creatures.

It's now recognized that the manufacture of PCBs accumulates difficulties inexorably downstream, and their manufacture is therefore banned in most countries, although they remain in use in existing electrical equipment. They are costly to destroy safely – incineration at very high temperatures is needed to avoid the release of other chlorine-containing toxic substances such as furans and dioxins. These two classes of compounds, which are also formed by other combustion processes, have become the focus of recent concern. They are released into the environment in car exhaust, by municipal incinerators and by wood burning. Like PCBs, they are toxic and potentially carcinogenic, and their various sources are still not fully quantified.

Regulations notwithstanding, it seems clear that these organochlorine chemicals continue to pose serious problems of water-borne contamination. In 1998 the US Environmental Protection Agency announced that around seven per cent of all US watersheds have sediments contaminated to a degree that poses potential health risks to people eating fish and wildlife from the areas. Most of this contamination is by mercury and by PCBs released by organic chemical manufacturers and pulp and paper plants.

Sewer talk

Water in nature has an impressive ability to clean itself. Moderate inputs of organic waste such as rotting vegetables, excrement and even crude oil can be biodegraded by bacteria: consumed, metabolized and passed along the biogeochemical cycles of the elements. For this reason, many of the components of domestic sewage may persist for only days to weeks in the environment.

But self-cleansing of natural water has its limits. The shores of the Mediterranean Sea are awash with human detritus. Sewage from Bombay issues untreated into the Arabian Sea, rendering the coastal waters around the city unsafe for swimming at all times. Human excrement harbours some of the most pathogenic bacteria and viruses, particularly those responsible for cholera and typhoid.

Most micro-organisms require oxygen, and the 'feeding frenzy' of microbes supplied with organic waste can deplete water of oxygen. The consequent crisis in oxygen availability can be fatal to other aquatic creatures such as fish – they are asphyxiated underwater. In buried groundwater there is very little dissolved oxygen, and most biodegradation is effected by anaerobic ('airless') micro-organisms, which do the job more slowly. In consequence, groundwater aquifers are much more susceptible than streams and lakes to contamination with organic waste.

Water pollution by sewage has endangered lives for centuries. Dumping of 'dung and filth' into England's rivers was outlawed in the fourteenth century, yet one can see how much effect that had from Michael Faraday's outburst five hundred years later. In characteristically lurid terms, French writer Joris Karl Huysmans made the same complaint of the River Bièvre, a tributary of the Seine, in the 1920s:

> That river in rags, that strange river, that dumping-ground for filth, that bilge which is the colour of slate and melted lead, frothing, here and there, with greenish eddies, starred with muddy spittle, which gurgles on the sluice gate and is lost, sobbing in the holes of a wall.[6]

Lack of proper sanitation and sewage facilities continues to be the major source of water-borne disease and death in developing countries. Sewage is now given at least some form of treatment in industrialized countries before being discharged into natural waters. The most rudimentary processing is called primary treatment, which involves little more than letting the sewage stand until suspended solids settle out, while greasy and oily wastes form a scum on top which can be skimmed off. Most sewage is also afforded some kind of secondary treatment in which organic substances such as faecal matter, food residues, detergents and shampoos are biodegraded. The biologically processed sediment is rich in nitrogen- and phosphorus-containing organic matter, and so can be used as a fertilizer. This practice is controversial, however, because the sludge can also contain high concentrations of potentially hazardous substances, such as viruses and toxic heavy metals, which may find their way into groundwater. Until recently, sludge from sewage treatment in Europe could be dumped at sea, and the UK discharged it into the North Sea via pipes and ships; but a directive of the European Commission prohibited this from 1998.

The outflow from secondary sewage treatment is not stuff you'd want to drink. It still contains dangerous micro-organisms, as well as soluble nitrogen and phosphorus compounds which can adversely affect aquatic ecosystems (see below). A further, costly round of treatment is needed to pick out specific substances that cause environmental problems.

It's hard to see how any industrialized society could fail to incur a considerable sewage burden, but we seem to have acquired a peculiar relationship with water on the one hand and with solid waste on the other that exacerbates this situation beyond reason. Water is considered the universal cleanser – even in India, whose waters are amongst the most contaminated in the world, the rivers are accorded an almost supernatural quality of purification. Perhaps this is why, even in rural areas where fertilizer is in demand, human excrement has to be flushed away with gallons of water rather than composted for its nutrients. We can spread animal muck on the land, but our own induces such revulsion that it must be washed away through shiny white porcelain. In the Middle Ages human waste wasn't waste at all, but a product exported to the countryside as fertilizer. Most people in the developed world would now be astonished to learn that human excreta can be composted to dark, crumbly fertilizer within a year. There seems to be no strong reason other than aesthetics why compost toilets are not standard items in many modern houses.

Surface or groundwater generally requires chemical treatment to render it safe for drinking, primarily to destroy pathogenic micro-organisms. Most commonly, this involves chlorination. Chlorine kills most kinds of microbes, but can also leave an undesirable taste.[7] More worryingly, it can introduce new and potentially harmful chemical compounds. In particular, chlorine reacts with dissolved methane (which is produced by anaerobic microbial processes) to generate chloroform, a potential carcinogen. Concerns of this sort have led to the introduction of ozone as a purifying agent. Ozone is highly toxic, and can destroy some micro-organisms – such as cryptosporidia parasites, which come from livestock and can induce diarrhoea, vomiting and even death – that chlorination cannot. But ozone may react with organic compounds in water to leave an unpleasant colour, taste and smell in the water. The city council of Tucson, Arizona, discovered this to their cost when they introduced an expensive ozone purification plant for the city's water supply in the 1990s. The public dissatisfaction at the resulting water quality forced the council to close the plant.

Fatal blooms

Not all water-borne contaminants are fundamentally poisonous. Nitrate, one of the most serious pollutants of surface water in developed countries, is an essential nutrient for plant growth and a crucial part of the biogeochemical nitrogen cycle. It is produced by bacteria from nitrogen in the atmosphere, and also through chemical weathering of rocks.

But as Paracelsus said, the poison is in the dose. In excess, nitrate can be harmful to humans and other animals. High nitrate levels in water are caused largely by the use of nitrate-rich fertilizers and manure. The problem is particularly acute in the Netherlands and Denmark, where pig farming is common. Nitrate is especially harmful to young babies: it is converted to nitrite in the digestive system of infants younger than ten weeks. Nitrite reduces the capacity of blood to take up oxygen, and this in turn lends the skin a bluish colour and leads to slow asphyxiation. This condition, called methaemoglobinaemia or 'blue baby syndrome', afflicts babies who are bottle-fed with water containing high nitrate concentrations. There is also some suggestion – not by any means confirmed – that excessive nitrate intake can induce stomach cancer. The European Commission recommends that nitrate concentrations in drinking water should not exceed fifty milligrams per litre; the World Health Organization recommends a limit of twice this. But no one really knows what is a 'safe' level of nitrate.

As a nutrient, nitrate promotes plant growth. Phosphates, components of many detergents, do likewise. In excess, these two nutrients can destroy the balance of aquatic ecosystems, with dire consequences. Lakes and coastal waters fed by rivers enriched in nitrate and phosphate experience rapid growth – 'blooms' – of algae. In lakes the thick algal mats prevent light from penetrating into deeper waters, and this depresses growth of bottom-dwelling plants. Because these plants supply the bottom waters with oxygen as they grow, their inhibition reduces oxygen availability in the lower lake waters. At the same time, organic matter from the algal mats rains down to the lake bed and decomposes, and this biodegradation consumes what oxygen there is at an accelerated pace. The result is severe depletion of oxygen, called anoxia, in the lake, which leads to the death of fish and other creatures. This process is called eutrophication, and it upsets many of the world's lakes.

In coastal regions, eutrophication creates unsightly algal mats which clog the shores and render the waters unsuitable for bathing. Some marine algae also produce highly toxic substances which kill fish or render them poisonous. Certain types of toxic marine algae have a red colour, and their blooms create the phenomenon of 'red tides' which have threatened coastal ecosystems as well as economies that depend on fishing or tourism. Red tides have also been seen in freshwater lakes, such as Lake Biwa, Japan's largest lake, in Shiga prefecture northeast of the Kyoto–Osaka–Kobe conurbation. Here eutrophication has led to the appearance of red tides since 1977. Since 1979, the sale and use of laundry products containing phosphorus has been banned in the prefecture in an effort to counter the problem.

Industrial strength

Industrial pollution is nothing new. Mining has for centuries laced rivers with toxic heavy metals, washed from exposed ore veins or released during the metal extraction process. In the Rio Tinto region of Spain mining has befouled the waters for at least 2000 years. When, as at Rio Tinto, the ore is a sulphide mineral, the problem does not stop once the ore vein is exhausted and the mine abandoned. Metals mined as sulphides – including copper, gold, silver, zinc, lead and uranium – tend to occur in conjunction with iron pyrite, which, when exposed to air, reacts with water and oxygen to generate sulphuric acid. The acidic effluent, called acid mine drainage, harms local ecosystems through both its acridity and its load of metal ions such as aluminium, zinc and manganese. The iron precipitates in the rust-red form of iron hydroxide, turning streams a vivid orange and coating rocks with a bright deposit called yellowboy in the United States. Acid mine drainage poses a serious environmental problem in many regions of intensive mining: it is estimated to affect around 20,000 kilometres of streams and rivers in Colorado alone, and in Canada the costs of remediating the problem by adding caustic soda or lime are estimated at over $3 billion.

Mercury pollution dates back at least to Roman times. Mercury poisoning of fish has been reported worldwide, and has been implicated in health problems amongst fishing communities from the Amazon (where it is released during gold mining) to Japan, where the fishing town of Minamata was devastated in the 1950s by mercury emissions from a local factory. As well as mercury, other toxic metals commonly released into waterways by industrial activities include lead, arsenic, zinc, copper, tin, cadmium and cobalt.

During the 1970s the sight of oil-laden sea birds struggling for their lives became one of the most familiar and most distressing images of water pollution. On average, around three and a half million tons of oil are spilt into the sea every year, mainly through accidents involving oil tankers. The grounding of the *Exxon Valdez* at Prince William Sound in Alaska in 1989 alone released eleven million gallons, killing around 400,000 sea birds.

Yet mercifully, oil spills are biodegradable. The oil is dispersed by evaporation from the sea surface and by settling to the sea bed, and is eventually broken down by micro-organisms. Some bacteria thrive on oil, and one approach to spillage clean-up involves seeding the afflicted area with these micro-organisms and adding nutrients to stimulate their growth. This method, called bioremediation, was used to some effect during the clean-up operations after the *Exxon Valdez* accident. But it is a messy business, labour-intensive and slow: by 1997 less than four per cent of the *Exxon Valdez* oil had been removed by natural and artificial means. In the meantime, the ecological effects can be appalling. We can all identify with sea-bird fatalities, but many of the most serious ecological effects of spills are less apparent: the deaths of fish and marine plants, including bottom dwellers, and the destruction of algae, a crucial part of the marine food web.

In parts of Northern Europe, Scandinavia and the USA, industrial emissions have produced rain that kills. When acid rain falls from the sky, freshwater fish are decimated and conifer trees shed their needles and die. Acid rain dissolves stonework, corrodes metal pipes and leaches toxic metals such as aluminium, copper and cadmium into the run-off waters. By the early 1990s, around half of the lakes in the Adirondack mountains of New York State, and twenty per cent of those in Sweden, no longer contained fish, while around half of the forests of West Germany, the Netherlands and Switzerland were damaged. In Germany they have a word for it: *Waldsterben*, meaning forest death.

Acid rain is the result of the release of nitrogen oxides and sulphur dioxide into the atmosphere by human activities, including fossil-fuel burning in power stations and vehicles, the burning of vegetation during forest clearing, and metal-ore smelting.[8] Chemical reactions in the lower atmosphere convert the oxides to nitric and sulphuric acid.

The injustice of it all is that the afflicted areas are not generally around the source of the pollution but hundreds of miles downwind, where the acids are eventually washed out of the air in rain. So it is the remote natural environments of Scandinavia that bear the brunt of the fumes from Northern Europe and the Eastern bloc countries, and the forests of Canada that suffer from the industrial activities of the more populated regions. The tendency towards building higher smokestacks in power plants has only increased the range over which the gases are transported.

Measures to ameliorate acid rain were instigated as early as 1982, when many industrialized nations pledged at a conference in Stockholm to reduce their emissions of acid-forming gases. In Ottawa in 1984 several European and Eastern bloc countries agreed to cut back sulphur dioxide emissions by thirty per cent over the ensuing decade – but the UK, France, West Germany and the USA resisted these strictures. Additional countries have now made similar pledges, both on sulphur and nitrogen oxides, although the USA has steadfastly refused to constrain its industries by such promises.

The main problem is cost. With current technologies, reducing global sulphur emissions by fifty per cent would cost thousands of millions of dollars annually. Nonetheless, encouraging results have already been seen throughout Europe: by the early 1990s, sulphur dioxide emissions in the Netherlands, Germany and France were reduced to half, or less, of their 1980 levels. Acidification of natural waters is one source of bitterness we can hope to put behind us.

The recycling barrier

Part of our struggles with issues of water quality derive from a failure to match questions of purity to the nature of usage. It makes little sense to have a single water supply, since that requires all the water to be of drinkable quality – something that may not always be achieved, but is hugely expensive even as an ideal. Why should drinking water also serve to cool industrial plants, to vivify crops and lawns, to wash cars and put out fires? While we retain the idea of water as being either pure or 'waste' – the mythical emblems of water as an elixir or as irredeemably corrupt and foul – we cannot ensure appropriate and efficient apportioning of our natural resources. Says Sandra Postel of the Global Water Policy Project in Massachusetts,

> By better matching supplies of varying quality to different uses, more value can be derived from each liter taken from a river, lake, or aquifer – and the economic and environmental costs of developing new freshwater sources can be lessened. The challenge, in a nutshell, is to take the 'waste' out of wastewater.[9]

In other words, we can use more than once the water that we draw from natural sources. Water discharged from municipalities is commonly rich in the substances, such as organic matter and phosphates, that would benefit crops after only minor treatment. 'Sewage farming' is well established in principle, yet has been little exploited because of fears about its safety. Water-impoverished Israel is one of the few countries to make extensive use of its sewage for agriculture: seventy per cent of the discharge is tapped for irrigation after preliminary treatment. This treatment is essential: in Chile and Mexico, where raw sewage is channelled onto fields, diseases such as typhoid have resulted from consumption of the crops. Moreover, industrial effluents contain contaminants such as heavy metals that render the water unfit for direct agricultural use.

Water reuse does not have to operate on large scales. It is relatively easy and cheap to convert domestic 'grey water', such as that from baths and washing, to a form that is suitable for garden use. Filtration requires little more than passage through successive gravel-filled tanks, within which growing plants can help the filtration process. If such inexpensive systems were to be required for all new housing, the effect on urban water use could be considerable. Again, the barriers seem to be mainly psychological rather than technological or economic. We need to rethink our relationship to water.

Freshen up

One dream for meeting the expanding demand for water is to expand the supply accordingly – to increase the total amount of fresh water available for human use. Most of nature's water is already contaminated – by salt. This renders it useless to land organisms, and even for most industrial purposes sea water is far too corrosive. The coastal populations of hot, dry countries can only gaze out to the horizon in the knowledge that the shimmering blue ocean is of no more practical value than the yellow deserts.

Bachelard understood what this means for our dreams of water: that 'fresh water is the true mythical water'. He says,

> It is a fact too long ignored by mythologists that sea water is an inhuman water, that it fails in the first *duty* of every revered element, which is to serve man *directly*.[10]

Can we redeem sea water, and make it serve us directly? Yes, but at a cost that is all but prohibitive. Removal of salt from ocean or brackish water – a process called desalination – supplies less than 0.2 per cent of the water used globally. It is conducted on a large scale only in countries whose desperate need for water is coupled to a relatively wealthy economy. The Middle Eastern countries, primarily Saudi Arabia, Israel, Kuwait and the United Arab Emirates, produce three quarters of the world's desalinated water, and the United States a further one tenth. The mighty desalination complex at al-Jubayl on the coast of the Persian Gulf supplies Saudi Arabia with over a billion gallons of fresh water a day.

The simplest and oldest method of desalination is distillation: heating sea water until it evaporates as steam, leaving the salt behind. The first ever desalination plant, built in 1869 at the port of Aden on the Red Sea to supply British colonial ships, operated on this principle which, with considerable methodological refinement, continues to provide half of the desalinated water in the world today. But a lot of energy is needed to evaporate water. The process can be made somewhat less expensive by operating distilling desalination plants in conjunction with power-generating facilities: the steam used to drive turbines for electricity generation is also used to heat the brine for distilling. All the same, desalinated sea water is hugely expensive – about $1–2 per cubic metre, typically between four and eight times the cost of urban supplies from surface or groundwater.

The other main approach to desalination – the use of plastic semipermeable membranes to separate water molecules from salt ions – provides drinking water at less than half this cost. But membrane-based methods are not so efficient at removing salt, and so can be employed only to purify brackish water. So long as the price remains high, desalination will never do more than provide for the water needs of a few wealthy coastal cities. It is no solution to an impending global crisis.

Water wars

Harbingers of this crisis can be found in almost any country in the world. Because of extensive tapping of their waters, several of the world's great waterways – the Colorado River, the Yangtze and Yellow River, the Ganges, and the Nile – no longer reach the sea for at least part of the year. As a result, their deltas are becoming increasingly silted. In Mexico City, seventy per cent of the water is extracted from aquifers, which have become salinized as a result. About eighty-eight per cent of Saudi Arabia's water comes from aquifers, extracted at a totally non-renewable rate that, if continued, will drain them dry within the first half of the twenty-first century. Israel has just two small main aquifers with a renewable capacity of about half the country's annual water consumption.

Meanwhile, demands for irrigation, flood control and hydroelectric power are leading to engineering of waterways on an unprecedented scale. In Turkey where the Tigris and Euphrates rivers begin their journey to the Persian Gulf, a massive damming project called the Great Anatolia Project, aimed at providing irrigation and hydroelectricity, threatens to cut off the flow of both rivers through Syria and Iraq for up to eight months each year. The damming of the Nile at the 111-metre-high Aswan Dam in Egypt has created one of the world's largest artificial lakes, Lake Nasser, nearly 500 kilometres long. Completed in 1971, the Aswan Dam provides about a quarter of Egypt's power by hydroelectricity, and has added 900,000 acres to the country's arable land. Concerns that the lake would exacerbate waterborne diseases such as malaria and bilharzia do not seem to have been borne out. But the damming of the Nile required the resettlement of 100,000 people, increased the salinity and erosion of soil by the river, and has depleted sardine stocks in the eastern Mediterranean Sea.

The potential benefits, drawbacks, risks and human costs of this sort of hydro-engineering are nowhere better illustrated than at the awesome Three Gorges Dam project on the Yangtze in central China. The first phase of this, the world's biggest hydroelectric project, was completed in November 1997, when the Yangtze was dammed and diverted so that it runs along a temporary man-made channel. On the dry river bed, construction is now underway of a dam wall 175 metres high and 2.3 kilometres wide which will eventually harness the Yangtze to generate 18,200 megawatts of power – fifty per cent more than the previous biggest hydroelectric project, the Itaipu Dam on the border of Brazil and Paraguay. The cost of the project is estimated at $30 billion (but such estimates have an almost inevitable tendency to rise), and will generate power from 2003, although completion will not be until 2009. On the one hand, the environmental pay-off could be substantial – the hydroelectric project will reduce China's coal consumption by forty million tonnes a year, and will produce the equivalent electrical output of a dozen nuclear power plants. It may also help to prevent the kind of flooding that, in 1954, killed 30,000 people along the middle and lower Yangtze. But the reservoir that the Three Gorges Dam will create – 630–690 kilometres long and 1.1 kilometres across – will inundate 13 cities, 140 towns and 1352 villages, forcing the relocation of around 1.2 million people. Peasant farmers may come off the worst, as they will be forced to till the higher, less fertile slopes of the Yangtze valley. And there are worries about the problem of sediment build-up in the dam, not least because this may cause upstream silting as far as Chongqing, the biggest river port in southwest China. Even if all goes according to plan, the human cost of the project will be severe, and it is hard to know how to weigh this against the benefits for China as a whole.

The inequality in distribution of water resources around the world means that for some water-rich countries it is becoming less a source of jokes about the local climate and more an internationally marketable commodity. The Canadian Global Water Corporation has a licence to ship fresh water from northwest Alaska to the city of Tianjin on the coast of China, while the Nordic Water Company in Norway has plans that verge on the surreal: to ship water to Europe in huge floating plastic bags.

As the scale of hydrological engineering increases, as burgeoning industrialization causes water pollution to intensify, and as water use expands in response to the needs of growing populations, water resources are becoming a central component of international politics. All of these issues boil down to a single question: who, if anyone, owns the water?

The common good?

Amongst the profound changes to the social landscape that inevitably accompany the process we call development, the issue of ownership looms large. Common land, where anyone had the right to grow crops and graze livestock, was already disappearing fast in England by the fourteenth century, through the practice of enclosure by landlords and the reservation of woodland for the use of the aristocracy and monasteries. The same pattern of acquisition was played out again in the colonial world: to take just one example, in Rhodesia almost all of the prime farming land was claimed by the European settlers, and remains commercial land in present-day Zimbabwe.

By 1994, just seven per cent of the irrigable land – and most Zimbabwean farming depends on irrigation – lies in communal areas or those designated for resettlement of native farmers.

Land is one thing: a map can be carved up however you will, walls and fences erected, ownership enforced. But rivers don't dally, they cross boundaries and national borders. So do aquifers: tapped at one location, their level may be lowered far away. How, then, can flowing water be owned?

In trying to apply our concept of ownership to a resource whose very nature runs contrary to the idea, we have a recipe for conflict. Damming of the course of a river denies this nature, and complex social factors enter the equation of relative benefits and drawbacks. Yes, we can confer fertility on famished lands and provide electricity to cities – but amongst the overheads are the displacement of populations, disease, pollution, catastrophic flooding and armed conflict.

The tensions that arise from the management of water provide a good litmus test for the prevailing social climate in which it is conducted. According to social scientists John Donahue and Barbara Rose Johnston, systems for controlling access to and use of water resources typically recreate and reproduce the inequities in the societies that generate them. Sociologist Karl Wittfogel goes further: he saw that 'it was as the arbiters of water that tyrannies anointed themselves as legitimate'.[11] I offer here just a few vignettes that flesh out these accusations; but examples are abundant the world over.

Sacred springs

Social scientist David Orr asks, 'What is the meaning of water?' The machinery of the developed world offers one answer: water has become a strategic commodity, apportioned largely between state and private ownership.

Orr, however, suggests another interpretation: 'One might as well ask, "What does it mean to be human?" The answer may be found in our relation to water, the mother of life.' That might sound like a strange conjunction to our ears; but to the Hopi native Americans of arid northeast Arizona, water is a defining essence of their culture. They have settled in this harsh, dry land from at least AD 800, and there is evidence that agriculture has been practised here for far longer. Water, largely drawn from springs fed by a vast aquifer, has never been taken for granted by the Hopi, knowing as they do that the springs have a precarious existence reliant on infrequent rainfall and the spring snow-melt. To entice the rain from the skies they perform religious rituals including the *kachina*, the rain dance, designed to 'evoke and celebrate the life-giving force of water'.

The sparse natural waters of the Hopi land are used to irrigate terraced gardens in which chillis, beans, corn, onions, radishes and fruit trees are grown. But the Hopi's prayer for rain is not just an appeal for adequate irrigation and drinking water; they recognize that their entire environment, their world, depends on it. 'We pray,' says one Hopi man, 'for rain so that all the animals, birds, insects and other life-forms will have enough to drink too.'[12] To the Hopi, every spring (*paahu*) is a sacred place, a shrine to Paaloqangw the Plumed Water Snake, who is the patron of water in heaven and earth. The Hopi believe there is a sort of magnetism between water and its imagery in song and ritual, so that by evoking the names of springs in their religious practices they are enticing water towards them. The sacredness of springs, and their practical importance, makes them cultural landmarks – they punctuate clan migration routes and define areas of clan rights.

The Peabody Western Coal Company, which mines for coal on the Black Mesa just north of the Hopi partitioned lands, sees water rather differently. The company draws it up from the aquifers below the Black Mesa to pump coal slurry along pipelines to Nevada, where the coal is burnt to generate electricity for Las Vegas, Phoenix and southern California. The Peabody Company is the largest private producer of coal in the world, and is a part of the British-owned Energy Group plc, a spin-off from the multinational Hanson Industries. Peabody's Black-Mesa-Kayenta mine uses over a billion gallons of drinking-quality water each year from the Black Mesa aquifers to transport the crushed coal.

The springs that serve the Hopi villages about thirty miles to the south of the mine are now drying up, and the Hopi say that the mining activity is to blame. Peabody denies this, and refuses to alter its methods. The company has maintained for several years that the mine draws water from a different, lower aquifer to that which supplies the Hopi springs. It suggests that the decline of the springs is due to recent drought conditions, as well as increased water usage in the Hopi communities themselves. Certainly, expansion and modernization (such as the introduction of indoor plumbing) both in the Hopi villages and in those of the adjacent Navajo partitioned lands has placed increasing demands on water sources. But there seems to be little hard evidence to support Peabody's assertion that the mining has not affected the Hopi water. The mining company has recently admitted to taking some water from other aquifers, including the main upper aquifer that supplies the Hopi springs. And reports from the US Geological Survey indicate that the hydrology of the region is too poorly understood to draw any definitive conclusions about how tapping the lower aquifer will affect the upper one. Data released by the USGS in 1995 indicate that as much as two thirds of the decline in water levels in the wells of the area might be due to the mining activity. The Survey's studies suggest that some of the major Hopi wells will be totally dry by 2011.

To the Hopi, arguments about aquifers and hydrology are beside the point. To them, all water is sacred, and to see it drawn off in such vast quantities to be mixed with pulverized coal and pumped to Nevada is an affront to their cultural beliefs. They consider that their deity Maasaw has entrusted to them the care of all the earth and its resources. They fear that neglect of this responsibility could bring catastrophe. When minor earth tremors were felt in the region due to mining-induced subsidence, some of the Hopi elders interpreted this as a sign of their spiritual negligence.

Yet their position is a delicate one, since, by Western standards, the Hopi tribe is impoverished, and coal royalties and water leases to Peabody account for eighty per cent of its annual revenue. With this in mind, the Hopi Tribal Council does not (or is not in a position to) oppose the mining in itself – its complaint is with the phenomenally wasteful way in which the mine uses the land's water. The slurry method is also opposed by the US Environmental Protection Agency, which is, however, powerless to intercede. A further complication is the way in which the local native American tribes, the Hopi and the Navajo, have been played off against each other – the Hopi feel that the Navajo have received greater concessions, leading the Navajo to be more favourably disposed towards the mine.

Ultimately at issue here are different views about the significance and appropriate use of water. What elevates the situation beyond a dispute between the rich and powerful and the impoverished and weak is the cultural dimension – the reverence with which the Hopi regard nature's waters. If water is removed from the Hopi land, their cultural identity seeps away with it. The same story can be found elsewhere in North America. On the coast of Washington State just below the Canadian border, the Lummi nation of native Americans has long been struggling to secure freshwater rights to the Nooksack river both as a guarantee of water quality and to safeguard the salmon fishing that is central to the Lummi culture. And the livelihood of the Cree people of James Bay in Quebec, traditional hunters in a riverine environment, is threatened by diversion of the Eastmain river to feed the massive James Bay Hydroelectric Project. In each case, the demands that a 'developed' lifestyle place on water resources seem to be incompatible with the need of fragile indigenous communities for clean natural waters and their concomitant ecosystems.

The long journey nowhere

Yet even when these cultural differences do not exist, conflicts between water users may lack any clear solution. The water-hungry states of the southwestern USA have been battling for over a century over who owns the water that courses through its rivers. The historian Donald Worster has encapsulated the problem:

> The American West can best be described as a modern hydraulic society, which is to say, a social order based on the intensive, large-scale manipulation of water and its products in an arid setting.[13]

And there are few settings more arid than Arizona, which for decades has simmered with resentment at the lushness and wealth of the state between it and the Pacific. Arizonians, it is said, suffer from 'California envy', from the perception that their state is just a hot, dry obstacle to the abundance of the West Coast.

Running like a fault line between these jittery neighbours is the Colorado River, a mighty artery that drains seven states: not just Arizona and California but Colorado itself as well as Wyoming, Utah, New Mexico and Nevada. All of these states foster cities greedy for the river's water, but none more so than Arizona and California. Who gets it?

The question is: how much water can be siphoned off by upstream users before those downstream are left with just a trickle? Arizona and California lie at the most downstream end of the river before it passes out of the USA into Mexico, and there discharges into the Gulf of Mexico.[14] But they are the biggest consumers, and the battle over the Colorado River has been mainly a two-state affair.

The river flows for almost half its length through Arizona; and the Californian watershed is only a minor contributor to the discharge. Yet California has always wielded the most political power, and to Arizonan eyes it has usually got its way – often at huge expense, and occasionally with alarming consequences.[15]

By the end of the nineteenth century, land in the arid West was considered worthless without water, and the catchphrase was that irrigation would 'make the desert bloom'. At the turn of the century, the engineer Charles Rockwood, financed by an investor named George Chaffey, initiated a bold scheme to transform the dry valley of the Salton Sink in California to a verdant paradise. Rockwood constructed the Alamo Canal to divert part of the Colorado River's flow into the valley; by 1904 it was being cultivated by over 7000 farmers, and Chaffey was proposing to rename it Imperial Valley. 'Water is king,' he said. 'Here is its kingdom.'

But in that year the canal silted up, cutting off the agricultural lifeline. Now short of cash after the massive expenditure on the canal, the speculators took the cheapest option: they built a second canal at a budget rate, gambling that there was no need to run to the expense of giving it a head gate to regulate the flow. They miscalculated. In the following year the Colorado River swelled with spring runoff, and the new canal flooded. Salton Valley indeed became water's kingdom: an inland sea seventy-two feet deep and covering an area of 150 square miles. This diversion of the river water created a waterfall in the river channel that began steadily to carve its way back upstream. By 1907 it was one hundred feet tall, and was threatening to reroute the lower part of the river permanently. Only by spending $3 million to dam the intake did the Southern Pacific Railroad company avert this catastrophe.

Wild schemes notwithstanding, the major issue at the beginning of the century was the growth of California, the state that contributed least to the river's water but demanded the most. In US water law, prior appropriation is what counts: the person who first puts the water to beneficial use secures the right to future use of that amount. When water is in short supply, the users with the oldest water rights get priority. In June 1922 the US Supreme Court ruled that this law applied across state boundaries, and other states along the Colorado River feared that rapidly growing California would acquire prior appropriation rights that would stymie later growth elsewhere.

But in late 1922 an historic agreement was reached that was

supposed to allay these fears. The Colorado River Compact apportioned half of the Colorado River water to the 'upper basin' states – Wyoming, Colorado, Utah and New Mexico – and half to the 'lower basin' states of California, Arizona and Nevada. The dividing point was at Lee's Ferry in northern Arizona. Yet Arizona was not convinced that the arrangement lay in their favour, and despite signing the Compact, the state refused to ratify it.

Arizonans were particularly concerned about plans to dam the Colorado River about halfway along its length. In principle this was intended to facilitate equable allocation of the lower-basin water; but to Arizonans, the dam project represented a victory for California, which would thereby be assisted in stealing Arizona's water. After fierce political debate, the construction of the dam began in 1930; it was ultimately to become the mighty Hoover Dam, named for President Herbert Hoover and representing at that time the highest dam in the world. At one point the tensions almost escalated into armed conflict. In 1934 Arizona's governor Benjamin Moeur called out the Arizona National Guard to prevent work on the Parker Dam, lower down the river at Lake Havasu, which he saw as a further violation of Arizona's water rights.

But by the 1940s pressure on Arizona was mounting to ratify the Colorado River Compact. The state needed a canal system to deliver its share of the water to the south-central region, where it was needed both by farmers and by the growing cities of Phoenix and Tucson. It was clear that support and goodwill for the project would be contingent on cooperation over the Compact. Arizona ratified the agreement in 1944, and plans for the Central Arizona Project (CAP) commenced.

The idea was to divert the river water from Lake Havasu by canal over 450 miles or so of desert to deliver water for irrigation and to quench the thirst of Arizona's two major cities. The Project could not take shape, however, until Arizona had settled its continuing dispute with California over its share of the lower-basin Colorado River water. Only after a protracted legal battle in the Supreme Court that began in 1952 and ended eleven years later was the plan ready to roll. Authorization for the Project was finally granted in 1968 by Lyndon Johnson, and construction began.

As the canal started to make its way across the desert, the costs began to escalate. The 1968 estimated cost of $832 million soared to a final figure of around $4.1 billion. Never free from controversy, the project was in frequent danger of grinding to a halt; Jimmy Carter's administration in the late 1970s was particularly keen to see it terminated. Further complications flowed from plans to impound the water from other Arizonan rivers into the CAP – a dam at the confluence of the Salt and Verde rivers, proposed as a flood control measure, would have inundated the Yavapai native American reservation and was vetoed by Carter.

As the price tag on the canal spiralled upwards, so did the cost of the water it supplied, until it seemed likely that CAP water would simply become uneconomic to those, such as Arizonan farmers, who were supposed to benefit from it. When water becomes an economic commodity, it no longer has universal value: each source enters a market where it competes with other sources. Even in sun-baked Arizona, CAP water costs more than that pumped from aquifers. The final irony came in 1993, when the Central Arizona Project at last reached Tucson across 335 miles of canal. Tucson's water situation has long been precarious – unlike Phoenix, it relies almost wholly on groundwater, a resource that has not been able to keep pace with the city's growth. Water usage in the city is unusually high – by the early 1970s it was around 200 gallons a day per capita, and in 1974 this high demand ran the city's pumps dry. Eagerly awaited throughout this period, however, the CAP ran spectacularly aground when it arrived at Tucson. Because of high salinity and contamination acquired en route from damaged and corroded pipes, the water was virtually undrinkable, and the residents of Tucson voted that it not be used for domestic supplies. After its long and arduous journey, the CAP found itself unwanted on arrival.

The potential for future dispute over Colorado River water is tremendous. Despite the development of water storage facilities along the river, a prolonged drought would surely resurrect bitter interstate disputes: it is not clear who, under such circumstances, gains priority. Ambiguities still remain about the water rights of native American tribes that relied on the river long before the states existed. And the USA must also negotiate water allocations with Mexico, through which the river passes before reaching the coast. Already the damming of the river means that it almost runs dry by the time it reaches the delta, where local ecosystems have been irreparably damaged.

Holy water

There can be little doubt that the protracted and agonizing conflict in the Middle East between Israel and its neighbours is a dispute primarily about land and culture. The carving up of Palestine by the United Nations in 1947 to create the state of Israel initiated a massive displacement of the largely agrarian Palestinian population that was bound to provoke hostilities. The fact that both the resettled Israeli Jews and the predominantly Muslim Palestinians have cultural and religious traditions that assure them they have each been allocated the land of Palestine by divine sanction contributes to the deadlock in negotiating borders and rights.

But increasingly over the past decades, control of water has emerged as a strong element in the conflict. Israel today represents a striking instance of what happens when Western aspirations and lifestyles run up against extremely limited water resources. In feeding the swimming pools and gardens of resettled Israelis bearing the legacy of expectation engendered in European and North American urban environments, the region's scant water resources are now in a parlous state. The coastal aquifers are increasingly afflicted by salinization; and their contamination, a consequence of lax pollution laws coupled to intensive chemical farming practices, has caused dysentery and kidney disease.

To many Israelis, Palestine is a desert wilderness that must be reclaimed by modern farming methods and intensive irrigation. Echoing the watchword of pioneers in the early American West, Zionist leader David Ben-Gurion proclaimed his ambition to make this desert bloom. To the Palestinian farmers who have lived there for centuries, however, most of the land has always been beautiful and abundant, supporting grain, fruit and olive trees. Nonetheless, even the Palestinians are making increasing use of synthetic chemicals in agriculture, including pesticides (such as DDT) that have been banned in the West. The Israeli law that permits the government to impound for resettlement land that is not being actively farmed obliges the Arab farmers to forego the traditional fallow periods, causing the soil to lose its fertility. There is then no alternative but to resort to inorganic fertilizers and the rest of the chemical panoply of Western-style farming.

Ever since the Zionist movement looked hungrily to Palestine for the creation of a Jewish homeland in the early twentieth century, it recognized the crucial role that water would play.[16] So when the state of Israel was sanctioned by the United Nations in 1947, the government wasted no time in restructuring the region's water supplies. A major objective was to bring water from the northern region around the Sea of Galilee, fed by rain and snow falling in the Golan Heights, to irrigate the Negev Desert in central Israel. By the 1980s, the pipeline of the Kinneret–Negev Conduit was bearing south 420–450 million cubic metres of water a year from the Sea of Galilee (also called Lake Kinneret) and the surrounding area. Lake Huleh, a freshwater lake in the Galilee region, was already drained dry by 1957. The marshlands around the Sea of Galilee, once inhabited by Arabs who used its reeds and grasses to weave baskets, were desiccated, and are now mostly resettled by Israeli Jews – even though the 1947 partition designated this area part of the Palestinian state. The Sea itself, famous from Biblical legend, is now increasingly saline and suffers from red-tide algal blooms twice a year owing to severe eutrophication. The drying of the Galilee region has also had adverse effects on the water supplies of surrounding countries: the drying-out has been felt in Lebanon, Jordan and Syria, and has diverted saline springs into Jordanian water.

But the most explosive issue in the region's water conflicts is the use of the Jordan River and its tributaries. Originating in the Golan Heights in Syria and Lebanon, the river runs south to the Sea of Galilee, from where it continues south to the Dead Sea. For much of its length the Jordan River runs along the border between Jordan and Israel, including the occupied West Bank. Israel now diverts most of the Jordan River's flow. But in the late 1950s Jordan planned a series of dams on tributaries of the Jordan River to support irrigation, and began construction of a dam on the Yarmuk River, a major tributary that runs from Syria along the Syria–Jordan and then Jordan–Israel borders. Concerned over the effect that this would have on its rights to the Yarmuk's waters and having been able to secure no guarantees about this from Jordan, Israel bombed the construction project in 1964 – the opening shot of the real Arab–Israeli water wars. In the Six Day War of 1967 Israel captured the proposed dam site when it occupied the Golan Heights in Syria. By also invading the Jordan's source region in southern Lebanon, Israel thereby acquired control over almost the entire watershed of the upper Jordan River. Jordan's restricted access to water today gives it the lowest per capita water use in this part of the Middle East – barely a third of that of Jewish Israelis.

Despite the progress of peace negotiations, access to water continues to be a source of tension in the region. After a particularly dry summer in 1990, King Hussein of Jordan was famously quoted as saying that the only reason which might bring Jordan to war again with Israel was water.[17] The Jordan–Israel Peace Treaty of 1994 went some way towards reconciling the two nations, however.[18] Palestinians, meanwhile, complain that Israeli laws on water use are discriminatory, providing Jewish Israelis with almost unlimited access while leaving others with barely sufficient supplies. Palestinian farmers are heavily taxed on water use for irrigation.

In urban areas in the occupied territories, Palestinian Arabs typically consume seven times less water than Jewish Israelis, and in some Palestinian cities per capita use falls much lower in the summer. In the West Bank, 1.2 million Palestinians were long granted access to only a quarter of the aquifer that provides the region's water, while the access of the 100,000 Israelis was unlimited. Again, recent developments such as the 1995 Israeli–Palestinian Interim Agreement have eased some of these inequalities, but in the light of impending water shortages the negotiations remain precarious.

The Palestinian complaints seem to highlight injustice by any standards; but they are felt particularly keenly by Muslims for whom water has a religious connotation. Water and its environmental manifestations – the sea, springs, rains, clouds – represent a major theme throughout the Qur'an. The cleansing power of water is a strong element of Islamic belief, reflected in the requirement of ritual cleansing before prayer. In consequence, pollution of or denial of access to water are punishable crimes in Islamic tradition. For those who cleave to such beliefs, the Israeli government is violating not just citizens' rights but holy commandments. Needless to say, all this is grist for the incendiary mill of fundamentalism.

How strong a role water plays in the Arab–Israeli conflict is much debated; some regard it merely as a complication that engineering could put right.[19] Such are the complications of the historical, cultural and political issues at stake that it is probably impossible to quantify causal roles. But in any event, there is to be found here a distillation of the water issues that seem destined to gain ever more prominence in the global arena – the question of ownership within and between nations, the complications of population growth in water-depleted regions, accountability for pollution, clashes of cultural attitudes towards water use, apportioning of the relative importance of development and environmental protection, and the extent to which technological fixes can solve water problems. There are no easy solutions to these matters, but a respect for natural water resources is surely an essential preliminary.

The meaning of water

For most of this book I have talked about what water is, and what it does. The issues that surround the use of water resources, meanwhile, demand that we ask a different question: what does water *mean*? Understanding water's physics, chemistry, geology and biology can provide only a part of the answer. Attempts like that of the Austrian Viktor Schauberger to reclaim water's *elemental* status (page 311) look absurd by scientific standards; but however misconceived they might be, they speak to a relationship with water that lies outside science. To human civilization water *is* elemental, in the sense that it forms an integral part, it is a rudimentary substance. This relationship imparts a value that we can ill afford to neglect. Ecologists will argue eloquently for the importance of keeping natural waters as untainted as we can manage, but if this is going to happen, first of all we need to care. There is an argument that the only way to protect the water that nature grants us is to commodify it, to give it an economic value – to put a price on it. In some ways that is indeed what needs to happen – only if water is costed, and at a price which at least matches that of delivering it, will there by incentive to use it sparingly. But this cannot be the sole principle of water management. When something is priced, its value will always be relative. If pollution is the most economical option even in the face of financial penalties, what is to stop that course being adopted? And inherent in all commodification is the potential for inequality: he who can pay, gets; he who cannot, goes without.

Sandra Postel expresses with admirable directness the drawback of regarding water as an 'economic good':

> The risk . . . is that water's economic functions will be elevated over its life-support functions, and that the three pillars of sustainability – efficiency, equity, and ecosystem protection – will not be given equal weight.[20]

To a large degree this is a matter of choosing what sort of a world we want to live in. But I believe it is also about understanding the basic nature of the substance at stake here. If I have persuaded you that we can take this substance apart, dissect it into its H's and O's, prise apart its molecular network and track it down across the cosmos, yet retain and, I hope, even enhance the reverence that we feel for its mythical embodiment – well, then, I'll have told a story worth the telling.

Notes

Preface

Epigraph: K. Sudhoff & W. Matthiessen (eds.) (1922–25): *Paracelsus. Samtliche Werke*, Vol. 9, O.W. Barth, Munich.

Chapter 1

Epigraph: G. Wald (1964): *Proceedings of the National Academy of Sciences*; T. Hughes: *Fire Eater*, Faber & Faber, London.

1 Genesis 1:1–9.
2 The *Rig Veda* 10.129, translated by Raimundo Panikkar in *The Vedic Experience – Mantra-manjari* (Motilal Banarasidas).
3 The fudge factor, called the cosmological constant, is currently enjoying a revival. It has been reinvoked to explain the extraordinary discovery in 1998 that the Universe seems to be not just expanding but accelerating. A cosmological constant implies the existence of some force that induces acceleration by acting, at enormous spatial scales, in opposition to the mutual attraction of galaxies.
4 The numbers here denote the total number of protons and neutrons in the nucleus – this is characteristic of each isotope.
5 Paracelsus (1537/38): *Astronomia Magna*, p. 169. In K. Sudhoff & W. Matthiessen (eds) (1928–33): *Paracelsus. Samtliche Werke*, R. Oldenbourg, Berlin.
6 This was not a new idea, as Prout acknowledged: Humphry Davy had suggested something similar in 1812.
7 Walt Whitman: *Leaves of Grass* (1855).
8 Stuart Ross Taylor, an expert on the formation of the solar system, suggests that Kant gets rather more credit than he deserves: 'It seems clear that the many contradictions in Kant's hypothesis do not accord with the general popular acclaim it has received. Perhaps this is due to his eminence as a philosopher.' (S.R. Taylor (1992): *Solar System Evolution*, Cambridge University Press.) It was Pierre-Simon Laplace who, in the late eighteenth century, put Kant's vague notions into a more rigorous form consistent with Newton's mechanics.
9 Recent evidence points to a snag in this line of argument, however. If the water that we drink began its journey on a comet, then we would expect the composition of sea water to be the same as that in recent comets. But isn't all water alike? In one sense, yes; but there is a tell-tale signature that allows us to compare planetary water with

cometary water, which resides in the amount of heavy hydrogen – the isotope deuterium – they contain. Around 0.0156 per cent of hydrogen in sea water is deuterium. But measurements of the deuterium-to-hydrogen ratio in the comets Halley (from 1984), Hyakutake (1996) and Hale-Bopp (1997) show that they contain about twice as much deuterium in their icy mantles. This implies that the oceans can't be made from cometary water alone.

10 Do we still receive these deliveries? Most scientists believe not: they estimate that, with the possible exception of very rare cometary impacts, the Earth now receives negligible amounts of water from space. But physicist Louis Frank at the University of Iowa has generated controversy in recent years by proposing that the upper atmosphere receives a steady and appreciable rain of tiny comets, loose snowballs that vaporize rapidly as they enter the atmosphere. Frank bases his idea on the fact that emission of ultraviolet light has been reported from the upper atmosphere, from an unknown source which could be the sign of vaporizing mini-comets. If Frank is right, the Earth could be receiving a million tons of water a day from this rain of ice – enough to fill the Mediterranean over ten million years.

11 J.E. Lovelock (1979): *Gaia, a New Look at Life on Earth*, Oxford University Press, p. 157.

Chapter 2

Epigraph: Titus Lucretius Carus: *De Rerum Natura*, transl. J. Evelyn (1656); Leonardo da Vinci: *The Notebooks of Leonardo da Vinci*, ed. and transl. E. MacCurdy, Reynal & Hitchcock, New York, 1938.

1 Leonardo da Vinci, *The Notebooks of Leonardo da Vinci* (transl. E. MacCurdy), Reynal & Hitchcock, New York, 1938. In H. Boynton (ed.) (1948): *The Beginnings of Modern Science*, Walter Black, New York.

2 We will see later that cycling of water through the deep Earth is driven by the planet's internal heat.

3 P. Perrault (1674): *L'Origine des Fontaines*. In H. Boynton (ed.) (1948): *The Beginnings of Modern Science*, Walter Black, New York.

4 'Physical' here is to be distinguished from 'chemical'. The stuff of the Earth is being transformed chemically all the time. Nitrogen in the air, for example, is transformed by microbes into nitrogen-containing compounds in the soil, such as nitrates.

5 The Coriolis force is the result of conservation of angular momentum. Just as an object travelling in a straight line has linear momentum – a tendency to preserve its motion unless opposed by a force – so a rotating object has so-called angular momentum, a tendency to preserve its rotation. As the object follows its trajectory, the total angular momentum must stay the same. This is what makes a spinning skater turn faster if she draws in her arms. An object moving

across the face of the spinning Earth possesses angular momentum which changes as the object moves, and the Coriolis force deflects the object so as to conserve angular momentum.

6 Shosammi Sueyoshi: *Shinkokinshu* (1205). In *Anthology of Japanese Literature* (1955), Penguin, London.

7 Actually the tidal bulge lags somewhat behind the zenith because of friction. The same is true of the bulge at the nadir.

8 S. Schama (1995): *Landscape and Memory*, HarperCollins, London, p. 247.

9 P. Claudel (1913): *Cinq grandes odes*, Paris.

10 S. Schama (1995), op. cit., p. 262.

11 There is no clear, formal distinction between the two terms.

12 Ki no Tsurayuki: *Tosa Nikki*. In *Anthology of Japanese Literature* (1955), Penguin, London.

13 Ibid.

Chapter 3

1 The ratio of these isotopes in the water molecules that evaporated from the sea and then fell as snow depends largely on the ambient temperature at the sea surface. A high ratio of oxygen-18 to oxygen-16 in the ice indicates relatively high temperatures when the water evaporated – see below.

2 The sedimentary cores consist partly of the calcium carbonate shells of microscopic sea creatures called foraminifera, or forams. The species of forams can be identified by inspecting their buried shells under a microscope, and so the ocean temperatures at the time of their deposition can be crudely estimated on the basis of whether they are warm- or cold-water forams. Better still, analysis of the ratios of oxygen's stable isotopes, oxygen-18 and oxygen-16, in the carbonate of the shells provides a more accurate indicator of climate, since it reflects the extent of the ice sheets at the time that the shells were formed. Water molecules containing oxygen-18, being heavier than those containing oxygen-16, are selectively retained in the sea during evaporation, while the ice sheets on which some of the water vapour subsequently falls get correspondingly enriched in oxygen-16. The bigger the ice sheets, the greater the oxygen-18 enrichment of sea water – and of the foram shells formed within it. (Conversely, recall that a high oxygen-18 content in polar ice cores implies higher global mean temperatures – smaller ice sheets.)

3 Vast tracts of sea ice melt around Antarctica every year, but this makes no difference to sea level because floating ice displaces exactly its own mass of water.

Chapter 4

1 H.S. Jones (1952): *Life on Other Worlds*, English Universities Press, London, p. 206.

2 I found a quaint acceptance of this idea, for example, in C. Hatcher (1961): *The Horizon Book of Science*, Paul Hamlyn, London.
3 P. Spudis (1996): *The Once and Future Moon*, Smithsonian Institution Press, Washington DC.
4 None, that is, beyond Earth orbit. Manned space stations and shuttle flights may yet foster good science, although the harvest from those that have been launched so far is arguably meagre in proportion to the expense. Perhaps one of the most useful was the repair mission to the Hubble space telescope: maintenance of an unmanned project.
5 These trophies were certainly valuable to planetary scientists; but unmanned lander missions could now provide a cheaper way to obtain them.
6 Percival Lowell (1896): *Mars*, Houghton Mifflin, Boston.
7 Water vapour *is* present in the Martian atmosphere, but at such low concentrations that it was not until the 1950s that the measuring techniques became sensitive enough to detect it.
8 B. Murray (1989): *Journey Into Space*, W.W. Norton, New York, p. 65.
9 D. McKay *et al.* (1996): *Science* **273**, 924–30.
10 K.S. Robinson: *Red Mars* (1992); *Green Mars* (1993); *Blue Mars* (1996), Bantam, New York.
11 C.P. McKay, O.B. Toon & J. Kasting (1991): *Nature* **352**, 489–96.
12 The details are in fact a little more complex, and reinforce the centrality of water in sealing Venus's fate. When water vapour condenses as clouds, as it would have done on the young Venus, heat is released. This is the 'latent heat' that a gas takes in as it evaporates from a liquid (see p. 146). So the formation of thick clouds warms up the atmosphere, and this in turn allows water vapour to ascend higher before it condenses. In other words, the capacity of the atmosphere for water vapour increases. You could say that water vapour pulls itself up through the atmosphere by its own bootstraps: the more that evaporates, the more the atmosphere can hold. As Venus's water rose higher and higher, it became increasingly susceptible to being split by the ultraviolet part of sunlight and then lost from the planet for good as the fragments escaped the planet's gravity.
13 J.I. Lunine & C.J. Lunine (1998): *Earth: Evolution of a Habitable World*, Cambridge University Press, p. 181.
14 D.H. Grinspoon (1997): *Venus Revealed*, Addison-Wesley, New York, p. 100.
15 The astronomer Simon Marius made the same discovery independently at much the same time.
16 T.V. Johnson (1981): The Galilean satellites. In *The New Solar System* (eds. J.K. Beatty, B. O'Leary & A. Chaikin) Book Club Associates, New York, p. 152.
17 Quoted in JPL press release 97–66 (9 April 1997).

18 Quoted in the *Independent on Sunday*, 1 March 1998, p. 41.
19 J.A. Jacobs (1998): An astro-tale, *Astrophysics & Geophysics* 39, 3.9.
20 Some will protest that it is very 'terracentric' to insist that life will be carbon-based and water-lubricated. It's a valid criticism, but at present it doesn't really take us anywhere else interesting. Chemists are ingenious folk, and more than happy to dip into non-aqueous solvents – but none has yet come up with an alternative medium for life that looks anything like plausible.
21 K. Vonnegut (1967): *The Sirens of Titan*, Coronet, London, p. 186.
22 Pluto's small size and odd orbit make its claim to true planetary status somewhat contentious. There was a proposal in early 1999 to relegate it to a minor, rather than major, planet; but this was turned down by the International Astronomical Union, showing that, if nothing else, astronomers have become rather attached to this distant, lonely world.

Chapter 5

Epigraph: A. De Morgan (1954): *A Budget of Paradoxes*, Dover, New York.

1 G. Bachelard (1983): *Water and Dreams*, Dallas Institute Publications, p. 138.
2 Lao Tzu: *Tao Te Ching*, (transl. G.-F. Feng & J. English, 1972), Wildwood House, London.
3 C. Merchant (1980): *The Death of Nature*, HarperCollins, New York.
4 Plato: *Timaeus*, in *The Dialogues of Plato* (trans. B. Jowett), Random House, New York, 1937, Vol. 2, p. 13.
5 G. Vidal (1981): *Creation*, Heinemann, London.
6 Titus Lucretius Carus: *De Rerum Natura* (transl. J. Evelyn, 1656).
7 Geber's writings are now thought to be not the product of one man but of an alchemical school. Writing under the name of another, generally a master or founder of a tradition, was common practice for alchemists.
8 H.M. Leicester (1956): *The Historical Background of Chemistry*, John Wiley, New York, p. 66.
9 Hero of Alexandria: *A Treatise on Pneumatics* (transl. Benet Woodcroft), Taylor, Walton and Maberley, London, 1851, pp. 6–7. Quoted in H.M. Leicester (1956): *The Historical Background of Chemistry*, John Wiley, New York, p. 51.
10 Yet even by 1624 the French parliament were still making arrests for the heretical notion of a five-element system, illustrating how the spurious authority of Aristotle was still conflated with that of the Church and State.
11 J. Baptist van Helmont: *Oriatrike, or Physick Refined* (transl. K.J. Chandler), London, 1662.
12 Plants get the carbon for their cellulose tissues from the carbon dioxide in the air, which they convert in the process of photosynthesis. But it is hardly surprising that van Helmont and his

contemporaries overlooked the possibility that something as corporeal as wood could be formed by a direct 'condensation' of tenuous 'air'.

13 Robert Boyle (1661): *The Skeptical Chymist.*

14 Mayow believed, however, that the gas represents the combustible part of air, which he called 'nitro-aerial'.

15 B. Jaffe (1976): *Crucibles: The Story of Chemistry*, Dover, New York, p. 47.

16 Contrast this with the situation today, when most scientists will publish as soon as they have a firm result, and occasionally before.

17 J. Read (1995): *From Alchemy to Chemistry*, Dover, New York, p. 138.

18 See Bachelard: *Water and Dreams*, p. 96, for more on the poetic union of water and fire. Bachelard misses, however, the chemical truth: water is born of fire, for few substances ignite more ferociously than hydrogen.

19 Henry Cavendish (1784): Experiments on Air, *Philosophical Transactions of the Royal Society.*

20 Antoine Lavoisier (1783): *Observations sur la Physique*, **23**, 452.

21 Ibid.

22 Jaffe (1976): *Crucibles: The Story of Chemistry*, Dover, New York, pp. 79–80.

23 In fact these two gentlemen were not the first to split water with electricity: the Dutchmen A. Paets van Troostwijk and J.R. Deimann accomplished it in 1789 using a 'static machine' in which electrical charge was accumulated by rubbing – the same principle by which a van der Graaf generator works. But they did not, apparently, notice the gases bubbling off from the two electrical poles of their apparatus.

24 I blush when recalling how, conducting an experiment in Grenoble that required liquid-nitrogen refrigeration, I enquired of the technicians after the whereabouts of *le nitrogène*. I had no idea I was reviving an old cross-Channel dispute. Fortunately, neither did they.

25 B. Jaffe (1976): *Crucibles: The Story of Chemistry*, Dover, New York, p. 97.

26 A further complication came with the discovery of hydrogen peroxide in 1815. This has the formula H_2O_2, and so a hydrogen-to-oxygen ratio of 1:1.

27 Certain British newspapers do not yet seem to have caught up with this change.

Chapter 6

Epigraph: T.H. Huxley (1869): On the physical basis of life, *The Fortnightly Review* **5**, 129.

1 Rev. Dr Brewer (1876): *A Guide to the Scientific Knowledge of Things Familiar* (or *Dr Brewer's Guide to Science*), Jarrold & Sons, London.

2 T. Stoppard (1996): *Arcadia*, Faber & Faber, London.

3 T. Andrews (1869): *Philosophical Transactions of the Royal Society* **159**, 575.

4 If I were to be stricter with the metaphor, I'd say rather that the city is a fabulous melange of Gaslanders and Liquidlanders mixed in every possible manner. Here there are households of either nationality living side by side; there, an entire street of Liquidlanders, a block of Gaslanders, districts and quarters of one or the other. There is, indeed, no unique scale to the segregation.

5 Such attractive forces operate in most solids; but in fact they are not *essential* for crystallization. In addition, not all solids are so orderly: plastics, for example, typically consist of tangled chains of atoms linked into long polymer molecules, while in window glass the constituent silicon and oxygen atoms are disorderly.

6 J.D. Bernal (1964): *Proceedings of the Royal Society* A **280**, 299. (From the Bakerian Lecture, 1962.)

7 For this reason, the weak forces of attraction are called van der Waals forces. The Dutchman couldn't say much about what they are like – how strong, how long a range they have – but he was the first to really acknowledge their importance.

8 Purists will condemn this statement on the grounds that hydrogen has several isotopes (p. 8). The angles of the kink in water molecules containing one or two deuterium atoms in place of hydrogen are slightly different.

9 It's a shame, really – if only Plato had associated water with the tetrahedron instead of the icosahedron, we could have accredited him with all kinds of spurious foresight.

10 By the time *The Nature of the Chemical Bond* was published, Pauling had modified this picture to include some element of electron-sharing in the hydrogen bond too. Although this dual nature of the hydrogen bond – part electrical attraction, part electron sharing – became broadly accepted in later years, it was not until 1999 that Pauling's view was confirmed experimentally. One suspects that Armstrong would not have approved.

11 I'm grateful to Zafra Lerman of Columbia College for the inspiration. She once told me how she taught schoolchildren the chemistry behind the destruction of the ozone layer by getting them to do the Dance of the Ozone Molecules. Groups of two and three dancers engage and disengage, whirling, pirouetting, and bringing literal truth to the Nobel laureate Roald Hoffmann's comment that 'It's a wild dance floor there at the molecular level.' I was even more impressed when Zafra told me that she had persuaded a conference full of awkward academics to do the dance too.

12 J.G. Ballard (1966): *The Crystal World*, Jonathan Cape, London.

13 Three years later, in 1895, Röntgen discovered X-rays, the searchlight beams with which the atomic and molecular structures of countless solids and liquids have since been illuminated.

14 Röntgen confessed that this idea was 'by no means new'. He may

have been alluding to the suggestion of H. Vernon in 1891 that the density maximum follows from the tendency of water molecules to aggregate into clusters, like Röntgen's ice fragments.

15 H.E. Armstrong (1925): *Nature* **115**, 85.

16 I recommend *A Bedside Nature* (ed. W. Gratzer), Macmillan, London, 1996, for a choice selection of Armstrong's vituperative broadsides.

Chapter 7

Epigraph: H. von Helmholtz (1865): Ice and glaciers. Lecture delivered in Frankfurt and Heidelberg; P. Høeg (1994): *Miss Smilla's Feeling for Snow*, Flamingo, London.

1 K. Vonnegut (1963): *Cat's Cradle*, Penguin, London, pp. 34–5.

2 T. Mann (1969): *The Magic Mountain* (transl. H.T. Lowe-Porter), Penguin, London, p. 480.

3 K. Vonnegut (1963): op. cit., p. 33.

4 Because no true crystal can have a repeating lattice with fivefold symmetry, this distinction is firmly enforced by the rules of geometry.

5 J. Kepler (1610): *S.C. Maiest. Mathematici Strena seu De nive sexangula.*

6 T.H. Huxley (1869): *The Fortnightly Review* **5**, 129.

7 J. Tyndall (1858): *Philosophical Transactions of the Royal Society* **148**, 211–27. I am indebted to Frank James of the Royal Institution for supplying this and related references, including letters between Faraday and William Thomson to be published in *The Correspondence of Michael Faraday* Vol. 4 (ed. F.A.J.L. James), Institution of Electrical Engineers, London, 1999.

8 Quoted in R.M. Hazen (1993): *The New Alchemists*, Time Books, New York, pp. 54–5.

9 If the pressure is released below 0°C, ice-VII reverts to ice-I. Only if it is kept very cold indeed – below about –170°C – does ice-VII survive at atmospheric pressure.

10 Window glass, for example, consists of amorphous silica (silicon dioxide), and is made by melting and rapidly cooling sand, which is predominantly the crystalline form of silica called quartz.

11 I. Langmuir (1943): *Nature* **151**, 268.

12 Cloud droplets can stay liquid until slightly lower temperatures because of the effect of size that I've just mentioned.

13 Water is not, however, completely unique in its candidacy for two liquid states. In particular, it has been proposed that carbon shares this unusual behaviour – but the idea is hard to verify, since carbon's melting point is so high. Semimetals such as gallium, selenium, tellurium and bismuth are also potential dual-liquid substances. See J.N. Glosli & F.H. Ree (1999): *Physical Review Letters* **82**, 4659; and M. Togaya (1997): *Physical Review Letters* **79**, 2474.

14 H. van Helmholtz (1865): Ice and glaciers. Lecture delivered in Frankfurt and Heidelberg.

Chapter 8

1 At least, this is my belief; it is not universally held.

2 C. Darwin (1871), quoted in I.S. Shklovskii & C. Sagan (1977): *Intelligent Life in the Universe*, Picador, London, p. 226. Contrast this with the view expressed by Darwin eight years earlier: 'It is mere rubbish, thinking at present of the origin of life' (ibid, p. 207).

3 A slight exaggeration; the 'equivalent' characters of DNA and RNA do actually differ in a very subtle way.

4 C. Chyba (1998): The stuff of life: why water? *Planetary Report* May/June, 16–17.

5 L. Margulis (1998): *The Symbiotic Planet*, Weidenfeld & Nicolson, London, p. 121.

6 This is, I fear, the third distinct use to which I've put this word in the book. I can't help it – atomic physicists, physical chemists and cell biologists have all tried to claim the term as their own. Cell biologists got there at least before the physicists, who consciously borrowed the term 'fusion' to describe the merging of their own version of nuclei.

7 Evolution has a way of accreting teleological metaphors, so it can't be overstressed that the watchmaker is blind. No organism ever directs its evolution towards an empty niche, however tempting it appears; if the species gets there at all, it is through the undirected drift of random mutation. My metaphor would work better if the Pilgrim Fathers had been out fishing and got blown across the Atlantic.

8 George Sandys: *A Relation of a Journey Begun in A.D. 1610* (London, 1637). Quoted in S. Schama (1995): *Landscape and Memory*, HarperCollins, London, p. 259.

9 Paul Claudel (1945): *Connaissance de l'Est*, Paris, p. 105.

10 William Harvey (1628): *On the Movement of the Heart and Blood in Animals*. In H.E. Boynton (ed.) (1948): *The Beginnings of Modern Science*, Walter Black, New York.

11 A salt is produced when an acid is reacted with an alkali. Sodium bicarbonate is the product of the reaction of carbonic acid with sodium hydroxide (caustic soda). It is nothing much more than the familiar baking powder.

12 The Galenic doctrine of the four humours is itself derived from the medical philosophy of Hippocrates, who lived six centuries earlier.

13 G. Bachelard (1983): *Water and Dreams*, Dallas Institute Publications, p. 118.

14 Christopher Fry: *The Lady's Not For Burning*.

Chapter 9

Epigraph: P. Claudel (1928–34): *Positions et Propositions*, Paris; P.M. Wiggins (1990): Role of water in some biological processes, *Microbiological Reviews* **54**, 432.

1 M. Gerstein & M. Levitt (1998): *Scientific American*, November, pp. 100–105.
2 To be more accurate, one should say that the affinity of ions and polar molecules for water resides in the latter's high *dielectric constant*, which is a measure of a substance's ability to attenuate an electric field – analogous, you might say, to the ability of some materials to damp out sound waves. Because water has a high dielectric constant, dissolved ions of opposite charge do not feel one another's electric field very strongly, and so have less tendency to gather together into a crystal.
3 Cellulose, however, which is nothing more than polymerized glucose, is highly insoluble and indigestible, because the long sugary molecules pack together in a tightly organized way, bound by hydrogen bonds. Sugar and sawdust are made up from pretty much the same stuff, but you'd never guess it from their solubilities.
4 Soap molecules have this same configuration, and work by embedding their hydrophobic tails in insoluble globules of grease.
5 This tendency to bury hydrophobic parts of the chain should not be overemphasized, however: typically about half of the protein's exposed surface remains hydrophobic. The consequences of exposed hydrophobic surfaces are still controversial, as we'll see.
6 I must make it clear that this is (yet again) an overly simplistic picture by several orders of magnitude. For one thing, enzymes are just as adept at helping bonds to break as to make. For another, they are not really so rigid, and wouldn't function properly if they were; most enzymes undergo small changes in structure as they bind their target molecules and guide the reaction. And the analogy obscures the important fact that the main role of any catalyst, be it an enzyme or otherwise, is not simply to increase the chances of the reactant molecules coming together with the right spatial disposition but to lower the energy barrier for the reaction, so that less energy is needed to form the unstable, ephemeral species that appear in the course of the reaction.
7 F. Franks (1983): *Water*, Royal Society of Chemistry, London.
8 Another almost-truth, I fear. Strictly speaking, the balance is between the change in entropy and the change in a quantity called *enthalpy*, which encompasses heat change due to the making or breaking of chemical bonds but also a contribution that results from changes in volume and pressure – as, for example, when a piston is pushed out as petroleum is ignited or water evaporates. The enthalpy change is

really a measure of the *work* that can be extracted from a process, and the motion of a piston (say) is clearly a part of such work.

9 H.S. Frank & M.W. Evans (1945): *Journal of Chemical Physics* **13**, 507.

10 See, for example, the article by J. Israelachvili & H. Wennerström (1996): *Nature* **379**, 219–25. This discusses the character of water close to hydrophilic surfaces, and in particular the tendency for such surfaces to repel one another when close together. Several explanations for this effect invoke the unique degree of structure in liquid water; but Israelachvili and Wennerström contest this. Their assertion that 'When confronted with unexpected experimental results water structure has commonly been used as a *deus ex machina* for explaining the observations' could serve almost as a watchword for Chapters 10 and 11. They suggest that the origins of the repulsive force may be more mundane – for example, 'hairs' of silica protruding from silica surfaces, which hold them apart. (Some of the measurements of a repulsive force were made for silica surfaces.) They conclude with the suggestion that 'as a solvent and suspending medium, water should be seen as an ordinary liquid', which has provoked outrage in some quarters (and which, in truth, is very hard to credit if interpreted too literally).

11 Knowing nothing of water's remarkable structure, Henderson had no inkling of just how exquisite water's fitness is. But he argued that properties already well known – its liquidity at most of Earth's surface temperatures, its large heat capacity (and so slow rate of warming and cooling), its dissolving power and its ability to participate in reactions of acids and bases – made water appear almost deliberately pre-adapted for life. Henderson promoted this as an 'argument by design' for the existence of God – quite the opposite of the conventional implication of evolutionary theory!

Chapter 10

Epigraph: I. Langmuir (1953): Pathological science. A lecture transcribed by R.N. Hall, reprinted in *Physics Today*, October 1989, p. 36.

1 The appropriate manner in which to conduct peer review is contentious, however. Most journals offer referees the option of remaining anonymous to the authors – an option that is adopted more often than not. But anonymous peer review introduces the potential for all sorts of mendacity, and requires careful judgement by the editor. The *Journal of the British Medical Association* announced in 1999 that it would henceforth conduct open peer review, with the referees' identities always being revealed to the authors. The journal's editor cites evidence that this does not result in a decline in quality, and so is preferable on ethical grounds. Some groups of scientists feel that peer review is outmoded and stifling. Many physicists now regard a paper as 'published' when they post an unrefereed copy to

the electronic archive maintained at Los Alamos National Laboratory. And printed, unreviewed journals do exist.

2 B. Appleyard (1993): *Understanding the Present: Science and the Soul of Modern Man*. Appleyard does not intend this as a compliment, but to my mind he makes a wonderful case for why science is worth listening to. Rarely has a book so convincingly defeated its own objective.

3 The quote marks are here to warn that there is no one true Scientific Method, only a broad consensus – uncodified, and uncodifiable – as to what constitutes good scientific practice.

4 I. Langmuir (1953): Pathological science. A lecture transcribed by R.N. Hall and reprinted in *Physics Today*, October 1989, 36–48.

5 Langmuir's analysis has been perceptively extended by Nicholas Turro, a chemist at Columbia University in New York, in the Columbia house magazine *21stC* (issue 3.4). Available at http://www.columbia.edu/cu/21stC/issue-3.4/turro.html.

6 See D. Bouwmeester *et al.* (1997): *Nature* **390**, 575–9.

7 But please do bear in mind that the latter group by no means excludes all, or even most, scientists.

8 I am immensely grateful to John Finney for providing me with a tape recording and transcript of this conversation.

9 E. Willis, G.K. Rennie, C. Smart & B.A. Pethica (1969): *Nature* **222**, 159–61.

10 We saw in Chapter 7 that symmetrical hydrogen bonds do seem to form in some types of ice, but only when the normal hydrogen bonds have been reduced in length by tremendous squeezing.

11 F.J. Donahoe (1969): *Nature* **224**, 198.

12 D.H. Everett, J.M. Haynes & P.J. McElroy (1969): *Nature* **224**, 394.

13 F. Franks (1981): *Polywater*, MIT Press, Cambridge MA.

14 D.L. Rousseau & S.P.S. Porto (1970): *Science* **167**, 1715–19.

15 R.E. Davis (1970): *Chemical & Engineering News*, 28 September.

Chapter 11

Epigraph: G. Bachelard (1983): *Water and Dreams*, Dallas Institute Publications; S. Bonini, E. Adriani & F. Balsano (1988): Evidence of non-reproducibility, *Nature* **334**, 559.

1 R. Rhodes (1986): *The Making of the Atomic Bomb*, Simon & Schuster, New York, p. 140.

2 S. Poliakoff (1996): *Blinded by the Sun*, Methuen, London.

3 All the same, the thirty-five deaths were caused by the fall, not the fire.

4 Although different isotopes of an element are on the whole chemically identical, deuterium furnishes some dramatic exceptions. In particular, heavy water D_2O is lethal: if we were to consume it in preference to H_2O, we'd be dead within a day. This is because the properties of

liquid water are acutely sensitive to the nature of the hydrogen atoms. Deuterium, being fully twice as heavy, just doesn't have what it takes to make water the stuff of life. Rumour has it that a plot was once hatched to assassinate the Shah of Iran by replacing his drinking water with heavy water. To normal chemical analyses, the switch would have been imperceptible.

Early studies of the biological effects of heavy water were conducted by American chemist Gilbert Lewis (see *Nature* **133**, 620; 1934). The stuff was at that stage only recently discovered, and still hard to come by. Physicist Ernest Lawrence was eager to use deuterium in nuclear physics experiments in his new particle accelerator, and was disgusted to find Lewis squandering it on mice – see *A Bedside Nature* (ed. W. Gratzer), Macmillan, London, 1996, p. 217.

5 Only later did it become clear that this energetic event happened one night when no one had actually witnessed it – they had found only its aftermath.

6 F. Close (1991): *Too Hot to Handle: The Race for Cold Fusion*, Princeton University Press.

7 R. Highfield (1989), *Daily Telegraph*, March 1989.

8 F. Close (1991): *Too Hot to Handle: The Race for Cold Fusion*, Princeton University Press, p. 144.

9 Editorial by J. Maddox (1989): *Nature*, **338**, 527.

10 Editorial by J. Maddox (1988): *Nature* **333**, 787.

11 In 1997 the National Institutes of Health gave a cautious endorsement of acupuncture: a panel of scientists concluded after four years of trials that it seemed to be effective against nausea and some causes of pain. The data, they said, were 'as strong as those for many accepted Western medical therapies'. This conclusion was met with controversy and fury from some quarters of the medical community. Others speculated that acupuncture might trigger the production of pain-killing endorphins or hormones that affect our emotional state. 'The challenge,' said panel chair David Ramsay, 'is to integrate the theory of Chinese medicine into the conventional Western biomedical research model.' See, for example, *Chemical & Engineering News*, 14 November 1997, p. 1231 and 7 December 1998, and *Nature* **383**, 285 (1996).

12 In fact there is no evidence that quinine really does this; but it is, at any rate, what Hahnemann seems to have experienced.

13 A. Campbell (1984): *The Two Faces of Homeopathy*, Robert Hale, London.

14 The quote marks are called for, since in most cases the 'solutions' contain no solute at all, so can't really be considered dilute in any sense.

15 These can be extraordinarily subtle. Sometimes a placebo effect is evident even when patients *are told* that the medicine they are being

given is not genuine. See W.A. Brown (1998): *Scientific American* **278** (January), 90–95.
16 D.T. Reilly, M.A. Taylor, C. McSharry & T. Aitchinson (1986): *Lancet* **ii**, 881–5.
17 See L. Wolpert (1992): *The Unnatural Nature of Science*, Faber & Faber, London.
18 Editorial by J. Maddox (1988): *Nature* **333**, 787.
19 J. Benveniste (1988): *Nature* **335**, 759.
20 Not for nothing is there still in circulation a magazine for cold-fusion enthusiasts called *Infinite Energy*.
21 Editorial footnote to E. Davenas *et al*. (1988): *Nature* **333**, 816–18.
22 E. Davenas *et al*., op. cit.
23 H. Metzger & S.C. Dreskin (1988): *Nature* **334**, 375.
24 S. Bonini, E. Adriani & F. Balsano (1988): *Nature* **334**, 559.
25 Editorial footnote to E. Davenas *et al*. (1988): *Nature* **333**, 816–18.
26 J. Maddox, J. Randi & W.W. Stewart (1988): *Nature* **334**, 287–90.
27 See D.J. Kevles (1998): *The Baltimore Case: A Trial of Politics, Science, and Character*, W.W. Norton, New York.
28 K. Snell (1988): *Nature* **334**, 559.
29 J. Benveniste (1988): *Nature* **334**, 291.
30 J. Maddox, J. Randi & W.W. Stewart (1988): *Nature* **334**, 287–90.
31 J. Benveniste (1988): *Nature* **334**, 291.
32 H. Metzger & S.C. Dreskin (1988): *Nature* **334**, 375.
33 In 1993 *Nature* published a very thorough investigation of the 'high-dilution' effect by a team from University College, London (S.J. Hirst, N.A. Hayes, J. Burridge, F.L. Pearce & J.C. Foreman (1993): *Nature* **366**, 525–7). The researchers attempted to reproduce Benveniste's experimental procedures as closely as possible. They reported some puzzling but minor variations in degranulation as dilution was increased, but no evidence for any pronounced, periodic or reproducible recurrence at high dilutions. By this stage, however, the Benveniste affair was no more than a distant memory to most scientists.
34 Editorial by J. Maddox (1989): *Nature* **340**, 82.
35 K. Linde *et al*. (1997): *Lancet* **350**, 834–43.
36 See G. Preparata (1995): *QED Coherence in Matter*. World Scientific, Singapore; R. Arani, I. Bono, E. Del Giudice & G. Preparata (1995): *International Journal of Modern Physics* B **9**, 1813.
37 See http://www.digibio.com/cgi-bin/concat.p1?nd=n3.
38 T. Schwenk (1996): *Sensitive Chaos*, Rudolf Steiner Press, London.
39 C. Coats (1996): *Living Energies*, Gateway Books, Bath. See also V. Schauberger (1998): *The Water Wizard*, ed. C. Coats, Gateway Books, Bath.

Epilogue

Epigraph: W. Stegner (1954): *Beyond the Hundredth Meridian: John Wesley Powell and the Second Opening of the West*, University of Nebraska Press, Lincoln, Nebraska; K. Ettenger (1998): 'A river that was once so strong and deep', in J.M. Donahue & B.R. Johnston (eds): *Water, Culture, & Power*, Island Press, Washington DC.

1 United Nations (1977). *Report of the United Nations Water Conference, Mar del Plata, 14–25 March 1977*. E/CONF.70/29.
2 This is not the same as saying that the world's arable land area is increasing by this rate, for such land is also being lost owing to salinization, desertification and other processes.
3 Such is the level of subsidy for agricultural water in California that during the 1980s, the average price of water from the canals that constitute the state's Central Valley Project fell to less than ten per cent of the cost of providing it.
4 Dumping of nuclear waste at sea was banned in the USA in 1970, and in Europe in 1982. But there is no global ban, and it still happens.
5 R. Gottlieb (1988): *A Life of its Own: The Politics and Power of Water*, Harcourt Brace, San Diego, p. 189.
6 J.K. Huysmans (1928–34): *Croquis Parisiens*.
7 Fluoride, the scourge of conspiracy theorists like General Ripper (page 227), is added to some water supplies not as a purification agent but because it is required for healthy teeth. But the difference between healthy and toxic doses is rather small.
8 Acid rain has at least one natural cause too: volcanic activity, which releases sulphur and nitrogen oxides, as well as chlorine which is converted to hydrochloric acid. But the acid-rain problem of the past few decades is unambiguously the result of human activities.
9 S. Postel (1992): *Last Oasis: Facing Water Scarcity*, W.W. Norton, New York, p. 127.
10 G. Bachelard (1983): *Water and Dreams*, Dallas Institute Publications, p. 152.
11 S. Schama (1995): *Landscape and Memory*, HarperCollins, London, p. 261. See also K. Wittfogel (1957): *Oriental Despotism: A Comparative Study of Total Power*, Yale University Press, New Haven.
12 P. Whiteley & V. Masayesva (1998): The use and abuse of aquifers. In *Water, Culture, and Power* (eds. J.M. Donahue & B.R. Johnston), Island Press, Washington DC, p. 18.
13 D. Worster (1985): *Rivers of Empire: Water, Aridity, and the Growth of the American West*, Pantheon Books, New York.
14 As I indicated earlier, so much of the river's waters are now withdrawn that it doesn't always reach the coast.
15 It is worth pointing out that California is by no means monolithic in its thirst. The state has its own internal water wars too, on which

much of Californian politics has hinged in the past several decades. The northern part of the state, where the Sacramento river reaches the Pacific in the vast delta of the Bay area, is relatively well supplied with water. This is coveted by both the huge corporate farms in the central state and the sprawling urbanization around Los Angeles in the south. Immense and vastly expensive canal systems, such as the State Water Project and the Central Valley Project, bring Bay water southwards to meet these needs. Almost invariably, these waterways were constructed in a climate of rancour between different regions and different interest groups. The battles are by no means over.

16 A.T. Wolf (1995): *Hydropolitics Along the Jordan River: Scarce Water and its Impact on the Arab–Israeli Conflict.* United Nations Press, Tokyo; S. Libiszewski (1996): Water Disputes in the Jordan Basin and Their Role in the Resolution of the Arab–Israeli Conflict. In *Environmental Degradation as a Cause of War – Kriegsursache Umweltzerstörung. Regional and Country Studies of Research Fellows – Regional- und Länderstudien von Projektmitarbeitern* (eds. G. Bächler & K.R. Spillmann) Vol. II. p. 337. Rüegger Verlag, Zürich. See also http://www.fsk.ethz.ch/encop/13/en13-ch7.htm.

17 *Independent*, 15 May 1990.

18 See P.H. Gleick (1998): *The World's Water*, Island Press, Washington DC, pp. 115–16.

19 See S. Libiszewski (1995): *Water Disputes in the Jordan Basin Region and Their Role in the Resolution of the Arab–Israeli Conflict*, Section 4.1, ENCOP Occasional Paper No. 13, Centre for Security Policy and Conflict Research, Zurich: http://www.fsk.ethz.ch/encop/13/en13-ch4.htm.

20 S. Postel (1997): *Last Oasis: Facing Water Scarcity*, W.W. Norton, New York, p. xxviii.

Bibliography

Part I: Cosmic Juice
M. Allaby (1992): *Water: Its Global Nature*, Facts on File, Oxford.
G. Bachelard (1983): *Water and Dreams*, Dallas Institute Publications.
V.R. Baker, M.H. Carr, V.C. Gulick, C.R. Williams & M.S. Marley (1992): Channels and valley networks. In *Mars* (eds. H.H. Kieffer, B.M. Jakosky, C.W. Snyder & M.S. Matthews), University of Arizona Press, Tucson, pp. 493–522.
J.K. Beatty, C.C. Petersen & A. Chaikin (1999): *The New Solar System*, 4th edn, Cambridge University Press.
L.. Bergeron (1997): Deep waters, *New Scientist*, 30 August, pp. 22–6.
G.R. Bigg (1996): *The Oceans and Climate*, Cambridge University Press.
P. Bond (1998): Prospector finds ten times more ice, *Astronomy Now*, October, 51–2.
H. Boynton, ed. (1948): *The Beginnings of Modern Science*, Walter J. Black, Roslyn NY.
M.H. Carr *et al*. (1998): Evidence for a subsurface ocean on Europa, *Nature* **391**, 363–5.
M.T. Chahine (1992): The hydrological cycle and its influence on climate, *Nature* **359**, 373–80.
W. Chorlton (1983): *Ice Ages*, Time-Life Books, Amsterdam.
C. Clark (1983): *Flood*, Time-Life Books, Amsterdam.
J.I. Drever (1997): *The Geochemistry of Natural Waters*, Prentice Hall, New Jersey.
D. Drewry (1996): Ice sheets, climate change and sea level, *Physics World*, January, p. 29.
A. Eliot (1990): *The Universal Myths*, Meridian, New York.
M. Elitzur (1992): Astronomical masers, *Annual Reviews of Astronomy and Astrophysics* **30**, 75–112.
W.C. Feldman *et al*. (1998): Fluxes of fast and epithermal neutrons from Lunar Prospector: evidence for water ice at the lunar poles, *Science* **281**, 1496–1500.
P.E. Geissler *et al*. (1998): Evidence for non-synchronous rotation of Europa, *Nature* **391**, 368–70.
D. Goldsmith (1997): Comet origin of oceans all wet? *Science* **277**, 318.
D.H. Grinspoon (1997): *Venus Revealed*, Addison Wesley, Reading MA.
M. Hanlon (1998): Life in the ocean wave? *Independent on Sunday*, 1 March, 40–1.
A. Henderson-Sellers (1983): *The Origin and Evolution of Planetary Atmospheres*, Adam Hilger, Bristol.

H.D. Holland (1984): *The Chemical Evolution of the Atmosphere and Oceans*, Princeton University Press.

B.F. Howell (1990): *An Introduction to Seismological Research: History and Development*, Cambridge University Press, Chapter 9.

R.J. Huggett (1997): *Environmental Change*, Routledge, London.

Intergovernmental Panel on Climate Change (1996): *Climate Change 1995* (eds. J.T. Houghton *et al.*), Cambridge University Press.

H. Kanamori & T.H. Heaton (1996): The wake of a legendary earthquake, *Nature* **379**, 203–4.

J.S. Kargel & R.G. Strom (1996): Global climatic change on Mars, *Scientific American* **275(5)**, 80–8.

J.S. Kargel (1998): The salt of Europa, *Science* **280**, 1211.

J.F. Kasting (1988): Runaway and moist greenhouse atmospheres and the evolution of Earth and Venus, *Icarus* **74**, 472–94.

J.F. Kasting (1998): The origins of water on Earth, *Scientific American Quarterly* **9(3)**, 16–22.

K.K. Khurana *et al.* (1998): Induced magnetic fields as evidence for subsurface oceans in Europa and Callisto, *Nature* **395**, 777–80.

J.I. Lunine & C.J. Lunine (1998): *Earth: Evolution of a Habitable World*, Cambridge University Press.

M.C. Malin & M.H. Carr (1999): Groundwater formation of martian valleys, *Nature* **397**, 589–91.

A.M. Mannion (1991): *Global Environmental Change*, Longman, Harlow.

S.F. Mason (1992): *Chemical Evolution*, Clarendon Press, Oxford.

A.S. McEwen (1986): Tidal reorientation and the fracturing of Jupiter's moon Europa, *Nature* **321**, 49–51.

C.P. McKay, R.L. Mancinelli, C.R. Stoker & R.A. Wharton, Jr. (1992): The possibility of life on Mars during a water-rich past. In *Mars* (eds. H.H. Kieffer, B.M. Jakosky, C.W. Snyder & M.S. Matthews), University of Arizona Press, Tucson, pp. 493–522.

C.P. McKay, O.B. Toon & J. Kasting (1991): Making Mars habitable, *Nature* **352**, 489–96.

H.Y. McSween (1997): *Fanfare for Earth*, St Martin's Press, New York.

R. Meier *et al.* (1998): A determination of the HDO/H_2O ratio in comet C/1995 O1 (Hale-Bopp), *Science* **279**, 842–7.

O. Morton (1998): Flatlands, *New Scientist*, 18 July, 36–9.

F. Neubauer (1998): Oceans inside Jupiter's moons, *Nature* **395**, 749–50.

T. Oka (1997): Water on the Sun: molecules everywhere, *Science* **277**, 328–9.

R.T. Pappalardo *et al.* (1998): Geological evidence for solid-state convection in Europa's ice shell, *Nature* **391**, 365–8.

W.S.B. Paterson (1994): *The Physics of Glaciers*, Pergamon.

E.C. Pielou (1998): *Fresh Water*, University of Chicago Press.

F. Press and R. Siever (1998): *Understanding Earth*, W.H. Freeman, New York.

G.W. Robinson (1994): *Minerals*, Simon & Schuster, New York.
M.N. Ross & G. Schubert (1987): Tidal heating in an internal ocean model of Europa, *Nature* **325**, 133–4.
K. Satake, K. Shimazaki, Y. Tsuji & K. Ueda (1996): Time and size of a giant earthquake in Cascadia inferred from Japanese tsunami records of January 1700, *Nature* **379**, 246–9.
S. Schama (1995): *Landscape and Memory*, HarperCollins, London.
W.H. Schlesinger (1991): *Biogeochemistry: An Analysis of Global Change*, Academic Press, San Diego.
D. Schneider (1998): The Rising Seas, *Scientific American Quarterly* **9(3)**, 28–35.
R. Scorer and A. Verkaik (1989): *Spacious Skies*, David & Charles, Newton Abbot.
I.S. Shklovskii & C. Sagan (1977): *Intelligent Life in the Universe*, Picador, London.
P.D. Spudis (1996): *The Once and Future Moon*, Smithsonian Institution Press, Washington DC.
S.W. Squyres (1984): The history of water on Mars, *Annual Review of Earth and Planetary Science* **12**, 83–106.
S.R. Taylor (1992): *Solar System Evolution*, Cambridge University Press.
A.B. Thompson (1992): Water in the Earth's upper mantle, *Nature* **358**, 295–301.
L. Wallace *et al.* (1995): Water on the Sun, *Science* **268**, 1155–8.
P.J. Webster & J.A. Curry (1998): The oceans and weather, *Scientific American Quarterly* **9(3)**, 38–43.
N. Wells (1997): *The Atmosphere and Ocean*, 2nd edn, John Wiley, Chichester.

Part II: Two Hands, Two Feet

A.W. Adamson (1990): *Physical Chemistry of Surfaces*, 5th edn, John Wiley, New York.
C.A. Angell (1988): Approaching the limits, *Nature* **331**, 206–7.
I. Asimov (1991): *Atom*, Penguin, London.
P.W. Atkins (1978): *Physical Chemistry*, Oxford University Press.
M.C. Bellissent-Funel (1998): Is there a liquid–liquid phase transition in supercooled water?, *Europhysics Letters* **42**, 161–6.
M. Benoit, D. Marx & M. Parrinello (1998): Tunnelling and zero-point motion in high-pressure ice, *Nature* **392**, 258–61.
J.D. Bernal & R.H. Fowler (1933): A theory of water and ionic solution, with particular reference to hydrogen and hydroxyl ions, *Journal of Chemical Physics* **1**, 515–48.
P.W. Bridgman (1935): The pressure–volume–temperature relations of the liquid, and the phase diagram of heavy water, *Journal of Chemical Physics* **3**, 597–605.
P.W. Bridgman (1911): Water, in the liquid and five solid forms, under

pressure, *Proceedings of the American Academy of Arts and Sciences* **47**, 441–558.
W.H. Brock (1992): *The Fontana History of Chemistry*, Fontana, London.
E.F. Burton & W.F. Oliver (1936): The crystal structure of ice at low temperatures, *Proceedings of the Royal Society of London*, Series A **153**, 166–72.
A.M. Buswell & W.H. Rodebush (1956): Water, *Scientific American* **194(4)**, 77–89.
C. Cobb & H. Goldwhite (1995): *Creations of Fire*, Plenum Press, New York.
J.C. Cooper (1990): *Chinese Alchemy*, Sterling, New York.
P.G. Debenedetti (1996): *Metastable Liquids*, Princeton University Press.
M.W. Denny (1993): *Air and Water: The Biology and Physics of Life's Media*, Princeton University Press.
A. Donovan (1993): *Antoine Lavoisier*, Cambridge University Press.
A. Doppenschmidt, M. Kappl & H.-J. Butt (1998): Surface properties of ice studied by atomic force microscopy, *Journal of Physical Chemistry* **102**, 7813–19.
F. Franks (1983): *Water*, Royal Society of Chemistry, London.
F. Franks (ed.) (1975): *Water – A Comprehensive Treatise* Vol. 1, Plenum Press, New York.
F. Franks (1975): The properties of ice. In F. Franks (ed.): *Water – A Comprehensive Treatise* Vol. 1, Plenum Press, New York.
H.S. Frank (1975): Structural models. In F. Franks (ed.): *Water – A Comprehensive Treatise* Vol. 1, Plenum Press, New York.
C.J.T. de Grotthuss (1806): Mémoire sur la décomposition de l'eau et des corps qu'elle tient en dissolution a l'aide de l'électricité galvanique, *Annales de Chimie* **LVIII**, 54–73.
J.-P. Hansen & I.R. McDonald (1990): *Theory of Simple Liquids* 2nd edn, Academic Press, London.
S. Harrington, R. Zhang, P.H. Poole, F. Sciortino & H.E. Stanley (1997): Liquid–liquid phase transition: evidence from simulations, *Physical Review Letters* **78**, 2409–12.
P.V. Hobbs (1974): *Ice Physics*, Clarendon Press, Oxford.
E.J. Holmyard (1957): *Alchemy*, Penguin, Harmondsworth.
W.B. Holzapfel (1972): On the symmetry of the hydrogen bonds in ice VII, *Journal of Chemical Physics* **56**, 712–15.
E.D. Isaacs *et al.* (1999): Covalency of the hydrogen bond in ice: a direct X-ray measurement, *Physical Review Letters* **82**, 600–603.
B. Jaffe (1976): *Crucibles: The Story of Chemistry*, Dover, New York.
J. Knight (1998): Life on ice, *New Scientist*, 2 May, 24–8.
H.M. Leicester (1956): *The Historical Background of Chemistry*, John Wiley.
C. Lobban, J.L. Finney & W.F. Kuhs (1998): The structure of a new phase of ice, *Nature* **391**, 268–70.
P. Loubeyre, R. LeToullec, E. Wolanin, M. Hanfland & D. Hausermann

(1999): Modulated phases and proton centring in ice observed by X-ray diffraction up to 170 GPa, *Nature* **397**, 503–6.

P.J. Marchand (1996): *Life in the Cold*, 3rd edn, University Press of New England, Hanover, New Hampshire.

O. Mishima, L.D. Calvert & E. Whalley (1984): 'Melting ice' I [sic] at 77 K and 10 kbar: a new method of making amorphous solids, *Nature* **310**, 393–5.

O. Mishima, L.D. Calvert & E. Whalley (1985): An apparently first-order transition between two amorphous phases of ice induced by pressure, *Nature* **314**, 76–8.

O. Mishima & H.E. Stanley (1998): Decompression-induced melting of ice IV and the liquid–liquid transition in water, *Nature* **392**, 164–8.

O. Mishima & H.E. Stanley (1998): The relationship between liquid, supercooled and glassy water, *Nature* **396**, 329–35.

R.P. Multhauf (1993): *The Origins of Chemistry*, Gordon and Breach, Langhorne, Pennsylvania.

J.N. Murrell & E.A. Boucher (1982): *Properties of Liquids and Solutions*, John Wiley, Chichester.

L. Pauling (1935): The structure and entropy of ice and of other crystals with some randomness of atomic arrangement, *Journal of the American Chemical Society* **57**, 2680–84.

P.H. Poole, F. Sciortino, U. Essmann & H.E. Stanley (1992): Phase behaviour of metastable water, *Nature* **360**, 324–8.

F.X. Prielmeier, E.W. Lang, H.-D. Ludeman & R.S. Speedy (1987): Diffusion in supercooled water to 300 MPa, *Physical Review Letters* **59**, 1128–32.

C.N.R. Rao (1975): Theory of Hydrogen Bonding in Water. In F. Franks (ed.): *Water – A Comprehensive Treatise* Vol. 1, Plenum Press, New York.

J. Read (1995): *From Alchemy to Chemistry*, Dover, New York.

R. Rhodes (1986): *The Making of the Atomic Bomb*, Simon & Schuster, New York.

J.S. Rowlinson (ed.) (1988): *J.D. van der Waals: On the Continuity of the Gaseous and Liquid States. Studies in Statistical Mechanics* Vol. XIV, North-Holland, Amsterdam.

K. Sassen, K.N. Liou, S. Kinne & M. Griffin (1985): Highly supercooled cirrus cloud water: confirmation and climatic implications, *Science* **227**, 411–13.

S. Sastry, P. Debenedetti, F. Sciortino & H.E. Stanley (1996): Singularity-free interpretation of the thermodynamics of supercooled water, *Physical Review* E **53**, 6144–54.

F. Sicheri & D.S.C. Yang (1995): Ice-binding structure and mechanism of an antifreeze protein from winter flounder, *Nature* **375**, 427–31.

A.K. Soper (1997): The quest for the structure of water and aqueous solutions, *Journal of Physics: Condensed Matter* **9**, 2717–30.

C.M. Sorensen (1992): Still waters run still deeper, *Nature* **360**, 303–304.

R.J. Speedy & C.A. Angell (1976): Isothermal compressibility of supercooled water and evidence for a thermodynamic singularity at –45°C, *Journal of Chemical Physics* **65**, 851–8.

R. Speedy (1982): Stability-limit conjecture. An interpretation of the properties of water, *Journal of Chemical Physics* **86**, 982–91.

H.E. Stanley *et al.* (1994): Is there a second critical point in liquid water? *Physical Review Letters* **205**, 122–39.

H.E. Stanley *et al.* (1997): Cooperative molecular motions in water: the liquid–liquid critical point hypothesis, *Physical Review Letters* **236**, 19–37.

H.E. Stanley *et al.* (1998): The puzzling statistical physics of liquid water, *Physical Review Letters* **257**, 213–32.

F.H. Stillinger (1980): Water revisited, *Science* **209**, 451–7.

H. Tanaka (1998): Simple physical explanation of the unusual thermodynamic behaviour of liquid water, *Physical Review Letters* **80**, 5750–3.

H.C. von Baeyer (1993): *Taming the Atom*, Penguin, London.

K. Vonnegut (1965): *Cat's Cradle*, Penguin, London.

Part III: Life's Matrix

P.W. Atkins (1991): *Atoms, Electrons, and Change*, W.H. Freeman, New York.

D.T. Bowron, A. Filipponi, M.A. Roberts & J. Finney (1998): Hydrophobic hydration and the formation of a clathrate hydrate, *Physical Review Letters* **81**, 4164–7.

A. Carambassis, L.C. Jonker, P. Attard & M.W. Rutland (1998): Forces measured between hydrophobic surfaces due to a submicroscopic bridging bubble, *Physical Review Letters* **80**, 5357–60.

C. Chyba (1998): The stuff of life: why water? *Planetary Report* May/June, 16–17.

P.H.K. de Jong, J.E. Wilson, G.W. Neilson & A.D. Buckingham (1997): Hydrophobic hydration of methane, *Molecular Physics* **91**, 99–103.

J.P. Ferris, A.R. Hill, R.-H. Liu & L.E. Orgel (1996): Synthesis of long prebiotic oligomers on mineral surfaces, *Nature* **381**, 59–61.

A. Filipponi, D.T. Bowron, C. Lobban & J.L. Finney (1997): Structural determination of the hydrophobic hydration shell of Kr, *Physical Review Letters* **79**, 1293.

S. Fox (1988): *The Emergence of Life*, Basic Books, New York.

H.S. Frank & M.W. Evans (1945): Free volume and entropy in condensed systems, *Journal of Chemical Physics* **13**, 507–32.

S.A. Galema, E. Howard, J.B.F.N. Engberts & J.R. Grigera (1994): The effect of stereochemistry upon carbohydrate hydration. A molecular-dynamics simulation of β-D-galactopyranose and (α, β)-D-talopyranose, *Carbohydrate Research* **265**, 215–25.

M. Gerstein & M. Levitt (1998): Simulating water and the molecules of life, *Scientific American* Nov., 100–105.

D.S. Goodsell (1993): *The Machinery of Life*, Springer-Verlag, New York.
D.S. Goodsell (1996): *Our Molecular Nature*, Springer-Verlag, New York.
J. Israelachvili & H. Wennerström (1996): Role of hydration and water structure in biological and colloidal interactions, *Nature* **379**, 219–25.
S.J. Lippard & J.M. Berg (1994): *Principles of Bioinorganic Chemistry*, University Science Books, Mill Valley, California.
K. Lum, D. Chandler & J.D. Weeks (1999): Hydrophobicity at small and large length scales, *Journal of Physical Chemistry* **103**, 4570–4577.
W.T. Pockman, J.S. Sperry & J.W. O'Leary (1995): Sustained and significant negative water pressure in xylem, *Nature* **378**, 715–16.
R. Pomes & B. Roux (1996): Structure and dynamics of a proton wire: a theoretical study of H^+ translocation along the single-file water chain in the gramicidin A channel, *Biophysical Journal* **71**, 19–39.
W.K. Purves, G.H. Orians, H.C. Heller & D. Sadava (1997): *Life: The Science of Biology*, W.H. Freeman, New York.
F.A. Quiocho, D.K. Wilson & N.K. Vyas (1989): Substrate specificity and affinity of a protein modulated by bound water molecules, *Nature* **340**, 404.
J.D. Rawn (1989): *Biochemistry*. Neil Patterson Publishers, Burlington, North Carolina.
W.E. Royer, Jr., A. Pardanani, Q.H. Gibson, E.S. Peterson & J.M. Friedman (1996): Ordered water molecules as key allosteric mediators in a cooperative dimeric hemoglobin, *Proceedings of the National Academy of Sciences USA* **93**, 14526.
J.W. Schopf (ed.) (1992): *Major Events in the History of Life*, Jones & Bartlett, Boston.
P.M. Wiggins (1990): Role of water in some biological processes, *Microbiological Reviews* **54**, 432–49.

Part IV: Strange Brew

Anon. (1988): When to believe the unbelievable, *Nature* **333**, 787.
Anon. (1989): Cold results from Utah, *Nature* **338**, 364.
Anon. (1989): Cold fusion causes frenzy but lacks confirmation, *Nature* **338**, 447.
Anon. (1989): Disorderly publication, *Nature* **338**, 527–8.
Anon. (1989): Can heresy be real? *Nature* **340**, 82.
Anon. (1997): Thumbs up for acupuncture, *Science* **278**, 1231.
J. Benveniste (1988): Benveniste on the Benveniste affair, *Nature* **335**, 759.
J. Benveniste (1988): Dr Jacques Benveniste replies, *Nature* **334**, 291.
J.D. Bernal, P. Barnes, I.A. Cherry & J.L. Finney (1969): 'Anomalous' water, *Nature* **224**, 393–4.
S. Bonini, E. Adriani & F. Balsano (1988): Evidence of non-reproducibility, *Nature* **334**, 559.
A. Campbell (1984): *The Two Faces of Homoeopathy*, Robert Hale, London.

A. Cherkin (1969): 'Anomalous' water: a silica dispersion?, *Nature* **224**, 1293.
F. Close (1991): *Too Hot to Handle: The Race for Cold Fusion*, Princeton University Press.
C. Coats (1996): *Living Energies*, Gateway Books, Bath.
P. Coles (1989): Benveniste under review, *Nature* **340**, 89.
P. Coles (1989): INSERM closes the file, *Nature* **340**, 178.
P. Coles (1990): Benveniste all-clear, *Nature* **343**, 197.
E. Davenas *et al.* (1988): Human basophil degranulation triggered by very dilute antiserum against IgE, *Nature* **333**, 816–18.
F.J. Donahoe (1969): 'Anomalous' water, *Nature* **224**, 198.
J. Donohue (1969): Structure of 'polywater', *Science* **166**, 1000–1.
M.J. Escribano (1988): Only the smile is left, *Nature* **334**, 376.
D.H. Everett, J.M. Haynes & P.J. McElroy (1969): 'Anomalous' water, *Nature* **224**, 394.
M. Fleischmann & S. Pons (1989): Electrochemically induced nuclear fusion of deuterium, *Journal of Electroanalytical Chemistry* **261**, 301–8.
M. Fleischmann, S. Pons, M. Hawkins & R.J. Hoffman (1989): Measurement of γ-rays from cold fusion, *Nature* **339**, 667.
F. Franks (1981): *Polywater*, MIT Press, Cambridge MA.
P.M. Gaylarde (1988): Only the smile is left, *Nature* **334**, 375.
J.L. Glick (1988): Only the smile is left, *Nature* **334**, 376.
S.J. Hirst, N.A. Hayes, J. Burridge, F.L. Pearce & J.C. Foreman (1993): Human basophil degranulation is not triggered by very dilute antiserum against human IgE, *Nature* **366**, 525–7.
K.S. Jayaraman (1996): 'Petrol from plants' claim baffles Indian scientists, *Nature* **383**, 112.
K.S. Jayaraman (1996): Production for 'herbal' petrol, *Nature* **383**, 292.
K.S. Jayaraman (1996): Indian hopes for 'herbal fuel' disappear into thin air, *Nature* **383**, 374.
S.E. Jones *et al.* (1989): Observation of cold nuclear fusion in condensed matter, *Nature* **338**, 737–40.
I. Langmuir (1953): Pathological science. A lecture transcribed by R.N. Hall and reprinted in *Physics Today*, October 1989, 36–48.
N.S. Lewis *et al.* (1989): Searches for low-temperature nuclear fusion of deuterium in palladium, *Nature* **340**, 525–30.
K. Linde *et al.* (1997): Are the clinical effects of homoeopathy placebo effects? A meta-analysis of placebo-controlled trials, *Lancet* **350**, 834.
E.R. Lippincott, R.R. Stromberg, W.H. Grant & G.L. Cessac (1969): Polywater, *Science* **164**, 1482–7.
J. Maddox (1989): What to say about cold fusion, *Nature* **338**, 701.
J. Maddox, J. Randi & W.W. Stewart (1988): 'High-dilution' experiments a delusion, *Nature* **334**, 287–90.
H. Metzger & S.C. Dreskin (1988): Only the smile is left, *Nature* **334**, 375.
F.D. Peat (1989): *Cold Fusion*, Contemporary Books, Chicago.

R.D. Petrasso *et al.* (1989): Measurement of γ-rays from cold fusion – Reply, *Nature* **339**, 667–9.
S. Poliakoff (1996): *Blinded by the Sun*, Methuen, London.
D.T. Reilly, M.A. Taylor, C. McSharry & T. Aitchison (1986): Is homoeopathy a placebo response? Controlled trial of homoeopathic potency, with pollen in hayfever as a model, *Lancet* **ii**, 881–5.
D.L. Rousseau & S.P.S. Porto (1970): Polywater: polymer or artifact?, *Science* **167**, 1715–19.
M. Schiff (1995): *The Memory of Water*, Thorsons, London.
T. Schwenk (1996): *Sensitive Chaos*, Rudolf Steiner Press, London.
J.-C. Seagrave (1988): Evidence of non-reproducibility, *Nature* **334**, 559.
K.S. Suslick (1988): Only the smile is left, *Nature* **334**, 375–6.
G. Taubes (1993): *Bad Science: The Short Life and Weird Times of Cold Fusion*, Random House, New York.
D.E. Williams *et al.* (1989): Upper bounds on 'cold fusion' in electrolytic cells, *Nature* **342**, 375–84.
E. Willis, G.K. Rennie, C. Smart & B.A. Pethica (1969): 'Anomalous' water, *Nature* **222**, 159–61.

Epilogue: Blue Gold

M. Allaby (1992): *Water: Its Global Nature*, Facts on File, Oxford.
M. Allaby (1996): *Basics of Environmental Science*, Routledge, London.
W.H. Baarschers (1996): *Eco-facts & Eco-fiction*, Routledge, London.
G. Bächler & K.R. Spillmann (1996): *Environmental Degradation as a Cause of War – Kriegsursache Umweltzerstörung. Regional and Country Studies of Research Fellows – Regional- und Länderstudien von Projektmitarbeitern* Vols II & III, Rüegger Verlag, Zürich.
D. Briggs, P. Smithson, K. Addison & K. Atkinson (1997): *Fundamentals of the Physical Environment*, 2nd edn, Routledge, London.
R. Carson (1994): *Silent Spring*, Houghton Mifflin, Boston.
R.B. Clark (1997): *Marine Pollution*, 4th edn, Oxford University Press.
J.M. Donahue & B.R. Johnston (eds.) (1998): *Water, Culture, and Power*, Island Press, Washington DC.
J. Gelt (1997): Sharing Colorado River Water: History, Public Policy and the Colorado River Compact. Available at http://ag.arizona.edu/azwater/arroyo/101comm.html.
P.H. Gleick (1993). *Water in Crisis: A Guide to the World's Fresh Water Resources*. Oxford University Press.
P.H. Gleick (1998). *The World's Water*. Island Press, Washington DC.
R. Gottlieb (1988): *A Life of its Own: The Politics and Power of Water*, Harcourt Brace, Orlando, Florida.
D. Hillel (1994): *Rivers of Eden: The Struggle for Water and the Quest for Peace in the Middle East*, Oxford University Press.
R.J. Huggett (1997): *Environmental Change*, Routledge, London.
X. Lei (1998): Going against the flow in China, *Science* **280**, 24–26.

A.M. Mannion (1991): *Global Environmental Change*, Longman, Harlow, Essex.

B.J. Mason (1992): *Acid Rain*, Oxford University Press.

J.W. Maurits la Rivière (1989): Threats to the world's water, *Scientific American*, September, 48–55.

K.T. Pickering & L.A. Owen (1994): *An Introduction to Global Environmental Issues*, Routledge, London.

S. Postel (1992): *Last Oasis: Facing Water Scarcity*, W.W. Norton, New York.

Z. Sardar (1995): Cruising for peace, *Nature* **373**, 483–4.

A. Stikker (1998): Water today and tomorrow, *Futures* **30**, 43.

G.J. Young, J.C.I. Dooge & J.C. Rodda (1994): *Global Water Resource Issues*, Cambridge University Press.

Index